PIC16F1847 Microcontroller-Based Programmable Logic Controller

PIC16F1847
Microcontroller-Based
Programmable Logic
Controller

Hardware and Basic Concepts

Murat Uzam

CRC Press
Taylor & Francis Group
Boca Raton London New York

CRC Press is an imprint of the
Taylor & Francis Group, an **informa** business

First edition published 2021

by CRC Press
6000 Broken Sound Parkway NW, Suite 300, Boca Raton, FL 33487-2742
and by CRC Press
2 Park Square, Milton Park, Abingdon, Oxon, OX14 4RN

First edition published by CRC Press 2021

CRC Press is an imprint of Taylor & Francis Group, LLC

ISBN: 978-0-367-50639-1 (hbk)
ISBN: 978-0-367-55605-1 (pbk)
ISBN: 978-1-003-05060-5 (ebk)

DOI: 10.1201/9781003050605

Typeset in Times
by Deanta Global Publishing Services, Chennai, India

Visit the Routledge website: https://www.routledge.com/9780367506391

To the memory of my beloved father, Mehmet Uzam (1937–2017)

to my mother Zeynep Uzam

to my family

who love and support me

and

to my teachers and students

who enriched my knowledge

Contents

Prologue

Think globally, act locally.

Never give up.

No pain, no gain.

Practice makes perfect.

If we hear, we forget
If we see, we remember
If we do, we understand.

Success is not an accident, excellence is not a coincidence.

Think out of the box.

Preface

Programmable Logic Controllers (PLC) have been extensively used in industry for the past five decades. PLC manufacturers offer different PLCs in terms of functions, program memories, and the number of inputs/outputs (I/O), ranging from a few to thousands of I/Os. The design and implementation of PLCs have long been a secret of the PLC manufacturers. A serious project was reported by the author of this book in his previous book, entitled *Building a Programmable Logic Controller with a PIC16F648A Microcontroller*, published by CRC Press in 2014, to describe a microcontroller-based implementation of a PLC. The current project, called *PIC16F1847 Microcontroller-Based Programmable Logic Controller*, is based on the improved version of the project reported in the above-mentioned book. The improvements include both hardware and software elements. The current project is reported in three books and a downloadable document explaining application examples:

1. *PIC16F1847 Microcontroller-Based Programmable Logic Controller: Hardware and Basic Concepts* (this book)
2. *PIC16F1847 Microcontroller-Based Programmable Logic Controller: Intermediate Concepts*
3. *PIC16F1847 Microcontroller-Based Programmable Logic Controller: Advanced Concepts*

The current project is presented for students attending the related departments of engineering or technology faculties, for practicing engineers, and for hobbyists who want to learn how to design and use a microcontroller-based PLC. The book assumes the reader has taken courses on digital logic design, microcontrollers, and PLCs. In addition, the reader is expected to be familiar with the PIC16F series of microcontrollers and to have been exposed to writing programs using PIC assembly language within the MPLAB integrated development environment.

The contents of this book may be used to construct two different courses. The first one may involve teaching the use of the PLC technology as described in this book. This course may well fit in the related departments of both engineering and technology faculties. The second one may involve teaching how to design the PLC technology. This second course may be taught in electrical and electronics engineering and computer engineering departments.

Source and example files defined for the basic concepts of the PIC16F1847-Based PLC project are downloadable from this book's webpage under the downloads section.

In addition, PCB files of the CPU and I/O extension boards of the PIC16F1847-Based
PLC can also be downloaded from the same link.

Prof. Dr. Murat UZAM
Yozgat Bozok Üniversitesi
Mühendislik-Mimarlık Fakültesi
Elektrik-Elektronik Mühendisliği Bölümü
Yozgat
Turkey

About the Author

 **Murat Uzam** was born in Söke, Turkey, in 1968. He received his B.Sc. and M.Sc. degrees from the Electrical Engineering Department, Yıldız Technical University, İstanbul, Turkey, in 1989 and 1991, respectively, and his Ph.D. degree from the University of Salford, Salford, U.K., in 1998. He was with Niğde University, Turkey, from 1993 to 2010 in the Department of Electrical and Electronics Engineering as a Research Assistant, Assistant Professor, Associate Professor, and Professor. He was a Professor in the Department of Electrical and Electronics Engineering at Melikşah University in Kayseri, Turkey, from 2011 to 2016. Since 15 April 2020, he has been serving as a Professor in the Department of Electrical and Electronics Engineering at Yozgat Bozok University in Yozgat, Turkey.

He was a Visiting Researcher with INRIA, University of Metz and University of Rennes, France, in 1999, with the University of Toronto, Toronto, ON, Canada, in 2003, and with Xidian University, Xi'an, China, in 2013, 2015, and 2019.

He has published 47 conference papers and 106 journal and magazine papers, 70 of which are indexed by Science Citation Index Expanded (SCIE). He has published two books in Turkish and five books in English by CRC Press (Taylor & Francis Group). According to Publons, his H-Index is 16 and his papers have been cited more than 1370 times by the papers indexed in the SCIE. Dr. Uzam has been serving as a reviewer for prestigious journals and conferences. According to Publons, the number of his verified reviews is 70. His current research interests include design and implementation of discrete event control systems modelled by Petri nets and, in particular, deadlock prevention/liveness enforcing in flexible manufacturing systems, programmable logic controllers (PLCs), microcontrollers (especially PIC microcontrollers), and the design of microcontroller-based PLCs. The details of his studies are accessible from his web page: https://pbs.bozok.edu.tr/goster.php?lookup=1074

Background and Use of the Book

This project has been completed during the search for an answer to the following question: "How could one design and implement a programmable logic controller (PLC)?". An answer to this question was provided by the author in his previous book entitled *Building a Programmable Logic Controller with a PIC16F648A Microcontroller*, published by CRC Press in 2014. This project is based on the improved version of the PLC project reported in the above-mentioned book. So many new features have been included within the PIC16F1847-Based PLC project to make it an almost perfect PLC. The reader should be aware of the fact that this project does not include a graphical interface PC software as in commercial PLCs for developing PLC programs. Rather, PLC programs are developed by using macros as done in the Instruction List (IL) PLC programming language. An interested and skilled reader could well (and is encouraged to) develop a graphical interface PC software for easy use of the PIC16F1847-Based PLC.

The improvements of the PLC project reported in this book (*Hardware and Basic Concepts*) compared with the previous version are summarized as follows.

1. The current version of the PLC explained in this book is based on the PIC16F1847 microcontroller with: 8,192 words of flash program memory, 1,024 bytes of SRAM data memory, 256 bytes of EEPROM data memory, the maximum operating speed of 32 MHz, a 16-level-deep hardware stack, and an enhanced instruction set consisting of 49 single-word instructions, while the previous one was based on the PIC16F648A microcontroller with: 4,096 words of flash program memory, 256 bytes of SRAM data memory, 256 bytes of EEPROM data memory, the maximum operating speed of 20 MHz, an 8-level-deep hardware stack, and an instruction set consisting of 35 single-word instructions.
2. The hardware explained in this book consists of 1 CPU board and 4 digital I/O extension boards, while the previous one consisted of 1 CPU board and 2 digital I/O extension boards.
3. The clock frequency is 32 MHz in the current version of PLC, while it was 20 MHz in the previous version.
4. The current version of the PLC supports up to 32 digital inputs and 32 digital outputs, while the previous one supported 16 digital inputs and 16 digital outputs.
5. The current version of the PLC supports up to 4 analog inputs and 1 analog output, while the previous one did not support analog inputs/outputs.
6. The current version of the PLC supports 1,024 internal relays (memory bits), while the previous one supported only 32 internal relays.

7. The current version of the PLC provides 30 contact and relay-based instructions (macros), while the previous version provided 18 contact and relay-based instructions.

8. The current version of the PLC provides 14 flip-flop instructions (macros), while the previous version provided 8 flip-flop instructions.

9. The current version of the PLC provides 80 timers in total. These timers can be chosen from on-delay timers (TON_8 or TON_16), retentive on-delay timers (RTO_8 or RTO_16), off-delay timers (TOF_8 or TOF_16), pulse timers (TP_8 or TP_16), extended pulse timers (TEP_8 or TEP_16), and oscillator timers (TOS_8 or TOS_16). The timers with the suffix "_8" have 8-bit resolution, i.e., they are based on 8-bit registers, while the timers with the suffix "_16" have 16-bit resolution, i.e., they are based on 16-bit registers. On the other hand, the previous version of the PLC provided 8 on-delay timers (TON_8), 8 off-delay timers (TOF_8), 8 pulse timers (TP_8), and 8 oscillator timers (TOS_8). All these timers had 8-bit resolution, i.e., they were based on 8-bit registers.

10. The current version of the PLC provides 80 counters in total. These counters can be chosen from up counters (CTU_8 or CTU_16), down counters (CTD_8 or CTD_16), up/down counters (CTUD_8 or CTUD_16), and generalized up/down counters (GCTUD_8 or GCTUD_16). The counters with the suffix "_8" have 8-bit resolution, i.e., they are based on 8-bit registers, while the counters with the suffix "_16" have 16-bit resolution, i.e., they are based on 16-bit registers. On the other hand, the previous version of the PLC provided in total only 8 counters (CTU8 or CTD8 or CTUD8). They had 8-bit resolution, i.e., they were based on 8-bit registers.

11. The current version of the PLC provides 30 comparison instructions (macros), while the previous version provided 12 comparison instructions.

12. Almost all macros are improved compared with the previous versions, in terms of flexibility. For example, there is no restriction on the SRAM Banks, i.e., Boolean variables, 8-bit variables, and 16-bit variables used as a parameter in an instruction can be in any Bank. This was not the case in the previous version.

13. Flowcharts are provided to help the understanding of macros (instructions).

In order to follow the topics explained in this book properly, it is expected that the reader will construct his/her own PIC16F1847-Based PLC consisting of the CPU board and 4 I/O extension boards using the PCB files provided on the book's webpage under the downloads section. In addition, the reader should also download and make use of the PLC project files from the book's webpage. In this project, as the PIC Assembly is used as the programming language within the MPLAB integrated development environment (IDE), the reader is referred to the homepage of Microchip (http://www.microchip.com/) to obtain the latest version of MPLAB IDE. References [R1 and R2] may be useful to understand some aspects of the PIC16F1847 microcontroller and MPASM™ Assembler, respectively.

The contents of this book's 7 chapters are explained briefly, as follows.

1. **Hardware of the PIC16F1847-Based PLC:** In this chapter, the hardware structure of the PIC16F1847-Based PLC, consisting of 32 discrete inputs,

32 discrete outputs, 4 analog inputs, 1 analog output, and 2 PWM outputs is explained in detail.

2. **Basic Software:** This chapter explains the basic software structure of the PIC16F1847-Based PLC. A PLC scan cycle includes the following: (1) obtain the inputs, (2) run the user program, (3) update the outputs. In addition, it is also necessary to define and initialize all variables used within a PLC. Necessary functions are all described as PIC Assembly macros to be used in the PIC16F1847-Based PLC. The source files of the PIC16F1847-Based PLC are as follows: "PICPLC_PIC16F1847_memory.inc" (the individual bits of 8-bit SRAM registers M0, M1, …, M127 are defined in this file), "PICPLC_PIC16F1847_main.asm" (processor-specific variable definitions, PICPLC definitions, the user program, and subroutines are included in the project by using this file), "PICPLC_PIC16F1847_user.inc" (this file contains two macros, namely "user_program_1" and "user_program_2", in order to accommodate user programs), "PICPLC_PIC16F1847_subr.inc" (this file contains the "subroutines" macro and it is defined to obtain time delays at the expense of CPU clocks; the "subroutines" macro contains two time delay–related subroutines: "pause_1ms" and "pause_10us"), and "PICPLC_PIC16F1847_macros.inc". The file "PICPLC_PIC16F1847_macros.inc" contains the following macros: "initialize" (for PLC initialization), "ISR" (interrupt service routines), "get_inputs" (for handling the inputs), "lpf_progs" (low-pass digital filter macros for analog inputs), and "send_outputs" (for sending the outputs).

3. **Contact and Relay-Based Macros:** The following contact and relay-based macros are described in this chapter: "ld" (load), "ld_not" (load_not), "not", "or", "or_not", "nor", "and", "and_not", "nand", "xor", "xor_not", "xnor", "out", "out_not", "mid_out" (midline output), "mid_out_not" (inverted midline output), "in_out", "inv_out", "_set", "_reset", "SR" (set–reset), "RS" (reset–set), "r_edge" (rising edge detector), "f_edge" (falling edge detector), "r_toggle" (output toggle with rising edge detector), "f_toggle" (output toggle with falling edge detector), "adrs_re" (Address rising edge detector), "adrs_fe" (Address falling edge detector), "setBF" (set bit field), and "resetBF" (reset bit field). These macros are defined to operate on 1-bit (Boolean) variables.

4. **Flip-Flop Macros:** The following flip-flop macros are described in this chapter: "latch1" (D latch with active high enable), "latch0" (D latch with active low enable), "dff_r" (rising edge–triggered D flip-flop), "dff_r_SR" (rising edge–triggered D flip-flop with active high preset [S] and clear [R] inputs), "dff_f" (falling edge–triggered D flip-flop), "dff_f_SR" (falling edge–triggered D flip-flop with active high preset [S] and clear [R] inputs), "tff_r" (rising edge–triggered T flip-flop), "tff_r_SR" (rising edge–triggered T flip-flop with active high preset [S] and clear [R] inputs), "tff_f" (falling edge–triggered T flip-flop), "tff_f_SR" (falling edge–triggered T flip-flop with active high preset [S] and clear [R] inputs), "jkff_r" (rising edge–triggered JK flip-flop), "jkff_r_SR" (rising edge–triggered JK flip-flop with active high preset [S] and clear [R] inputs), "jkff_f" (falling

edge–triggered JK flip-flop), and "jkff_f_SR" (falling edge–triggered JK flip-flop with active high preset [S] and clear [R] inputs). Flip-flop macros are defined to operate on Boolean (1-bit) variables. 21 examples are provided to show the applications of these flip-flop macros, including the implementation of asynchronous and synchronous counters, and shift registers constructed by using the flip-flop macros.

5. **Timer Macros:** The following timer macros are described in this chapter: "TON_8" (8-bit on-delay timer), "TON_16" (16-bit on-delay timer), "RTO_8" (8-bit retentive on-delay timer), "RTO_16" (16-bit retentive on-delay timer), "TOF_8" (8-bit off-delay timer), "TOF_16" (16-bit off-delay timer), "TP_8" (8-bit pulse timer), "TP_16" (16-bit pulse timer), "TEP_8" (8-bit extended pulse timer), "TEP_16" (16-bit extended pulse timer), "TOS_8" (8-bit oscillator timer), and "TOS_16" (16-bit oscillator timer).

6. **Counter Macros:** The following counter macros are described in this chapter: "CTU_8" (8-bit up counter), "CTU_16" (16-bit up counter), "CTD_8" (8-bit down counter), "CTD_16" (16-bit down counter), "CTUD_8" (8-bit up/down counter), "CTUD_16" (8-bit up/down counter), "GCTUD_8" (8-bit generalized up/down counter), and "GCTUD_16" (16-bit generalized up/down counter).

7. **Comparison Macros:** In this chapter, the majority of the comparison macros are described according to the following notation: **GT** (**G**reater **T**han—">"), **GE** (**G**reater than or **E**qual to—"≥"), **EQ** (**EQ**ual to—"="), **LT** (**L**ess **T**han—"<"), **LE** (**L**ess than or **E**qual to—"≤"), or **NE** (**N**ot **E**qual to—"≠"). The contents of two 8-bit registers (R1 and R2) are compared with the following comparison macros: "R1_GT_R2" (Is R1 greater than R2?), "R1_GE_R2" (Is R1 greater than or equal to R2?), "R1_EQ_R2" (Is R1 equal to R2?), "R1_LT_R2" (Is R1 less than R2?), "R1_LE_R2" (Is R1 less than or equal to R2?), and "R1_NE_R2" (Is R1 not equal to R2?). Similar comparison macros are also described for comparing the contents of an 8-bit register (R) with an 8-bit constant (K): "R_GT_K" (Is R greater than K?), "R_GE_K" (Is R greater than or equal to K?), "R_EQ_K" (Is R equal to K?), "R_LT_K" (Is R less than K?), "R_LE_K" (Is R less than or equal to K?), and "R_NE_K" (Is R not equal to K?).

The contents of two 16-bit registers (R1 and R2) are compared with the following comparison macros: "R1_GT_R2_16" (Is R1 greater than R2?), "R1_GE_R2_16" (Is R1 greater than or equal to R2?), "R1_EQ_R2_16" (Is R1 equal to R2?), "R1_LT_R2_16" (Is R1 less than R2?), "R1_LE_R2_16" (Is R1 less than or equal to R2?), and "R1_NE_R2_16" (Is R1 not equal to R2?). Similar comparison macros are also described for comparing the contents of a 16-bit register (R) with a 16-bit constant (K): "R_GT_K_16" (Is R greater than K?), "R_GE_K_16" (Is R greater than or equal to K?), "R_EQ_K_16" (Is R equal to K?), "R_LT_K_16" (Is R less than K?), "R_LE_K_16" (Is R less than or equal to K?), and "R_NE_K_16" (Is R not equal to K?). In addition, the following comparison macros are also provided: "in_RANGE" (Is the value within the given range?), "in_RANGE_16" (Is the value within the given range?), "out_RANGE" (Is the value out of the given range?),

TABLE 1
General Characteristics of the PIC16F1847-Based PLC

Inputs/Outputs/Functions	Byte addresses/Related bytes	Bit addresses or function names/numbers
32 discrete inputs	I0,	I0.0, I0.1, ..., I0.7
(external inputs: 5 or 24V DC)	I1,	I1.0, I1.1, ..., I1.7
	I2,	I2.0, I2.1, ..., I2.7
	I3	I3.0, I3.1, ..., I3.7
32 discrete outputs	Q0,	Q0.0, Q0.1, ..., Q0.7
(relay-type outputs)	Q1,	Q1.0, Q1.1, ..., Q1.7
	Q2,	Q2.0, Q2.1, ..., Q2.7
	Q3	Q3.0, Q3.1, ..., Q3.7
4 analog inputs	AI0H:AI0L,	AI0H,1, AI0H,0, ..., AI0L,0
	AI1H:AI1L,	AI1H,1, AI1H,0, ..., AI1L,0
	AI2H:AI2L,	AI2H,1, AI2H,0, ..., AI2L,0
	AI3H:AI3L	AI3H,1, AI3H,0, ..., AI3L,0
1 analog output	-	RA2
1 high speed counter input	-	RB6
2 PWM outputs	-	RA4 & RA7
Drum sequencer instruction with up to 16 steps and 16 outputs on each step	Details are available in Chapter 4 of Volume III - *Advanced Concepts*	Details are available in Chapter 4 of Volume III - *Advanced Concepts*
1,024 internal relays	M0,	M0.0, M0.1, ..., M0.7
(memory bits)	M1,	M1.0, M1.1, ..., M1.7
	.	.
	.	
	M127	M127.0, M127.1, ..., M127.7
80 8-bit on-delay timers (TON_8)	TV_L, TV_L+1, ..., TV_L+79 T_Q0, T_Q1, ..., T_Q9	TQ0, TQ1, ..., TQ79
80 8-bit retentive on-delay timers (RTO_8)	TV_L, TV_L+1, ..., TV_L+79 T_Q0, T_Q1, ..., T_Q9	TQ0, TQ1, ..., TQ79
80 8-bit off-delay timers (TOF_8)	TV_L, TV_L+1, ..., TV_L+79 T_Q0, T_Q1, ..., T_Q9	TQ0, TQ1, ..., TQ79
80 8-bit pulse timers (TP_8)	TV_L, TV_L+1, ..., TV_L+79 T_Q0, T_Q1, ..., T_Q9	TQ0, TQ1, ..., TQ79
80 8-bit extended pulse timers (TEP_8)	TV_L, TV_L+1, ..., TV_L+79 T_Q0, T_Q1, ..., T_Q9	TQ0, TQ1, ..., TQ79
80 8-bit oscillator timers (TOS_8)	TV_L, TV_L+1, ..., TV_L+79 T_Q0, T_Q1, ..., T_Q9	TQ0, TQ1, ..., TQ79
80 16-bit on-delay timers (TON_16)	TV_L, TV_L+1, ..., TV_L+79 TV_H, TV_H+1, ..., TV_H+79 T_Q0, T_Q1, ..., T_Q9	TQ0, TQ1, ..., TQ79

(Continued)

TABLE 1 (CONTINUED)
General Characteristics of the PIC16F1847-Based PLC

Inputs/Outputs/Functions	Byte addresses/Related bytes	Bit addresses or function names/numbers
80 16-bit retentive on-delay timers (RTO_16)	TV_L, TV_L+1, ..., TV_L+79 TV_H, TV_H+1, ..., TV_H+79 T_Q0, T_Q1, ..., T_Q9	TQ0, TQ1, ..., TQ79
80 16-bit off-delay timers (TOF_16)	TV_L, TV_L+1, ..., TV_L+79 TV_H, TV_H+1, ..., TV_H+79 T_Q0, T_Q1, ..., T_Q9	TQ0, TQ1, ..., TQ79
80 16-bit pulse timers (TP_16)	TV_L, TV_L+1, ..., TV_L+79 TV_H, TV_H+1, ..., TV_H+79 T_Q0, T_Q1, ..., T_Q9	TQ0, TQ1, ..., TQ79
80 16-bit extended pulse timers (TEP_16)	TV_L, TV_L+1, ..., TV_L+79 TV_H, TV_H+1, ..., TV_H+79 T_Q0, T_Q1, ..., T_Q9	TQ0, TQ1, ..., TQ79
80 16-bit oscillator timers (TOS_16)	TV_L, TV_L+1, ..., TV_L+79 TV_H, TV_H+1, ..., TV_H+79 T_Q0, T_Q1, ..., T_Q9	TQ0, TQ1, ..., TQ79
80 8-bit up counters (CTU_8)	CV_L, CV_L+1, ..., CV_L+79 C_Q0, C_Q1, ..., C_Q9	CQ0, CQ1, ..., CQ79
80 8-bit down counters (CTD_8)	CV_L, CV_L+1, ..., CV_L+79 C_Q0, C_Q1, ..., C_Q9	CQ0, CQ1, ..., CQ79
80 8-bit up/down counters (CTUD_8)	CV_L, CV_L+1, ..., CV_L+79 C_Q0, C_Q1, ..., C_Q9	CQ0, CQ1, ..., CQ79
80 8-bit generalized up/down counters (GCTUD_8)	CV_L, CV_L+1, ..., CV_L+79 C_Q0, C_Q1, ..., C_Q9	CQ0, CQ1, ..., CQ79
80 16-bit up counters (CTU_16)	CV_L, CV_L+1, ..., CV_L+79 CV_H, CV_H+1, ..., CV_H+79 C_Q0, C_Q1, ..., C_Q9	CQ0, CQ1, ..., CQ79
80 16-bit down counters (CTD_16)	CV_L, CV_L+1, ..., CV_L+79 CV_H, CV_H+1, ..., CV_H+79 C_Q0, C_Q1, ..., C_Q9	CQ0, CQ1, ..., CQ79
80 16-bit up/down counters (CTUD_16)	CV_L, CV_L+1, ..., CV_L+79 CV_H, CV_H+1, ..., CV_H+79 C_Q0, C_Q1, ..., C_Q9	CQ0, CQ1, ..., CQ79
80 16-bit generalized up/down counters (GCTUD_16)	CV_L, CV_L+1, ..., CV_L+79 CV_H, CV_H+1, ..., CV_H+79 C_Q0, C_Q1, ..., C_Q9	CQ0, CQ1, ..., CQ79

"out_RANGE_16" (Is the value out of the given range?), "Hbit_CaC" (high bit count and compare), and "Lbit_CaC" (low bit count and compare). Note that this chapter is provided as downloadable ancillary material.

Application Examples: In total there are 20 application examples considered. For some application examples, more than one solution is provided in order to point out how different methods can be used for controlling the same problem. When the three books are purchased separately, application examples 1–9 (or 10–11 and 13–18; 7–12 and 20, respectively) are provided as downloadable ancillary material for the book *PIC16F1847 Microcontroller-Based Programmable Logic Controller: Hardware and Basic Concepts* (*Intermediate Concepts*; *Advanced Concepts*, respectively). On the other hand, when the three books are purchased as a set, all application examples are provided as a single ancillary material.

Appendix A: The list of components for all boards and modules developed in this project as reported in this book, together with the photographs of all components, are provided in Appendix A.

Table 1 shows the general characteristics of the PIC16F1847-Based PLC.

IMPORTANT NOTES

1. At any time, at most 80 different timers can be used. A unique timer number from 0 to 79 can be assigned to only one of the macros "TP_8", "TEP_8", "TOS_8", "TON_16", "RTO_16", "TOF_16", "TP_16", "TEP_16", and "TOS_16".

2. At any time, at most 80 different counters can be used. A unique counter number from 0 to 79 can be assigned to only one of the macros "CTU_8", "CTD_8", "CTUD_8", "GCTUD_8", "CTU_16", "CTD_16", "CTUD_16", and "GCTUD_16".

REFERENCES

R1. PIC16(L)F1847 Data Sheet, DS40001453F, 2011–2017, Microchip Technology Inc. http://ww1.microchip.com/downloads/en/DeviceDoc/40001453F.pdf

R2. MPASM™ Assembler, MPLINK™ Object Linker, MPLIB™ Object Librarian User's Guide DS33014J, 2005, Microchip Technology Inc. http://ww1.microchip.com/downloads/en/devicedoc/33014j.pdf

1 Hardware of the PIC16F1847-Based PLC

The hardware of the PIC16F1847-Based PLC consists of mainly two parts: the *CPU board* and the *I/O extension board*. The schematic diagram and the photograph of the PIC16F1847-Based PLC CPU board are shown in Figures 1.1 and 1.2, respectively. The CPU board contains mainly *three sections*: power, programming, and CPU (central processing unit).

The **power section** accepts 12V DC input used as the operating voltage of relays. 5V DC is also used for ICs, inputs, etc. An adjustable LM2596 step-down voltage regulator module is used to obtain 5V DC voltage from the 12V DC input voltage. It has the following specifications—conversion efficiency: up to 92%; switching frequency: 150 KHz; rectifier: nonsynchronous rectification; module properties: non-isolated step-down module (buck); operating temperature: industrial grade (–40 to +85); load regulation: ± 0.5%; voltage regulation: ± 2.5%; dynamic response speed: 5% 200 µs; input voltage: 3–40V; output voltage: 1.5–35V (adjustable); output current: maximum 3A; size: 43mm*21mm*14mm (length*width*height).

It is important to note that the output voltage (OUT+) of the adjustable LM2596 step-down voltage regulator module must be set to 5.00V by adjusting the potentiometer on the module before inserting the CPU. 12V DC input voltage can be subjected to electric surge or electrostatic discharge on the external terminal connections. The TVS (transient voltage suppressor) 1.5KE13A shown in the circuit provides highly effective protection against such discharges. It is also used to protect the circuit from accidental reverse polarity of the DC input voltage. For a proper operation of the PIC16F1847-Based PLC make sure that the DC input voltage < 13V DC.

The **programming section** deals with the programming of the PIC16F1847 microcontroller. For programming the PIC16F1847 in circuit, it is necessary to use a PIC programmer hardware and a software with ICSP (in-circuit serial programming) capability. In this project, Microchip's PICkit 3 In-Circuit Debugger/Programmer (www.microchip.com/PICkit3) is used as the PIC programmer hardware. MPLAB X IDE software (www.microchip.com/mplab/mplab-x-ide), freely available by Microchip (www.microchip.com), is used for the program development and for programming the PIC16F1847 microcontroller. The ICSP connector takes the lines VPP (MCLR), VDD, VSS (GND), DATA (RB7), and CLOCK (RB6) from the PIC programmer hardware through a properly prepared cable and it connects them to a 4PDT (four pole double throw) switch. There are two positions of the 4PDT switch. As seen from Figure 1.1, in the PROG position of the 4PDT switch, PIC16F1847 is ready to be programmed and in the RUN position, the loaded program is run. For programming the PIC16F1847 properly by means of a PIC programmer and the 4PDT switch, it is also a necessity to *switch off* the power switch. The **CPU section** consists of the PIC16F1847 microcontroller. In the project reported in

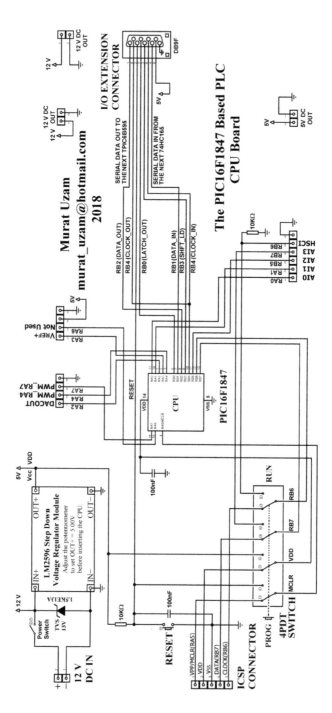

FIGURE 1.1 Schematic diagram of the PIC16F1847-Based PLC CPU board.

FIGURE 1.2 Photograph of the CPU board.

this book, the PLC is fixed to run at 32 MHz with an internal oscillator (oscillator frequency = 8 MHz and PLL = 4). This frequency is fixed because time delays are calculated based on this speed. RB1, RB3, and RB4 pins are all reserved to be used for 8-bit parallel-to-serial converter registers 74HC/LS165. Through these three pins and with added 74HC/LS165 registers we can describe as many inputs as necessary. RB1, RB3, and RB4 are the "data in", the "shift/load", and the "clock in" pins, respectively. Similarly, the RB2, RB4, and RB0 pins are all reserved to be used for 8-bit serial-to-parallel converter register/drivers TPIC6B595. Through these three pins and with added TPIC6B595 registers we can describe as many outputs as necessary. RB2, RB4, and RB0 are the "data out", the "clock out", and the "latch out" pins, respectively.

The RA0, RA1, RB5, and RB7 pins are described and used as analog inputs. They are called AI0, AI1, AI2, and AI3, respectively. The RA2 pin is used as an analog output and it is called DACOUT. The RA3 pin is used as VREF+ (ADC voltage reference input). The RB6 pin is used as the clock input of the high speed counter and it is called HSCI. The RA4 and RA7 pins are used as PWM (pulse width modulation) outputs. Therefore, they are called PWM_RA4 and PWM_RA7, respectively. The RA6 pin is not used. The PIC16F1847 provides the following—flash program memory (words): 8K; SRAM data memory (bytes): 1,024; and EEPROM data memory (bytes): 256. The PIC16F1847-Based PLC macros make use of registers defined in SRAM data memory.

Figures 1.3 and 1.4 show the schematic diagram and the photograph of the *I/O extension board*, respectively. The I/O extension board contains mainly *two sections*: 8 digital inputs and 8 digital outputs. The I/O extension connector DB9M, seen on the left, connects the I/O extension board to the CPU board or to a previous

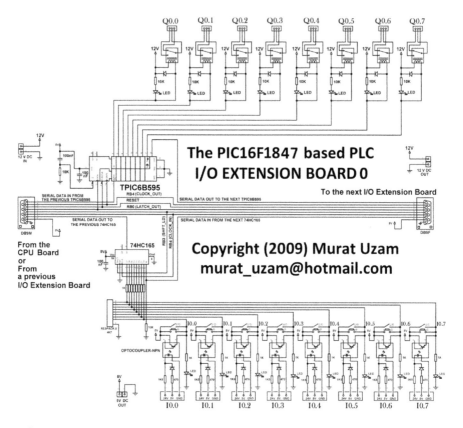

FIGURE 1.3 Schematic diagram of the I/O extension board.

I/O extension board. Similarly, the I/O extension connector DB9F, seen on the right, connects the I/O extension board to a next I/O extension board. In this way we can connect as many I/O extension boards as necessary. 5V DC and 12V DC are taken from the CPU board or from a previous I/O extension board and they are passed to the next I/O extension boards. All I/O data are sent to and taken from all the connected extension I/O boards by means of I/O extension connectors DB9M and DB9F.

The **inputs section** of each I/O extension board introduces 8 digital inputs for the PIC16F1847-Based PLC (called I0.0, I0.1, ..., I0.7 for the first I/O extension board, called I1.0, I1.1, ..., I1.7 for the second I/O extension board, called I2.0, I2.1, ..., I2.7 for the third I/O extension board, and called I3.0, I3.1, ..., I3.7 for the fourth and last I/O extension board). 5V DC or 24V DC input signals can be accepted by each input. These external input signals are isolated from the other parts of the hardware by using NPN-type optocouplers (e.g., 4N25). For simulating input signals, one can use on-board push buttons as temporary inputs and slide switches as permanent inputs. In the beginning of each PLC scan cycle (get_inputs), the 74HC/LS165 of each I/O extension board is loaded (RB3 [shift/load] = 0) with the level of 8 inputs, and then these data are serially clocked in (when RB3 = 1, through the RB1 "data in" and RB4

FIGURE 1.4   Photograph of the I/O extension board.

"clock in" pins). If there is only one I/O extension board used, then 8 clock_in signals are enough to get the 8 input signals. For each additional I/O extension board, 8 more clock_in signals are necessary. The serial data coming from the I/O extension board(s) are taken from the "SI" input of the 74HC/LS165.

The **outputs section** of each I/O extension board introduces 8 discrete relay outputs for the PIC16F1847-Based PLC (called Q0.0, Q0.1, ..., Q0.7 for the first I/O extension board, called Q1.0, Q1.1, ..., Q1.7 for the second I/O extension board, called Q2.0, Q2.1, ..., Q0.7 for the third I/O extension board, and called Q3.0, Q3.1, ..., Q3.7 for the fourth and last I/O extension board). Each relay operates with 12V DC and driven by an 8-bit serial-to-parallel converter register/driver TPIC6B595. Relays have SPDT (single pole double throw) contacts with C (common), NC (normally closed), and NO (normally open) terminals. At the end of each PLC scan cycle (send_outputs), the output data are serially clocked out (through the RB4 "clock out" and RB2 "data out" pins) and finally latched within the TPIC6B595. If there is only one I/O extension board used, then 8 clock_out signals are enough to send the 8 output signals. For each additional I/O extension board, 8 more clock_out signals are necessary. The serial data going to the I/O extension board(s) are sent out from the "SER OUT" (pin 18) of the TPIC6B595.

The PCB Gerber files of both the CPU board and the I/O extension board are downloadable from this book's webpage under the downloads section. Note that in the PCB design of the CPU board and the I/O extension board, some lines of I/O extension connectors DB9M and DB9F are different from the ones shown in Figures 1.1 and 1.3.

The project reported in this book makes use of a CPU board and four I/O extension boards. Thus, in total there are 32 digital inputs and 32 digital outputs. Figure 1.5 shows the PIC16F1847-Based PLC consisting of a CPU board, four I/O extension boards, a 12V DC adapter, and a PICkit 3 PIC programmer.

In addition to the CPU board and I/O extension boards, in this section let us briefly consider some additional input and output modules to be used with the PIC16F1847-Based PLC, as shown in Figure 1.6. The following is the list of these additional input and output modules:

1. Analog input modules
2. Analog output modules
3. RC low-pass filters module
4. 5.00V voltage reference module
5. Voltage regulator module

Analog input modules designed within this project are as follows:

1. 0V to 5V Analog Input Module 1
2. 0V to 5V Analog Input Module 2

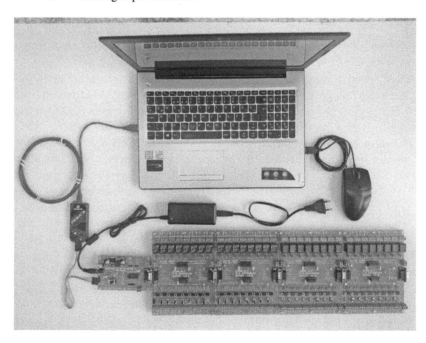

FIGURE 1.5 Photograph of the CPU board plus four I/O extension boards and a PICkit 3 PIC programmer.

FIGURE 1.6 Photograph of the CPU board together with 13 analog input modules and 7 analog output modules.

3. 0V to 5V Analog Input Module 3
4. 0V to 5V Analog Input Module 4
5. 0V to 5V Analog Input Module 5
6. 0–10V to 0–5V Signal Converter—Analog Input Module 1
7. 0–10V to 0–5V Signal Converter—Analog Input Module 2
8. –5V – +5V to 0–5V Signal Converter—Analog Input Module 1
9. –5V – +5V to 0–5V Signal Converter—Analog Input Module 2
10. –10V – +10V to 0–5V Signal Converter—Analog Input Module 1
11. –10V – +10V to 0–5V Signal Converter—Analog Input Module 2
12. 0–5V or 4–20mA to 0–5V Signal Converter—Analog Input Module 1
13. 0–5V or 4–20mA to 0–5V Signal Converter—Analog Input Module 2

Analog output modules designed within this project are as follows:

1. 0V to 5V Analog Output Module
2. 0–5V to 0–10V Signal Converter—Analog Output Module
3. 0–5V to –5V – +5V Signal Converter—Analog Output Module 1
4. 0–5V to –5V – +5V Signal Converter—Analog Output Module 2

5. 0–5V to –10V – +10V Signal Converter—Analog Output Module 1
6. 0–5V to –10V – +10V Signal Converter—Analog Output Module 2
7. 0–5V to 4–20mA Signal Converter—Analog Output Module

These analog input and analog output modules are explained in detail in Chapter 6 of the *Advanced Concepts* book.

An RC low-pass filter is a filter circuit, composed of a resistor and a capacitor, which passes low-frequency signals and blocks high-frequency signals. When a resistor is placed in series with the power source and a capacitor is placed parallel to that same power source, this type of circuit forms a low-pass filter. Figure 1.7 depicts the schematic diagram of RC low-pass filters constructed for analog inputs AI0, AI1, AI2, and AI3, with the cut-off frequency of 48Hz.

An external 5.00V voltage reference is necessary to be used with the analog-to-digital converter (ADC) module and the digital-to-analog converter (DAC) module of the PIC16F1847. To satisfy this requirement, a low-cost solution is obtained by using the REF02 voltage reference from Analog Devices. Figure 1.8(a) shows the schematic diagram of the 5.00V voltage reference REF02 with a trim adjustment circuit consisting of R1, R2, and POT, while Figure 1.8(b) depicts the photograph of the 5.00V voltage reference module.

In analog input modules and analog output modules (see Chapter 6 of the *Advanced Concepts* book) +5.00V and +6.26V power supplies are necessary, and in the DC motor control examples with an L298N dual full-bridge driver (see "Application Examples"), a +6.00V power supply is necessary. As considered before, LM2596 step-down voltage regulators can be used to obtain these DC voltages from the 12V DC input voltage. To address this need, a voltage regulator module is designed. Figure 1.9(a) shows the schematic diagram of the voltage regulator module, consisting of three LM2596 step-down voltage regulators, while Figure 1.9(b) depicts the photograph of the voltage regulator module. By using this voltage regulator module,

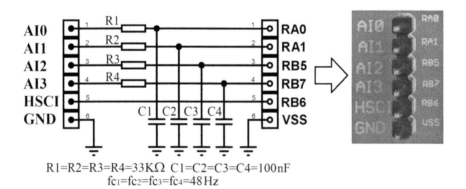

$$R1=R2=R3=R4=33K\Omega \quad C1=C2=C3=C4=100nF$$
$$fc_1=fc_2=fc_3=fc_4=48\,Hz$$

FIGURE 1.7 Schematic diagram of RC low-pass filters for analog inputs AI0, AI1, AI2, and AI3.

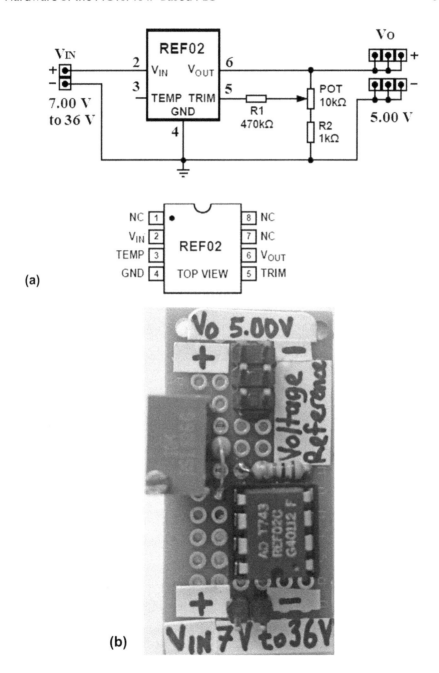

(a)

(b)

FIGURE 1.8 (a) Schematic diagram of the 5.00V voltage reference REF02 with a trim adjustment circuit consisting of R1, R2, and POT; (b) Photograph of the 5.00V voltage reference module.

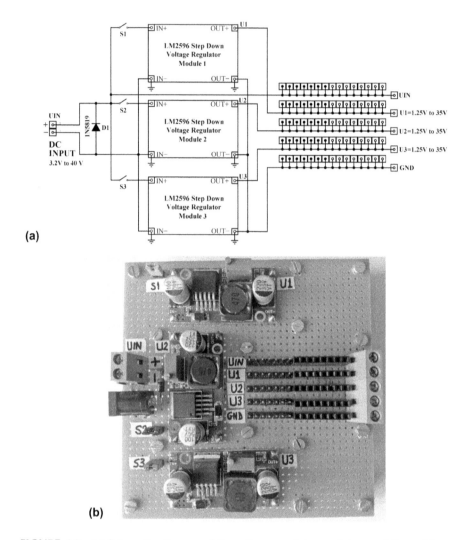

FIGURE 1.9 (a) Schematic diagram of the voltage regulator module, consisting of three LM2596 step-down voltage regulators; (b) Photograph of the voltage regulator module.

three independent voltage values can be adjusted and used. D1 is used to make sure that the polarity of the DC input voltage is correct. Switches S1, S2, and S3 (implemented by using jumpers) are used to turn on or off the LM2596S voltage regulators 1, 2, and 3, respectively.

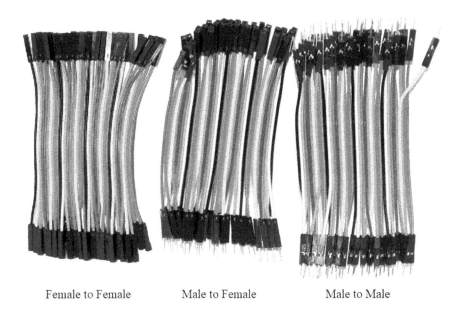

Female to Female   Male to Female   Male to Male

FIGURE 1.10   Three types of Dupont cables used in the project described in this book.

Last but not least, in order to connect the above-mentioned input and output modules with the PIC16F1847-Based PLC input/output terminals, it is necessary to use some cables. For this purpose, three types of Dupont cables, shown in Figure 1.10, are used.

2 Basic Software

INTRODUCTION

In this chapter, the basic software of the PIC16F1847-Based PLC is explained. A PLC scan cycle includes the following: obtain the inputs, run the user program, and update the outputs. It is also necessary to define and initialize all variables used within a PLC. Necessary functions are all described as PIC Assembly macros to be used in the PIC16F1847-Based PLC. As can be seen from Figure 2.1, the source files and their macros developed in the PICPLC_PIC16F1847 project file are as follows:

1. PICPLC_PIC16F1847_memory.inc
2. PICPLC_PIC16F1847_main.asm
3. PICPLC_PIC16F1847_user_Bsc.inc
4. PICPLC_PIC16F1847_subr.inc
5. PICPLC_PIC16F1847_macros_Bsc.inc
    5.1 initialize (for PLC initialization)
    5.2 ISR (interrupt service routines)
    5.3 get_inputs (for handling the inputs)
    5.4 lpf_progs (low-pass digital filter macros for analog inputs)
    5.5 send_outputs (for sending the outputs)

The basic software of the PIC16F1847-Based PLC makes use of general-purpose 8-bit registers (GPR) of SRAM data memory of the PIC16F1847 microcontroller. 1,024 SRAM bytes of PIC16F1847 are allocated in 13 banks, namely Bank0, Bank1, ..., Bank12. In this PLC project, 695 SRAM bytes are defined and reserved to be used within the PLC functions. GPRs in banks Bank0, Bank1, Bank2, and Bank3 are intentionally left unused for general use. Thus there are 329 GPRs ready to be used. The directory called "PICPLC_PIC16F1847_Bsc", downloadable from this book's webpage under the downloads section, contains all project files, macros, definitions, and examples necessary for the PIC16F1847-Based PLC project explained in this book (*Hardware and Basic Concepts*).

Note that files "PICPLC_PIC16F1847_macros_Bsc.inc" and "PICPLC_PIC16F1847_user_Bsc.inc" refer to the macros and user program files of the basic concepts developed in the PIC16F1847-Based PLC project, respectively. They do not contain files related to the intermediate and advanced concepts. These files are intended for the readers who purchased this book as a standalone book. On the other hand, when this book is purchased as a part of the set of three books, all project files including basic, intermediate, and advanced concepts are put in the same directory and the reader is entitled to download and use the whole of the project files in one directory, the name of which becomes "PICPLC_PIC16F1847" instead of "PICPLC_PIC16F1847_Bsc". Therefore, in the second case, the name of the file

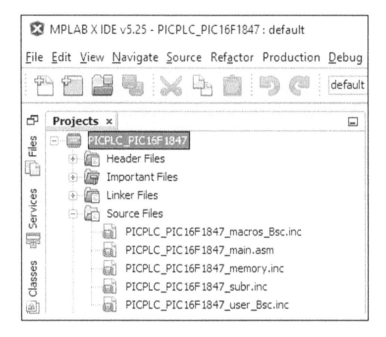

FIGURE 2.1   Screenshot of the "PICPLC_PIC16F1847" project, showing the five source files developed and used in the project.

"PICPLC_PIC16F1847_macros_Bsc.inc"   (and   PICPLC_PIC16F1847_user_Bsc .inc, respectively) becomes "PICPLC_PIC16F1847_macros.inc" (and PICPLC_ PIC16F1847_user.inc, respectively).

In this section the contents of the source files depicted in Figure 2.1 are explained. In addition, the concept of a "contact bouncing" problem and how it is solved in the PIC16F1847-Based PLC are explained in detail.

2.1   DEFINITION AND ALLOCATION OF VARIABLES

The definitions of all 8-bit variables to be used for the PIC16F1847-Based PLC project and their allocation in SRAM data memory are shown in Figures 2.2 and 2.3, respectively. These definitions are placed in the "PICPLC_PIC16F1847_macros _Bsc.inc" file. Although detailed explanations for these variables are provided in the related sections of this book, let us now briefly consider these 8-bit variables. In this project, we define four 8-bit registers (I0, I1, I2, and I3) to hold the debounced state of physical digital input registers (74HC/LS165) and four 8-bit registers (Q0, Q1, Q2, and Q3) to hold the state of physical digital output registers. Temp_1 and Temp_2 are general temporary registers declared to be used in some macros. SMB1 is declared to be used for obtaining special memory bits. SMB2 is declared to be used for obtaining reference timing signals.

It is well known that digital inputs taken from contacts always suffer from "contact bouncing". To circumvent this problem, we define a "debouncing" mechanism for the digital inputs, and this will be explained later. In the "get_inputs" stage of the

```
;---------------------------------------------------------------------
;       VARIABLE DEFINITIONS                                         ;
;---------------------------------------------------------------------
;------------------ beginning of BANK0 -------------------------------
        cblock 0x020    ; There are 80 8-bit GPRs available in BANK0.

        endc            ;
;------------------ end of BANK0 -------------------------------------
;------------------ beginning of common RAM memory ------------------
        cblock 0x70
        I0, I1, I2, I3, Q0, Q1, Q2, Q3, Temp_1, Temp_2, SMB1, SMB2
        endc
;------------------ end of common RAM memory ------------------------

;------------------ beginning of BANK1 ------------------------------
        cblock 0x0A0    ; There are 80 8-bit GPRs available in BANK1.

        endc            ;
;------------------ end of BANK1 ------------------------------------
;------------------ beginning of BANK2 ------------------------------
        cblock 0x120    ; There are 80 8-bit GPRs available in BANK2.

        endc            ;
;------------------ end of BANK2 ------------------------------------
;------------------ beginning of BANK3 ------------------------------
        cblock 0x1A0    ;There are 80 8-bit GPRs available in BANK3.

        endc            ;
;------------------ end of BANK3 ------------------------------------
;------------------ beginning of BANK4 ------------------------------
        cblock 0x220    ; 80 8-bit-variables are defined to hold
        TV_L            ; low byte timing values
        endc            ; TV_L, TV_L+1, ..., TV_L+79
;------------------ end of BANK4 ------------------------------------
;------------------ beginning of BANK5 ------------------------------
        cblock 0x2A0    ; 80 8-bit-variables are defined to hold
        TV_H            ; high byte timing values
        endc            ; TV_H, TV_H+1, ..., TV_H+79
;------------------ end of BANK5 ------------------------------------
;
;------------------ beginning of BANK6 ------------------------------
        cblock 0x320    ; 80 8-bit-variables are defined to hold
        CV_L            ; low byte count values
        endc            ; CV_L, CV_L+1, ..., CV_L+79
;------------------ end of BANK6 ------------------------------------
;
;------------------ beginning of BANK7 ------------------------------
        cblock 0x3A0    ; 80 8-bit-variables are defined to hold
        CV_H            ; high byte count values
        endc            ; CV_H, CV_H+1, ..., CV_H+79
;------------------ end of BANK7 ------------------------------------
```

FIGURE 2.2 (*1 of 5*) Definition of 8-bit variables.

PLC scan cycle, digital input signals are serially taken from the related 74HC/LS165 registers and stored in the SRAM registers. As a result, bI0, bI1, bI2, and bI3 will hold these bouncing digital input signals. After applying the debouncing mechanism to the bouncing digital input signals bI0, bI1, bI2, and bI3, we obtain "debounced" input signals and they are stored in SRAM registers I0, I1, I2, and I3 respectively. In the "send_outputs" stage of the PLC scan cycle, the output information stored in

```
;-------------------- beginning of BANK8 --------------------------------
        cblock 0x420   ; 8 Memory bytes, 8x8=64 Memory bits (Internal Relays)
        M0, M1, M2, M3, M4, M5, M6, M7
        endc          ;
        cblock 0x428   ; 8 Memory bytes, 8x8=64 Memory bits (Internal Relays)
        M8, M9, M10, M11, M12, M13, M14, M15
        endc          ;
        cblock 0x430   ; 8 Memory bytes, 8x8=64 Memory bits (Internal Relays)
        M16, M17, M18, M19, M20, M21, M22, M23
        endc          ;
        cblock 0x438   ; 8 Memory bytes, 8x8=64 Memory bits (Internal Relays)
        M24, M25, M26, M27, M28, M29, M30, M31
        endc          ;
        cblock 0x440   ; 8 Memory bytes, 8x8=64 Memory bits (Internal Relays)
        M32, M33, M34, M35, M36, M37, M38, M39
        endc          ;
        cblock 0x448   ; 8 Memory bytes, 8x8=64 Memory bits (Internal Relays)
        M40, M41, M42, M43, M44, M45, M46, M47
        endc          ;
        cblock 0x450   ; 8 Memory bytes, 8x8=64 Memory bits (Internal Relays)
        M48, M49, M50, M51, M52, M53, M54, M55
        endc          ;
        cblock 0x458   ; 8 Memory bytes, 8x8=64 Memory bits (Internal Relays)
        M56, M57, M58, M59, M60, M61, M62, M63
        endc          ;
        cblock 0x460   ; 8 Memory bytes, 8x8=64 Memory bits (Internal Relays)
        M64, M65, M66, M67, M68, M69, M70, M71
        endc          ;
        cblock 0x468   ; 8 Memory bytes, 8x8=64 Memory bits (Internal Relays)
        M72, M73, M74, M75, M76, M77, M78, M79
        endc             ; In BANK8 80 Memory bytes (640 Memory bits) are defined
;-------------------- end of BANK8 -----------------------------------------
```

FIGURE 2.2   Continued

the 8-bit SRAM registers Q0, Q1, Q2, and Q3 is serially sent out to and stored in the related TPIC6B595 registers. This means that the Q0, Q1, Q2, and Q3 registers will hold output information and their contents will be copied into the TPIC6B595 registers at the end of each PLC scan cycle.

160 8-bit registers, namely TV_L, TV_L+1, ..., TV_L+79 and TV_H, TV_H+1, ..., TV_H+79, are defined to be used in timer macros (see Chapter 5 of this book) for holding current timing values of timers. Ten 8-bit registers, namely T_Q0, T_Q1, ..., T_Q9 are defined to be used in timer macros for holding timer status bits (timer outputs). 160 8-bit registers, namely CV_L, CV_L+1, ..., CV_L+79 and CV_H, CV_H+1, ..., CV_H+79, are defined to be used in counter macros (see Chapter 6 of this book) for holding current count values of counters. 20 8-bit registers, namely C_Q0, C_Q1, ..., C_Q9 and C_QD0, C_QD1, ..., C_QD9, are defined to be used in counter macros for holding counter status bits (counter outputs). 128 8-bit registers, namely M0, M1, ..., M127, are defined for obtaining 1,024 memory bits (internal relays, in PLC jargon). The following 43 8-bit registers are defined to be used in drum sequencer instruction: drum_TVL, drum_TVL+1, ..., drum_TVL+15, drum_ TVH, drum_TVH+1, ..., drum_TVH+15, drum_TQL, drum_TQH, drum_stepsL, drum_stepsH, drum_eventsL, drum_eventsH, drum_QL, drum_QH, drum_tmp, drum_tmpL, and drum_tmpH. The following 54 8-bit registers are defined to be

```
          ;
          ;--------------- beginning of BANK9 ----------------------------------
          cblock 0x4A0   ; 8 Memory bytes,  8x8=64 Memory bits (Internal Relays)
          M80, M81, M82, M83, M84, M85, M86, M87
          endc           ;
          cblock 0x4A8   ; 8 Memory bytes,  8x8=64 Memory bits (Internal Relays)
          M88, M89, M90, M91, M92, M93, M94, M95
          endc           ;
          cblock 0x4B0   ; 8 Memory bytes,  8x8=64 Memory bits (Internal Relays)
          M96, M97, M98, M99, M100, M101, M102, M103
          endc           ;
          cblock 0x4B8   ; 8 Memory bytes,  8x8=64 Memory bits (Internal Relays)
          M104, M105, M106, M107, M108, M109, M110, M111
          endc           ;
          cblock 0x4C0   ; 8 Memory bytes,  8x8=64 Memory bits (Internal Relays)
          M112, M113, M114, M115, M116, M117, M118, M119
          endc           ;
          cblock 0x4C8   ; 8 Memory bytes,  8x8=64 Memory bits (Internal Relays)
          M120, M121, M122, M123, M124, M125, M126, M127
          endc           ; In BANK9 48 Memory bytes (384 Memory bits) are defined.
          ;-------------------------------------------------------------
          ; In BANK8 and BANK9, 128 Memory bytes (1024 Memory bits) are defined.
          ; M0, M1, ..., M127
          ;-------------------------------------------------------------
          cblock 0x4D0   ; Timer status registers
          T_Q0, T_Q1, T_Q2, T_Q3, T_Q4, T_Q5, T_Q6, T_Q7, T_Q8, T_Q9
          endc
          cblock 0x4DA   ; Counter status registers:
          C_Q0, C_Q1, C_Q2, C_Q3, C_Q4, C_Q5, C_Q6, C_Q7, C_Q8, C_Q9
          endc
          cblock 0x4E4   ; Down Counter status registers:
          C_QD0,C_QD1,C_QD2,C_QD3,C_QD4,C_QD5,C_QD6,C_QD7,C_QD8,C_QD9
          endc           ;
          cblock 0x4EE   ;
          ; --------- 2 RAM locations in BANK9: 4EEh & 4EFh are not used
          endc           ;
          ;--------------- end of BANK9 ----------------------------------
```

FIGURE 2.2   Continued

used in SFC (sequential function charts)-related macros (see Chapter 5 of *Advanced Concepts*): step_1.TL, step_1.TL+1, …, step_1.TL+24, step_1.TH, step_1.TH+1, …, step_1.TH+24, SF0, SF1, SF2, MB0, MB1, and MB2. 40 8-bit registers, namely LPF, LPF+1, …, LPF+39, are defined to be used in low-pass digital filter macros for holding current timing values of low-pass digital filters. The following eight 8-bit registers hold four 10-bit noisy digital values for 4 analog inputs: nAI0L, nAI0H, nAI1L, nAI1H, nAI2L, nAI2H, nAI3L, and nAI3H. The following eight 8-bit registers hold four 10-bit filtered digital values for 4 analog inputs: AI0L, AI0H, AI1L, AI1H, AI2L, AI2H, AI3L, and AI3H. Registers HSC_B2 and HSC_B3 are defined to be used in the HSC_RB6 macro (see Chapter 2 of *Advanced Concepts*) to hold the most significant two bytes of 32-bit count values. 32 8-bit registers, namely DBNCR, DBNCR+1, …, DBNCR+31, are defined to be used in the debouncer macro "dbncrN" for holding current timing values of debouncer macros. 8-bit registers CNT1, CNT2, and CNT3 are defined to be used in the "ISR" macro in order to obtain reference timing signals T_2ms, T_10ms, T_100ms, and T_1s. 8-bit registers TenK, Thou, Hund, Tens, and Ones are defined to be used in the following macros: "Conv_UInt_2_BCD_P",

```
;-------------------- beginning of BANK10 ---------------------------------------
cblock 0x520   ; 16 8-bit-variables are defined for d_TON16
drum_TVL       ; to hold low byte timing values
endc           ; drum_TVL, drum_TVL +1, ..., drum_TVL +15
cblock 0x530   ; 16 8-bit-variables are defined for d_TON16
drum_TVH       ; to hold high byte timing values
endc           ; drum_TVH, drum_TVH +1, ..., drum_TVH +15
cblock 0x540   ;16 Status bits for 16 d_TON16
drum_TQL,drum_TQH
endc
cblock 0x542   ;16 Steps for Drum Sequencer Instruction
drum_stepsL,drum_stepsH
endc
cblock 0x544   ;16 drum events for Drum Sequencer Instruction
drum_eventsL,drum_eventsH
endc
cblock 0x546
drum_QL,drum_QH;16 final drum outputs
endc
cblock 0x548   ;These 3 registers are used in Drum Sequencer Instruction.
drum_tmp,drum_tmpL,drum_tmpH
endc
cblock 0x54B   ;
SF0,SF1,SF2    ;24 step flags defined for SFC
endc           ;
cblock 0x54E   ;
MB0,MB1,MB2    ;24 Memory bits defined for SFC
endc           ;
cblock 0x551   ;24 Memory words defined for SFC to be used in elapsed times
step_1.TL,step_1.TH,step_2.TL,step_2.TH,step_3.TL,step_3.TH,step_4.TL,step_4.TH
endc           ;
cblock 0x559   ;24 Memory words defined for SFC to be used in elapsed times
step_5.TL,step_5.TH,step_6.TL,step_6.TH,step_7.TL,step_7.TH,step_8.TL,step_8.TH
endc           ;
cblock 0x561   ;24 Memory words defined for SFC to be used in elapsed times
step_9.TL,step_9.TH,step_10.TL,step_10.TH,step_11.TL,step_11.TH,step_12.TL,step_12.TH
endc           ;
cblock 0x569   ;24 Memory words defined for SFC to be used in elapsed times
step_13.TL,step_13.TH,step_14.TL,step_14.TH,step_15.TL,step_15.TH,step_16.TL
endc           ;
;------------------------------------------------------------------------
;-------------------- end of BANK10 -----------------------------------------
;
```

FIGURE 2.2   Continued

"Conv_BCD_U_2_Uint", "Conv_BCD_P_2_Uint", "Conv_UsInt_2_BCD_U", and "Conv_UsInt_2_BCD_P". The 8-bit register "STP_bits" is defined to be used in the PWM macros and the HSC macro. 8-bit registers i, j, and k are defined to be used in the selection macros (see Chapter 2 of *Advanced Concepts*).

The individual bits (1-bit variables) of 8-bit SRAM registers M0, M1, M2, ..., M127 are all considered in the next section. The definitions of 1-bit (Boolean) variables are placed in the "PICPLC_PIC16F1847_macros_Bsc.inc" file. The definitions of 32 bouncing digital input signals bI0.0, bI0.1, ..., bI3.7 by using all bits of 8-bit SRAM registers bI0, bI1, bI2, and bI3 are shown in Figure 2.4.

The allocation of individual bits (1-bit variables) of 8-bit SRAM registers bI0, bI1, bI2, and bI3 is shown in Table 2.1.

```
;--------------------- beginning of BANK11 ----------------------------------------------
;--------------------- LPF Variables are in BANK11 ----------------------------------
        cblock 0x5A0   ; 40 8-bit-variables are defined for low pass digital filters
        LPF            ; LPF, LPF+1, ..., LPF+39
        endc           ;
        cblock 0x5C8   ; 4 noisy Digital Values for 4 Analog inputs
        nAI0L, nAI0H, nAI1L, nAI1H, nAI2L, nAI2H, nAI3L, nAI3H
        endc           ; are stored in these registers
        cblock 0x5D0   ; Filtered Digital Values for 4 Analog inputs
        AI0L, AI0H, AI1L, AI1H, AI2L, AI2H, AI3L, AI3H
        endc           ; are stored in these registers
        cblock 0x5D8   ;
        step_16.TH     ;
        endc           ;
        cblock 0x5D9   ;24 Memory words defined for SFC to be used in elapsed times
        step_17.TL,step_17.TH,step_18.TL,step_18.TH,step_19.TL,step_19.TH,step_20.TL,step_20.TH
        endc           ;
        cblock 0x5E1   ;24 Memory words defined for SFC to be used in elapsed times
        step_21.TL,step_21.TH,step_22.TL,step_22.TH,step_23.TL,step_23.TH,step_24.TL,step_24.TH
        endc           ;
        cblock 0x5E9   ; HSC_B2 and HSC_B3 registers are used in the HSC_RB6 macro.
        HSC_B2,HSC_B3
        endc           ;
        cblock 0x5EB   ;
                       ;5 Bytes are available.
        endc           ;
;--------------------- end of BANK11 ----------------------------------------------
;
;--------------------- beginning of BANK12 ----------------------------------------------
;--------------------- Debouncer Variables are in BANK12 -----------------------
        cblock 0x620        ; 32 8-bit-variables are defined to hold timing values
        DBNCR               ; DBNCR, DBNCR+1, ..., DBNCR+31
        endc                ;
        cblock 0x640        ; 32 bouncing digital inputs are stored in these four registers
        bI0, bI1, bI2, bI3  ; bI0.0, bI0.1, ..., bI3.7
        endc                ;
        cblock 0x644        ;
        CNT1, CNT2, CNT3    ;These 3 registers are used in the ISR.
        endc                ;
        cblock 0x647        ;These five temporary registers are defined for
        TenK,Thou,Hund,Tens,Ones;"Conv_UInt_2_BCD_P", "Conv_BCD_U_2_UInt"
        endc                ;"Conv_BCD_P_2_UInt", "Conv_UsInt_2_BCD_U",
                            ;and "Conv_UsInt_2_BCD_P" macros.
        cblock 0x64C        ;
        STP_bits            ;This setup register is used in
        endc                ;PWM macros and HSC_RB6 macro.
        cblock 0x64D        ;
        i,j,k               ;These registers are used in the selection macros.
        endc                ;
;--------------------- end of BANK12 ----------------------------------------------
```

FIGURE 2.2   Continued

The definitions of 32 debounced digital input signals I0.0, I0.1, …, I3.7 by using all bits of 8-bit SRAM registers I0, I1, I2, and I3 are shown in Figure 2.5.

The allocation of individual bits (1-bit variables) of 8-bit SRAM registers I0, I1, I2, and I3 is shown in Table 2.2.

The definitions of 32 digital output signals Q0.0, Q0.1, …, Q3.7 by using all bits of 8-bit SRAM registers Q0, Q1, Q2, and Q3 are shown in Figure 2.6.

Addr		Addr		Addr		Addr	
020h		050h		0A0h		0D0h	
021h		051h		0A1h		0D1h	
022h		052h		0A2h		0D2h	
023h		053h		0A3h		0D3h	
024h		054h		0A4h		0D4h	
025h		055h		0A5h		0D5h	
026h		056h		0A6h		0D6h	
027h		057h		0A7h		0D7h	
028h		058h		0A8h		0D8h	
029h		059h		0A9h		0D9h	
02Ah		05Ah		0AAh		0DAh	
02Bh		05Bh		0ABh		0DBh	
02Ch		05Ch		0ACh		0DCh	
02Dh		05Dh		0ADh		0DDh	
02Eh		05Eh		0AEh		0DEh	
02Fh		05Fh		0AFh		0DFh	
030h		060h		0B0h		0E0h	
031h		061h		0B1h		0E1h	
032h		062h		0B2h		0E2h	
033h		063h		0B3h		0E3h	
034h		064h		0B4h		0E4h	
035h		065h		0B5h		0E5h	
036h		066h		0B6h		0E6h	
037h		067h		0B7h		0E7h	
038h		068h		0B8h		0E8h	
039h		069h		0B9h		0E9h	
03Ah		06Ah		0BAh		0EAh	
03Bh		06Bh		0BBh		0EBh	
03Ch		06Ch		0BCh		0ECh	
03Dh		06Dh		0BDh		0EDh	
03Eh		06Eh		0BEh		0EEh	
03Fh		06Fh		0BFh		0EFh	
040h		070h	I0	0C0h		0F0h	I0
041h		071h	I1	0C1h		0F1h	I1
042h		072h	I2	0C2h		0F2h	I2
043h		073h	I3	0C3h		0F3h	I3
044h		074h	Q0	0C4h		0F4h	Q0
045h		075h	Q1	0C5h		0F5h	Q1
046h		076h	Q2	0C6h		0F6h	Q2
047h		077h	Q3	0C7h		0F7h	Q3
048h		078h	Temp_1	0C8h		0F8h	Temp_1
049h		079h	Temp_2	0C9h		0F9h	Temp_2
04Ah		07Ah	SMB1	0CAh		0FAh	SMB1
04Bh		07Bh	SMB2	0CBh		0FBh	SMB2
04Ch		07Ch		0CCh		0FCh	
04Dh		07Dh		0CDh		0FDh	
04Eh		07Eh		0CEh		0FEh	
04Fh		07Fh		0CFh		0FFh	
	Bank 0				Bank 1		

FIGURE 2.3 (*1 of 7*) Allocation of 8-bit variables in SRAM data memory.

The allocation of individual bits (1-bit variables) of 8-bit SRAM registers Q0, Q1, Q2, and Q3 is shown in Table 2.3.

The definitions of special memory bits and for 74HC165 and TPIC6B595 ICs are depicted in Figure 2.7(a) and (b), respectively. Tables 2.4 and 2.5 show the allocation of individual bits of the SMB1 register and SMB2 register, respectively.

Addr	Value	Addr	Value	Addr	Value	Addr	Value
120h		150h		1A0h		1D0h	
121h		151h		1A1h		1D1h	
122h		152h		1A2h		1D2h	
123h		153h		1A3h		1D3h	
124h		154h		1A4h		1D4h	
125h		155h		1A5h		1D5h	
126h		156h		1A6h		1D6h	
127h		157h		1A7h		1D7h	
128h		158h		1A8h		1D8h	
129h		159h		1A9h		1D9h	
12Ah		15Ah		1AAh		1DAh	
12Bh		15Bh		1ABh		1DBh	
12Ch		15Ch		1ACh		1DCh	
12Dh		15Dh		1ADh		1DDh	
12Eh		15Eh		1AEh		1DEh	
12Fh		15Fh		1AFh		1DFh	
130h		160h		1B0h		1E0h	
131h		161h		1B1h		1E1h	
132h		162h		1B2h		1E2h	
133h		163h		1B3h		1E3h	
134h		164h		1B4h		1E4h	
135h		165h		1B5h		1E5h	
136h		166h		1B6h		1E6h	
137h		167h		1B7h		1E7h	
138h		168h		1B8h		1E8h	
139h		169h		1B9h		1E9h	
13Ah		16Ah		1BAh		1EAh	
13Bh		16Bh		1BBh		1EBh	
13Ch		16Ch		1BCh		1ECh	
13Dh		16Dh		1BDh		1EDh	
13Eh		16Eh		1BEh		1EEh	
13Fh		16Fh		1BFh		1EFh	
140h		**170h**	**I0**	1C0h		**1F0h**	**I0**
141h		**171h**	**I1**	1C1h		**1F1h**	**I1**
142h		**172h**	**I2**	1C2h		**1F2h**	**I2**
143h		**173h**	**I3**	1C3h		**1F3h**	**I3**
144h		**174h**	**Q0**	1C4h		**1F4h**	**Q0**
145h		**175h**	**Q1**	1C5h		**1F5h**	**Q1**
146h		**176h**	**Q2**	1C6h		**1F6h**	**Q2**
147h		**177h**	**Q3**	1C7h		**1F7h**	**Q3**
148h		**178h**	**Temp_1**	1C8h		**1F8h**	**Temp_1**
149h		**179h**	**Temp_2**	1C9h		**1F9h**	**Temp_2**
14Ah		**17Ah**	**SMB1**	1CAh		**1FAh**	**SMB1**
14Bh		**17Bh**	**SMB2**	1CBh		**1FBh**	**SMB2**
14Ch		**17Ch**		1CCh		**1FCh**	
14Dh		**17Dh**		1CDh		**1FDh**	
14Eh		**17Eh**		1CEh		**1FEh**	
14Fh		**17Fh**		1CFh		**1FFh**	

Bank 2	Bank 3

FIGURE 2.3   Continued

The variable "LOGIC0" is defined to hold a logic "0" value throughout the PLC operation. At the initialization stage it is deposited with this value. Similarly, the variable "LOGIC1" is defined to hold a logic "1" value throughout the PLC operation. At the initialization stage it is deposited with this value. The special memory bit "FRSTSCN" is arranged to hold the value of "1" at the first PLC scan cycle only.

Addr	Value	Addr	Value	Addr	Value	Addr	Value
220h	TV_L	250h	TV_L+48	2A0h	TV_H	2D0h	TV_H+48
221h	TV_L+1	251h	TV_L+49	2A1h	TV_H+1	2D1h	TV_H+49
222h	TV_L+2	252h	TV_L+50	2A2h	TV_H+2	2D2h	TV_H+50
223h	TV_L+3	253h	TV_L+51	2A3h	TV_H+3	2D3h	TV_H+51
224h	TV_L+4	254h	TV_L+52	2A4h	TV_H+4	2D4h	TV_H+52
225h	TV_L+5	255h	TV_L+53	2A5h	TV_H+5	2D5h	TV_H+53
226h	TV_L+6	256h	TV_L+54	2A6h	TV_H+6	2D6h	TV_H+54
227h	TV_L+7	257h	TV_L+55	2A7h	TV_H+7	2D7h	TV_H+55
228h	TV_L+8	258h	TV_L+56	2A8h	TV_H+8	2D8h	TV_H+56
229h	TV_L+9	259h	TV_L+57	2A9h	TV_H+9	2D9h	TV_H+57
22Ah	TV_L+10	25Ah	TV_L+58	2AAh	TV_H+10	2DAh	TV_H+58
22Bh	TV_L+11	25Bh	TV_L+59	2ABh	TV_H+11	2DBh	TV_H+59
22Ch	TV_L+12	25Ch	TV_L+60	2ACh	TV_H+12	2DCh	TV_H+60
22Dh	TV_L+13	25Dh	TV_L+61	2ADh	TV_H+13	2DDh	TV_H+61
22Eh	TV_L+14	25Eh	TV_L+62	2AEh	TV_H+14	2DEh	TV_H+62
22Fh	TV_L+15	25Fh	TV_L+63	2AFh	TV_H+15	2DFh	TV_H+63
230h	TV_L+16	260h	TV_L+64	2B0h	TV_H+16	2E0h	TV_H+64
231h	TV_L+17	261h	TV_L+65	2B1h	TV_H+17	2E1h	TV_H+65
232h	TV_L+18	262h	TV_L+66	2B2h	TV_H+18	2E2h	TV_H+66
233h	TV_L+19	263h	TV_L+67	2B3h	TV_H+19	2E3h	TV_H+67
234h	TV_L+20	264h	TV_L+68	2B4h	TV_H+20	2E4h	TV_H+68
235h	TV_L+21	265h	TV_L+69	2B5h	TV_H+21	2E5h	TV_H+69
236h	TV_L+22	266h	TV_L+70	2B6h	TV_H+22	2E6h	TV_H+70
237h	TV_L+23	267h	TV_L+71	2B7h	TV_H+23	2E7h	TV_H+71
238h	TV_L+24	268h	TV_L+72	2B8h	TV_H+24	2E8h	TV_H+72
239h	TV_L+25	269h	TV_L+73	2B9h	TV_H+25	2E9h	TV_H+73
23Ah	TV_L+26	26Ah	TV_L+74	2BAh	TV_H+26	2EAh	TV_H+74
23Bh	TV_L+27	26Bh	TV_L+75	2BBh	TV_H+27	2EBh	TV_H+75
23Ch	TV_L+28	26Ch	TV_L+76	2BCh	TV_H+28	2ECh	TV_H+76
23Dh	TV_L+29	26Dh	TV_L+77	2BDh	TV_H+29	2EDh	TV_H+77
23Eh	TV_L+30	26Eh	TV_L+78	2BEh	TV_H+30	2EEh	TV_H+78
23Fh	TV_L+31	26Fh	TV_L+79	2BFh	TV_H+31	2EFh	TV_H+79
240h	TV_L+32	270h	I0	2C0h	TV_H+32	2F0h	I0
241h	TV_L+33	271h	I1	2C1h	TV_H+33	2F1h	I1
242h	TV_L+34	272h	I2	2C2h	TV_H+34	2F2h	I2
243h	TV_L+35	273h	I3	2C3h	TV_H+35	2F3h	I3
244h	TV_L+36	274h	Q0	2C4h	TV_H+36	2F4h	Q0
245h	TV_L+37	275h	Q1	2C5h	TV_H+37	2F5h	Q1
246h	TV_L+38	276h	Q2	2C6h	TV_H+38	2F6h	Q2
247h	TV_L+39	277h	Q3	2C7h	TV_H+39	2F7h	Q3
248h	TV_L+40	278h	Temp_1	2C8h	TV_H+40	2F8h	Temp_1
249h	TV_L+41	279h	Temp_2	2C9h	TV_H+41	2F9h	Temp_2
24Ah	TV_L+42	27Ah	SMB1	2CAh	TV_H+42	2FAh	SMB1
24Bh	TV_L+43	27Bh	SMB2	2CBh	TV_H+43	2FBh	SMB2
24Ch	TV_L+44	27Ch		2CCh	TV_H+44	2FCh	
24Dh	TV_L+45	27Dh		2CDh	TV_H+45	2FDh	
24Eh	TV_L+46	27Eh		2CEh	TV_H+46	2FEh	
24Fh	TV_L+47	27Fh		2CFh	TV_H+47	2FFh	

Bank 4   Bank 5

FIGURE 2.3 Continued

In the other PLC scan cycles following the first one, it is reset. The special memory bit "SCNOSC" is arranged to work as a "scan oscillator". This means that in one PLC scan cycle this special bit will hold the value of "0", in the next one the value of "1", in the next one the value of "0", and so on. This will keep on going for every PLC scan cycle.

320h	CV_L	350h	CV_L+48	3A0h	CV_H	3D0h	CV_H+48
321h	CV_L+1	351h	CV_L+49	3A1h	CV_H+1	3D1h	CV_H+49
322h	CV_L+2	352h	CV_L+50	3A2h	CV_H+2	3D2h	CV_H+50
323h	CV_L+3	353h	CV_L+51	3A3h	CV_H+3	3D3h	CV_H+51
324h	CV_L+4	354h	CV_L+52	3A4h	CV_H+4	3D4h	CV_H+52
325h	CV_L+5	355h	CV_L+53	3A5h	CV_H+5	3D5h	CV_H+53
326h	CV_L+6	356h	CV_L+54	3A6h	CV_H+6	3D6h	CV_H+54
327h	CV_L+7	357h	CV_L+55	3A7h	CV_H+7	3D7h	CV_H+55
328h	CV_L+8	358h	CV_L+56	3A8h	CV_H+8	3D8h	CV_H+56
329h	CV_L+9	359h	CV_L+57	3A9h	CV_H+9	3D9h	CV_H+57
32Ah	CV_L+10	35Ah	CV_L+58	3AAh	CV_H+10	3DAh	CV_H+58
32Bh	CV_L+11	35Bh	CV_L+59	3ABh	CV_H+11	3DBh	CV_H+59
32Ch	CV_L+12	35Ch	CV_L+60	3ACh	CV_H+12	3DCh	CV_H+60
32Dh	CV_L+13	35Dh	CV_L+61	3ADh	CV_H+13	3DDh	CV_H+61
32Eh	CV_L+14	35Eh	CV_L+62	3AEh	CV_H+14	3DEh	CV_H+62
32Fh	CV_L+15	35Fh	CV_L+63	3AFh	CV_H+15	3DFh	CV_H+63
330h	CV_L+16	360h	CV_L+64	3B0h	CV_H+16	3E0h	CV_H+64
331h	CV_L+17	361h	CV_L+65	3B1h	CV_H+17	3E1h	CV_H+65
332h	CV_L+18	362h	CV_L+66	3B2h	CV_H+18	3E2h	CV_H+66
333h	CV_L+19	363h	CV_L+67	3B3h	CV_H+19	3E3h	CV_H+67
334h	CV_L+20	364h	CV_L+68	3B4h	CV_H+20	3E4h	CV_H+68
335h	CV_L+21	365h	CV_L+69	3B5h	CV_H+21	3E5h	CV_H+69
336h	CV_L+22	366h	CV_L+70	3B6h	CV_H+22	3E6h	CV_H+70
337h	CV_L+23	367h	CV_L+71	3B7h	CV_H+23	3E7h	CV_H+71
338h	CV_L+24	368h	CV_L+72	3B8h	CV_H+24	3E8h	CV_H+72
339h	CV_L+25	369h	CV_L+73	3B9h	CV_H+25	3E9h	CV_H+73
33Ah	CV_L+26	36Ah	CV_L+74	3BAh	CV_H+26	3EAh	CV_H+74
33Bh	CV_L+27	36Bh	CV_L+75	3BBh	CV_H+27	3EBh	CV_H+75
33Ch	CV_L+28	36Ch	CV_L+76	3BCh	CV_H+28	3ECh	CV_H+76
33Dh	CV_L+29	36Dh	CV_L+77	3BDh	CV_H+29	3EDh	CV_H+77
33Eh	CV_L+30	36Eh	CV_L+78	3BEh	CV_H+30	3EEh	CV_H+78
33Fh	CV_L+31	36Fh	CV_L+79	3BFh	CV_H+31	3EFh	CV_H+79
340h	CV_L+32	370h	I0	3C0h	CV_H+32	3F0h	I0
341h	CV_L+33	371h	I1	3C1h	CV_H+33	3F1h	I1
342h	CV_L+34	372h	I2	3C2h	CV_H+34	3F2h	I2
343h	CV_L+35	373h	I3	3C3h	CV_H+35	3F3h	I3
344h	CV_L+36	374h	Q0	3C4h	CV_H+36	3F4h	Q0
345h	CV_L+37	375h	Q1	3C5h	CV_H+37	3F5h	Q1
346h	CV_L+38	376h	Q2	3C6h	CV_H+38	3F6h	Q2
347h	CV_L+39	377h	Q3	3C7h	CV_H+39	3F7h	Q3
348h	CV_L+40	378h	Temp_1	3C8h	CV_H+40	3F8h	Temp_1
349h	CV_L+41	379h	Temp_2	3C9h	CV_H+41	3F9h	Temp_2
34Ah	CV_L+42	37Ah	SMB1	3CAh	CV_H+42	3FAh	SMB1
34Bh	CV_L+43	37Bh	SMB2	3CBh	CV_H+43	3FBh	SMB2
34Ch	CV_L+44	37Ch		3CCh	CV_H+44	3FCh	
34Dh	CV_L+45	37Dh		3CDh	CV_H+45	3FDh	
34Eh	CV_L+46	37Eh		3CEh	CV_H+46	3FEh	
34Fh	CV_L+47	37Fh		3CFh	CV_H+47	3FFh	
		Bank 6				Bank 7	

FIGURE 2.3 Continued

Let us now consider the four reference timing signals, namely T_2ms, T_10ms, T_100ms, and T_1s. As will be explained later, timer TMR6 of PIC16F1847 is set up to count ¼ of the 32-MHz oscillator signal, i.e., 8 MHz with a prescaler arranged to divide the signal to 64. Then the TMR6 interrupt flag, i.e., TMR6IF, will be set at every 1 ms. When TMR6IF is set, Boolean variables T_2ms, T_10ms, T_100ms, and T_1s will be processed within the "ISR" to obtain timing signals with periods

Addr	Value	Addr	Value
420h	M0	450h	M48
421h	M1	451h	M49
422h	M2	452h	M50
423h	M3	453h	M51
424h	M4	454h	M52
425h	M5	455h	M53
426h	M6	456h	M54
427h	M7	457h	M55
428h	M8	458h	M56
429h	M9	459h	M57
42Ah	M10	45Ah	M58
42Bh	M11	45Bh	M59
42Ch	M12	45Ch	M60
42Dh	M13	45Dh	M61
42Eh	M14	45Eh	M62
42Fh	M15	45Fh	M63
430h	M16	460h	M64
431h	M17	461h	M65
432h	M18	462h	M66
433h	M19	463h	M67
434h	M20	464h	M68
435h	M21	465h	M69
436h	M22	466h	M70
437h	M23	467h	M71
438h	M24	468h	M72
439h	M25	469h	M73
43Ah	M26	46Ah	M74
43Bh	M27	46Bh	M75
43Ch	M28	46Ch	M76
43Dh	M29	46Dh	M77
43Eh	M30	46Eh	M78
43Fh	M31	46Fh	M79
440h	M32	470h	I0
441h	M33	471h	I1
442h	M34	472h	I2
443h	M35	473h	I3
444h	M36	474h	Q0
445h	M37	475h	Q1
446h	M38	476h	Q2
447h	M39	477h	Q3
448h	M40	478h	Temp_1
449h	M41	479h	Temp_2
44Ah	M42	47Ah	SMB1
44Bh	M43	47Bh	SMB2
44Ch	M44	47Ch	
44Dh	M45	47Dh	
44Eh	M46	47Eh	
44Fh	M47	47Fh	

Bank 8

Addr	Value	Addr	Value
4A0h	M80	4D0h	T_Q0
4A1h	M81	4D1h	T_Q1
4A2h	M82	4D2h	T_Q2
4A3h	M83	4D3h	T_Q3
4A4h	M84	4D4h	T_Q4
4A5h	M85	4D5h	T_Q5
4A6h	M86	4D6h	T_Q6
4A7h	M87	4D7h	T_Q7
4A8h	M88	4D8h	T_Q8
4A9h	M89	4D9h	T_Q9
4AAh	M90	4DAh	C_Q0
4ABh	M91	4DBh	C_Q1
4ACh	M92	4DCh	C_Q2
4ADh	M93	4DDh	C_Q3
4AEh	M94	4DEh	C_Q4
4AFh	M95	4DFh	C_Q5
4B0h	M96	4E0h	C_Q6
4B1h	M97	4E1h	C_Q7
4B2h	M98	4E2h	C_Q8
4B3h	M99	4E3h	C_Q9
4B4h	M100	4E4h	C_QD0
4B5h	M101	4E5h	C_QD1
4B6h	M102	4E6h	C_QD2
4B7h	M103	4E7h	C_QD3
4B8h	M104	4E8h	C_QD4
4B9h	M105	4E9h	C_QD5
4BAh	M106	4EAh	C_QD6
4BBh	M107	4EBh	C_QD7
4BCh	M108	4ECh	C_QD8
4BDh	M109	4EDh	C_QD9
4BEh	M110	4EEh	
4BFh	M111	4EFh	
4C0h	M112	4F0h	I0
4C1h	M113	4F1h	I1
4C2h	M114	4F2h	I2
4C3h	M115	4F3h	I3
4C4h	M116	4F4h	Q0
4C5h	M117	4F5h	Q1
4C6h	M118	4F6h	Q2
4C7h	M119	4F7h	Q3
4C8h	M120	4F8h	Temp_1
4C9h	M121	4F9h	Temp_2
4CAh	M122	4FAh	SMB1
4CBh	M123	4FBh	SMB2
4CCh	M124	4FCh	
4CDh	M125	4FDh	
4CEh	M126	4FEh	
4CFh	M127	4FFh	

Bank 9

FIGURE 2.3 Continued

of 2 milliseconds, 10 milliseconds, 100 milliseconds, and 1 second, respectively. Timing diagrams of the reference timing signals T_2ms, T_10ms, T_100ms, and T_1s are depicted in Figure 2.8. Note that the evaluation of TMR6 is independent from PLC scan cycles. When the PLC is switched on, four reference timing signals (clock pulses), namely T_2ms, T_10ms, T_100ms, and T_1s, will start their operation automatically as shown in Figure 2.8.

Addr	Bank 10	Addr	Bank 10	Addr	Bank 11	Addr	Bank 11
520h	drum_TVL	550h	MB2	5A0h	LPF	5D0h	AI0L
521h	drum_TVL+1	551h	step_1.TL	5A1h	LPF+1	5D1h	AI0H
522h	drum_TVL+2	552h	step_1.TH	5A2h	LPF+2	5D2h	AI1L
523h	drum_TVL+3	553h	step_2.TL	5A3h	LPF+3	5D3h	AI1H
524h	drum_TVL+4	554h	step_2.TH	5A4h	LPF+4	5D4h	AI2L
525h	drum_TVL+5	555h	step_3.TL	5A5h	LPF+5	5D5h	AI2H
526h	drum_TVL+6	556h	step_3.TH	5A6h	LPF+6	5D6h	AI3L
527h	drum_TVL+7	557h	step_4.TL	5A7h	LPF+7	5D7h	AI3H
528h	drum_TVL+8	558h	step_4.TH	5A8h	LPF+8	5D8h	step_16.TH
529h	drum_TVL+9	559h	step_5.TL	5A9h	LPF+9	5D9h	step_17.TL
52Ah	drum_TVL+10	55Ah	step_5.TH	5AAh	LPF+10	5DAh	step_17.TH
52Bh	drum_TVL+11	55Bh	step_6.TL	5ABh	LPF+11	5DBh	step_18.TL
52Ch	drum_TVL+12	55Ch	step_6.TH	5ACh	LPF+12	5DCh	step_18.TH
52Dh	drum_TVL+13	55Dh	step_7.TL	5ADh	LPF+13	5DDh	step_19.TL
52Eh	drum_TVL+14	55Eh	step_7.TH	5AEh	LPF+14	5DEh	step_19.TH
52Fh	drum_TVL+15	55Fh	step_8.TL	5AFh	LPF+15	5DFh	step_20.TL
530h	drum_TVH	560h	step_8.TH	5B0h	LPF+16	5E0h	step_20.TH
531h	drum_TVH+1	561h	step_9.TL	5B1h	LPF+17	5E1h	step_21.TL
532h	drum_TVH+2	562h	step_9.TH	5B2h	LPF+18	5E2h	step_21.TH
533h	drum_TVH+3	563h	step_10.TL	5B3h	LPF+19	5E3h	step_22.TL
534h	drum_TVH+4	564h	step_10.TH	5B4h	LPF+20	5E4h	step_22.TH
535h	drum_TVH+5	565h	step_11.TL	5B5h	LPF+21	5E5h	step_23.TL
536h	drum_TVH+6	566h	step_11.TH	5B6h	LPF+22	5E6h	step_23.TH
537h	drum_TVH+7	567h	step_12.TL	5B7h	LPF+23	5E7h	step_24.TL
538h	drum_TVH+8	568h	step_12.TH	5B8h	LPF+24	5E8h	step_24.TH
539h	drum_TVH+9	569h	step_13.TL	5B9h	LPF+25	5E9h	HSC_B2
53Ah	drum_TVH+10	56Ah	step_13.TH	5BAh	LPF+26	5EAh	HSC_B3
53Bh	drum_TVH+11	56Bh	step_14.TL	5BBh	LPF+27	5EBh	
53Ch	drum_TVH+12	56Ch	step_14.TH	5BCh	LPF+28	5ECh	
53Dh	drum_TVH+13	56Dh	step_15.TL	5BDh	LPF+29	5EDh	
53Eh	drum_TVH+14	56Eh	step_15.TH	5BEh	LPF+30	5EEh	
53Fh	drum_TVH+15	56Fh	step_16.TL	5BFh	LPF+31	5EFh	
540h	drum_TQL	570h	I0	5C0h	LPF+32	5F0h	I0
541h	drum_TQH	571h	I1	5C1h	LPF+33	5F1h	I1
542h	drum_stepsL	572h	I2	5C2h	LPF+34	5F2h	I2
543h	drum_stepsH	573h	I3	5C3h	LPF+35	5F3h	I3
544h	drum_eventsL	574h	Q0	5C4h	LPF+36	5F4h	Q0
545h	drum_eventsH	575h	Q1	5C5h	LPF+37	5F5h	Q1
546h	drum_QL	576h	Q2	5C6h	LPF+38	5F6h	Q2
547h	drum_QH	577h	Q3	5C7h	LPF+39	5F7h	Q3
548h	drum_tmp	578h	Temp_1	5C8h	nAI0L	5F8h	Temp_1
549h	drum_tmpL	579h	Temp_2	5C9h	nAI0H	5F9h	Temp_2
54Ah	drum_tmpH	57Ah	SMB1	5CAh	nAI1L	5FAh	SMB1
54Bh	SF0	57Bh	SMB2	5CBh	nAI1H	5FBh	SMB2
54Ch	SF1	57Ch		5CCh	nAI2L	5FCh	
54Dh	SF2	57Dh		5CDh	nAI2H	5FDh	
54Eh	MB0	57Eh		5CEh	nAI3L	5FEh	
54Fh	MB1	57Fh		5CFh	nAI3H	5FFh	

Bank 10   Bank 11

FIGURE 2.3 Continued

Time delays are obtained by using one of these four reference timing signals. For example, if, say, we need 5 seconds' time delay, we can obtain it by counting the T_10ms signal 500 times (10 ms × 500 = 5,000 ms = 5 s) or by counting the T_100ms signal 50 times (100 ms × 50 = 5000 ms = 5 s). The counting process is carried out by using the rising edge signals instead of using the original reference timing signals. The time interval from one rising edge of a reference timing signal to the

Addr	Label	Addr	Label
620h	DBNCR	650h	
621h	DBNCR+1		
622h	DBNCR+2		
623h	DBNCR+3		
624h	DBNCR+4		
625h	DBNCR+5		
626h	DBNCR+6		
627h	DBNCR+7		
628h	DBNCR+8		
629h	DBNCR+9		
62Ah	DBNCR+10		
62Bh	DBNCR+11		
62Ch	DBNCR+12		
62Dh	DBNCR+13		
62Eh	DBNCR+14		
62Fh	DBNCR+15		Unimplemented
630h	DBNCR+16		Read as '0'
631h	DBNCR+17		
632h	DBNCR+18		
633h	DBNCR+19		
634h	DBNCR+20		
635h	DBNCR+21		
636h	DBNCR+22		
637h	DBNCR+23		
638h	DBNCR+24		
639h	DBNCR+25		
63Ah	DBNCR+26		
63Bh	DBNCR+27		
63Ch	DBNCR+28		
63Dh	DBNCR+29		
63Eh	DBNCR+30		
63Fh	DBNCR+31	66Fh	
640h	bl0	670h	I0
641h	bl1	671h	I1
642h	bl2	672h	I2
643h	bl3	673h	I3
644h	CNT1	674h	Q0
645h	CNT2	675h	Q1
646h	CNT3	676h	Q2
647h	TenK	677h	Q3
648h	Thou	678h	Temp_1
649h	Hund	679h	Temp_2
64Ah	Tens	67Ah	SMB1
64Bh	Ones	67Bh	SMB2
64Ch	STP_bits	67Ch	
64Dh	i	67Dh	
64Eh	j	67Eh	
64Fh	k	67Fh	

Bank 12

FIGURE 2.3 Continued

next one is equal to the period of that signal. As a result, in this project, rising edge signals re_T2ms, re_T10ms, re_T100ms, and re_T_1s are obtained from reference timing signals T_2ms, T_10ms, T_100ms, and T_1s, respectively, to be used in timing-related functions. Figure 2.9 shows timing diagrams of a reference timing signal (RTS) (T[period] = 2 ms, 10 ms, 100 ms, 1 s) and the rising edge signal of the RTS.

```
;------------------------------------------
#define bI0.0  bI0,0     ;b:bouncing
#define bI0.1  bI0,1
#define bI0.2  bI0,2
#define bI0.3  bI0,3
#define bI0.4  bI0,4
#define bI0.5  bI0,5
#define bI0.6  bI0,6
#define bI0.7  bI0,7
;------------------------------------------
#define bI1.0  bI1,0     ;b:bouncing
#define bI1.1  bI1,1
#define bI1.2  bI1,2
#define bI1.3  bI1,3
#define bI1.4  bI1,4
#define bI1.5  bI1,5
#define bI1.6  bI1,6
#define bI1.7  bI1,7
;------------------------------------------
#define bI2.0  bI2,0     ;b:bouncing
#define bI2.1  bI2,1
#define bI2.2  bI2,2
#define bI2.3  bI2,3
#define bI2.4  bI2,4
#define bI2.5  bI2,5
#define bI2.6  bI2,6
#define bI2.7  bI2,7
;------------------------------------------
#define bI3.0  bI3,0     ;b:bouncing
#define bI3.1  bI3,1
#define bI3.2  bI3,2
#define bI3.3  bI3,3
#define bI3.4  bI3,4
#define bI3.5  bI3,5
#define bI3.6  bI3,6
#define bI3.7  bI3,7
;------------------------------------------
```

FIGURE 2.4   Definition of 32 bouncing digital input signals by using all bits of 8-bit SRAM registers bI0, bI1, bI2, and bI3.

TABLE 2.1
Allocation of Individual Bits of 8-Bit SRAM Registers bI0, bI1, bI2, and bI3

Address	Name	Bit 7	Bit 6	Bit 5	Bit 4	Bit 3	Bit 2	Bit 1	Bit 0
640h	bI0	bI0.7	bI0.6	bI0.5	bI0.4	bI0.3	bI0.2	bI0.1	bI0.0
641h	bI1	bI1.7	bI1.6	bI1.5	bI1.4	bI1.3	bI1.2	bI1.1	bI1.0
642h	bI2	bI2.7	bI2.6	bI2.5	bI2.4	bI2.3	bI2.2	bI2.1	bI2.0
643h	bI3	bI3.7	bI3.6	bI3.5	bI3.4	bI3.3	bI3.2	bI3.1	bI3.0

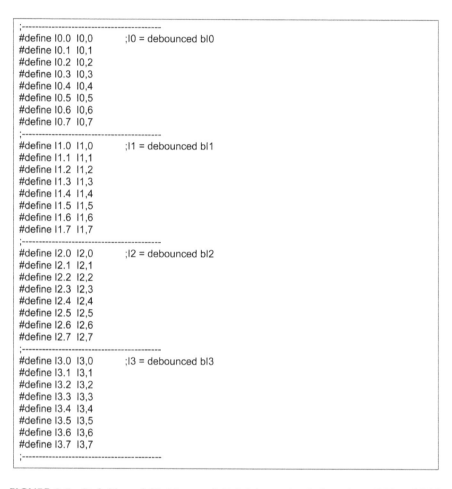

```
;-------------------------------------
#define I0.0 I0,0        ;I0 = debounced bI0
#define I0.1 I0,1
#define I0.2 I0,2
#define I0.3 I0,3
#define I0.4 I0,4
#define I0.5 I0,5
#define I0.6 I0,6
#define I0.7 I0,7
;-------------------------------------
#define I1.0 I1,0        ;I1 = debounced bI1
#define I1.1 I1,1
#define I1.2 I1,2
#define I1.3 I1,3
#define I1.4 I1,4
#define I1.5 I1,5
#define I1.6 I1,6
#define I1.7 I1,7
;-------------------------------------
#define I2.0 I2,0        ;I2 = debounced bI2
#define I2.1 I2,1
#define I2.2 I2,2
#define I2.3 I2,3
#define I2.4 I2,4
#define I2.5 I2,5
#define I2.6 I2,6
#define I2.7 I2,7
;-------------------------------------
#define I3.0 I3,0        ;I3 = debounced bI3
#define I3.1 I3,1
#define I3.2 I3,2
#define I3.3 I3,3
#define I3.4 I3,4
#define I3.5 I3,5
#define I3.6 I3,6
#define I3.7 I3,7
;-------------------------------------
```

FIGURE 2.5 Definition of 32 debounced digital input signals by using all bits of 8-bit SRAM registers I0, I1, I2, and I3.

TABLE 2.2

Allocation of Individual Bits of 8-Bit SRAM Registers I0, I1, I2, and I3

Address	Name	Bit 7	Bit 6	Bit 5	Bit 4	Bit 3	Bit 2	Bit 1	Bit 0
070h	I0	I0.7	I0.6	I0.5	I0.4	I0.3	I0.2	I0.1	I0.0
071h	I1	I1.7	I1.6	I1.5	I1.4	I1.3	I1.2	I1.1	I1.0
072h	I2	I2.7	I2.6	I2.5	I2.4	I2.3	I2.2	I2.1	I2.0
073h	I3	I3.7	I3.6	I3.5	I3.4	I3.3	I3.2	I3.1	I3.0

```
#define Q0.0  Q0,0
#define Q0.1  Q0,1
#define Q0.2  Q0,2
#define Q0.3  Q0,3
#define Q0.4  Q0,4
#define Q0.5  Q0,5
#define Q0.6  Q0,6
#define Q0.7  Q0,7
;-------------------------------------------
#define Q1.0  Q1,0
#define Q1.1  Q1,1
#define Q1.2  Q1,2
#define Q1.3  Q1,3
#define Q1.4  Q1,4
#define Q1.5  Q1,5
#define Q1.6  Q1,6
#define Q1.7  Q1,7
;-------------------------------------------
#define Q2.0  Q2,0
#define Q2.1  Q2,1
#define Q2.2  Q2,2
#define Q2.3  Q2,3
#define Q2.4  Q2,4
#define Q2.5  Q2,5
#define Q2.6  Q2,6
#define Q2.7  Q2,7
;-------------------------------------------
#define Q3.0  Q3,0
#define Q3.1  Q3,1
#define Q3.2  Q3,2
#define Q3.3  Q3,3
#define Q3.4  Q3,4
#define Q3.5  Q3,5
#define Q3.6  Q3,6
#define Q3.7  Q3,7
```

FIGURE 2.6 Definition of 32 digital output signals by using all bits of 8-bit SRAM registers Q0, Q1, Q2, and Q3.

TABLE 2.3
Allocation of Individual Bits of 8-Bit SRAM Registers Q0, Q1, Q2, and Q3

Address	Name	Bit 7	Bit 6	Bit 5	Bit 4	Bit 3	Bit 2	Bit 1	Bit 0
074h	Q0	Q0.7	Q0.6	Q0.5	Q0.4	Q0.3	Q0.2	Q0.1	Q0.0
075h	Q1	Q1.7	Q1.6	Q1.5	Q1.4	Q1.3	Q1.2	Q1.1	Q1.0
076h	Q2	Q2.7	Q2.6	Q2.5	Q2.4	Q2.3	Q2.2	Q2.1	Q2.0
077h	Q3	Q3.7	Q3.6	Q3.5	Q3.4	Q3.3	Q3.2	Q3.1	Q3.0

2.2 CONTENTS OF THE FILE "PICPLC_PIC16F1847_MEMORY.INC"

The individual bits (1-bit variables) of 8-bit SRAM registers M0, M1, …, M127 are defined in the file "PICPLC_PIC16F1847_memory.inc" as shown in Figure 2.10. Table 2.6 shows the allocation of individual bits of 8-bit SRAM registers M0, M1, …, M127.

```
;----- SPECIAL MEMORY BITS -----------
#define      LOGIC0      SMB1,0      ; Logic 0
#define      LOGIC1      SMB1,1      ; Logic 1
#define      FRSTSCN     SMB1,2      ; First scan bit
#define      SCNOSC      SMB1,3      ; Scan oscillator
#define      T_2ms       SMB2,0      ; T(period) of T_2ms is 2 ms
#define      T_10ms      SMB2,1      ; T(period) of T_10ms is 10 ms
#define      T_100ms     SMB2,2      ; T(period) of T_100ms is 100 ms
#define      T_1s        SMB2,3      ; T(period) of T_1s is 1000 ms = 1 s
#define      re_T2ms     SMB2,4      ; rising edge signal of T_2ms
#define      re_T10ms    SMB2,5      ; rising edge signal of T_10ms
#define      re_T100ms   SMB2,6      ; rising edge signal of T_100ms
#define      re_T1s      SMB2,7      ; rising edge signal of T_1s
#define      HSC_Q       STP_bits,7  ; status bit of High Speed Counter (HSC)
; ----------------------------------------------------
```

(a)

```
;-Definition for 74HC165---
#define       shft_ld           PORTB,RB3
;---------------------------
;-Definition for TPIC6B595-
#define       latch_out         PORTB,RB0
;---------------------------
```

(b)

FIGURE 2.7 (a) Definition of special memory bits, (b) Definitions for 74HC165 and TPIC6B595 ICs.

TABLE 2.4

Allocation of Individual Bits of the SMB1 Register

Address	Name	Bit 7	Bit 6	Bit 5	Bit 4	Bit 3	Bit 2	Bit 1	Bit 0
07Ah	SMB1	R	R	R	R	SCNOSC	FRSTSCN	LOGIC1	LOGIC0

LOGIC0   : set to 0 after the first scan.
LOGIC1   : set to 1 after the first scan.
FRSTSCN   : is set to 1 during the first scan and set to 0 after the first scan.
SCNOSC   : is toggled between 0 and 1 at each scan.
R   : Reserved and used in the "re_RTS" macro.

2.3 CONTENTS OF THE FILE "PICPLC_PIC16F1847_MAIN.ASM"

In this section, the contents of the file "PICPLC_PIC16F1847_main.asm" are explained. Processor-specific variable definitions, PICPLC definitions, user program, and subroutines are included to the project through the include files p16F1847 .inc, PICPLC_PIC16F1847_macros_Bsc.inc, PICPLC_PIC16F1847_user_Bsc.inc, and PICPLC_PIC16F1847_subr.inc, respectively. As is well known, a PLC scan cycle includes the following: (1) obtain the inputs, (2) run the user program, and (3) update the outputs. This cycle is always repeated as long as the PLC runs. Before getting into these endless PLC scan cycles, the initial conditions of the PLC are set

TABLE 2.5
Allocation of Individual Bits of the SMB2 Register

Address	Name	Bit 7	Bit 6	Bit 5	Bit 4	Bit 3	Bit 2	Bit 1	Bit 0
07Bh	SMB2	re_T1s	re_T100ms	re_T10ms	re_T2ms	T_1s	T_100ms	T_10ms	T_2ms

T_2ms	: reference timing signal with 2-millisecond period.
T_10ms	: reference timing signal with 10-millisecond period.
T_100ms	: reference timing signal with 100-millisecond period.
T_1s	: reference timing signal with 1-second period.
re_T2ms	: rising edge signal of T_2ms.
re_T10ms	: rising edge signal of T_10ms.
re_T100ms	: rising edge signal of T_100ms.
re_T1s	: rising edge signal of T_1s.

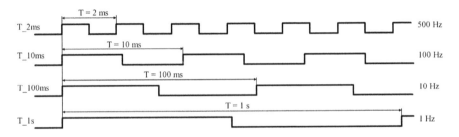

FIGURE 2.8 Timing diagrams of reference timing signals (RTS) T_2ms, T_10ms, T_100ms, T_1s (not to scale).

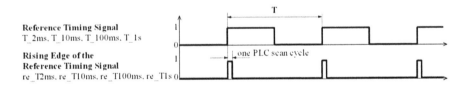

FIGURE 2.9 Timing diagrams of a free-running reference timing signal (RTS) (T[period] = 2 ms, 10 ms, 100 ms, 1 s) and the rising edge signal of the RTS (not to scale).

up in the initialization stage. These main steps can be seen from Figure 2.11, where "initialize" is the macro for setting up initial conditions, "get_inputs" is the macro for getting and handling inputs, and "send_outputs" is the macro for updating outputs. The user PLC program is included through the file "PICPLC_PIC16F1847_user _Bsc.inc" and allocated between "get_inputs" and "send_outputs" macros. The user PLC program is split into two separate parts to be included into two macros, namely "user_program_1" and "user_program_2". Endless PLC scan cycles are obtained by means of the label "PLC_scan" and the instruction "goto PLC_scan". The "ISR" (interrupt service routines) macro is located in the flash program memory (FPM) after the interrupt vector 0x04. The macro "lpf_progs" defines low-pass digital filters

```
;--------------------------------------------------------------;
;              GENERAL DATA SRAM DEFINITIONS          ;
;--------------------------------------------------------------;
;
;--- 1024 Memory Bits (Internal Relays) ----------
#define M0.0  M0,0
#define M0.1  M0,1
#define M0.2  M0,2
#define M0.3  M0,3
#define M0.4  M0,4
#define M0.5  M0,5
#define M0.6  M0,6
#define M0.7  M0,7

#define M1.0  M1,0
#define M1.1  M1,1
#define M1.2  M1,2
#define M1.3  M1,3
#define M1.4  M1,4
#define M1.5  M1,5
#define M1.6  M1,6
#define M1.7  M1,7

#define M2.0  M2,0
#define M2.1  M2,1
#define M2.2  M2,2
#define M2.3  M2,3
#define M2.4  M2,4
#define M2.5  M2,5
#define M2.6  M2,6
#define M2.7  M2,7
;.                                  ;In total there are 1024 definitions.
;
#define M126.0  M126,0
#define M126.1  M126,1
#define M126.2  M126,2
#define M126.3  M126,3
#define M126.4  M126,4
#define M126.5  M126,5
#define M126.6  M126,6
#define M126.7  M126,7

#define M127.0  M127,0
#define M127.1  M127,1
#define M127.2  M127,2
#define M127.3  M127,3
#define M127.4  M127,4
#define M127.5  M127,5
#define M127.6  M127,6
#define M127.7  M127,7
```

FIGURE 2.10 Definition of individual bits of 8-bit SRAM registers M0, M1, ..., M127 in the file "PICPLC_PIC16F1847_memory.inc".

TABLE 2.6
Allocation of Individual Bits of 8-Bit SRAM Registers M0, M1, ..., M127

Address	Name	Bit 7	Bit 6	Bit 5	Bit 4	Bit 3	Bit 2	Bit 1	Bit 0
420h	M0	M0.7	M0.6	M0.5	M0.4	M0.3	M0.2	M0.1	M0.0
421h	M1	M1.7	M1.6	M1.5	M1.4	M1.3	M1.2	M1.1	M1.0
422h	M2	M2.7	M2.6	M2.5	M2.4	M2.3	M2.2	M2.1	M2.0
423h	M3	M3.7	M3.6	M3.5	M3.4	M3.3	M3.2	M3.1	M3.0
424h	M4	M4.7	M4.6	M4.5	M4.4	M4.3	M4.2	M4.1	M4.0
425h	M5	M5.7	M5.6	M5.5	M5.4	M5.3	M5.2	M5.1	M5.0
426h	M6	M6.7	M6.6	M6.5	M6.4	M6.3	M6.2	M6.1	M6.0
427h	M7	M7.7	M7.6	M7.5	M7.4	M7.3	M7.2	M7.1	M7.0
428h	M8	M8.7	M8.6	M8.5	M8.4	M8.3	M8.2	M8.1	M8.0
429h	M9	M9.7	M9.6	M9.5	M9.4	M9.3	M9.2	M9.1	M9.0
42Ah	M10	M10.7	M10.6	M10.5	M10.4	M10.3	M10.2	M10.1	M10.0
42Bh	M11	M11.7	M11.6	M11.5	M11.4	M11.3	M11.2	M11.1	M11.0
42Ch	M12	M12.7	M12.6	M12.5	M12.4	M12.3	M12.2	M12.1	M12.0
42Dh	M13	M13.7	M13.6	M13.5	M13.4	M13.3	M13.2	M13.1	M13.0
42Eh	M14	M14.7	M14.6	M14.5	M14.4	M14.3	M14.2	M14.1	M14.0
42Fh	M15	M15.7	M15.6	M15.5	M15.4	M15.3	M15.2	M15.1	M15.0
430h	M16	M16.7	M16.6	M16.5	M16.4	M16.3	M16.2	M16.1	M16.0
431h	M17	M17.7	M17.6	M17.5	M17.4	M17.3	M17.2	M17.1	M17.0
432h	M18	M18.7	M18.6	M18.5	M18.4	M18.3	M18.2	M18.1	M18.0
433h	M19	M19.7	M19.6	M19.5	M19.4	M19.3	M19.2	M19.1	M19.0
434h	M20	M20.7	M20.6	M20.5	M20.4	M20.3	M20.2	M20.1	M20.0
435h	M21	M21.7	M21.6	M21.5	M21.4	M21.3	M21.2	M21.1	M21.0
436h	M22	M22.7	M22.6	M22.5	M22.4	M22.3	M22.2	M22.1	M22.0
437h	M23	M23.7	M23.6	M23.5	M23.4	M23.3	M23.2	M23.1	M23.0
438h	M24	M24.7	M24.6	M24.5	M24.4	M24.3	M24.2	M24.1	M24.0
439h	M25	M25.7	M25.6	M25.5	M25.4	M25.3	M25.2	M25.1	M25.0
43Ah	M26	M26.7	M26.6	M26.5	M26.4	M26.3	M26.2	M26.1	M26.0
43Bh	M27	M27.7	M27.6	M27.5	M27.4	M27.3	M27.2	M27.1	M27.0
43Ch	M28	M28.7	M28.6	M28.5	M28.4	M28.3	M28.2	M28.1	M28.0
43Dh	M29	M29.7	M29.6	M29.5	M29.4	M29.3	M29.2	M29.1	M29.0
43Eh	M30	M30.7	M30.6	M30.5	M30.4	M30.3	M30.2	M30.1	M30.0
43Fh	M31	M31.7	M31.6	M31.5	M31.4	M31.3	M31.2	M31.1	M31.0
440h	M32	M32.7	M32.6	M32.5	M32.4	M32.3	M32.2	M32.1	M32.0
441h	M33	M33.7	M33.6	M33.5	M33.4	M33.3	M33.2	M33.1	M33.0
442h	M34	M34.7	M34.6	M34.5	M34.4	M34.3	M34.2	M34.1	M34.0
443h	M35	M35.7	M35.6	M35.5	M35.4	M35.3	M35.2	M35.1	M35.0
444h	M36	M36.7	M36.6	M36.5	M36.4	M36.3	M36.2	M36.1	M36.0
445h	M37	M37.7	M37.6	M37.5	M37.4	M37.3	M37.2	M37.1	M37.0
446h	M38	M38.7	M38.6	M38.5	M38.4	M38.3	M38.2	M38.1	M38.0
447h	M39	M39.7	M39.6	M39.5	M39.4	M39.3	M39.2	M39.1	M39.0
448h	M40	M40.7	M40.6	M40.5	M40.4	M40.3	M40.2	M40.1	M40.0

(*Continued*)

TABLE 2.6 (CONTINUED)
Allocation of Individual Bits of 8-Bit SRAM Registers M0, M1, …, M127

Address	Name	Bit 7	Bit 6	Bit 5	Bit 4	Bit 3	Bit 2	Bit 1	Bit 0
449h	M41	M41.7	M41.6	M41.5	M41.4	M41.3	M41.2	M41.1	M41.0
44Ah	M42	M42.7	M42.6	M42.5	M42.4	M42.3	M42.2	M42.1	M42.0
44Bh	M43	M43.7	M43.6	M43.5	M43.4	M43.3	M43.2	M43.1	M43.0
44Ch	M44	M44.7	M44.6	M44.5	M44.4	M44.3	M44.2	M44.1	M44.0
44Dh	M45	M45.7	M45.6	M45.5	M45.4	M45.3	M45.2	M45.1	M45.0
44Eh	M46	M46.7	M46.6	M46.5	M46.4	M46.3	M46.2	M46.1	M46.0
44Fh	M47	M47.7	M47.6	M47.5	M47.4	M47.3	M47.2	M47.1	M47.0
450h	M48	M48.7	M48.6	M48.5	M48.4	M48.3	M48.2	M48.1	M48.0
451h	M49	M49.7	M49.6	M49.5	M49.4	M49.3	M49.2	M49.1	M49.0
452h	M50	M50.7	M50.6	M50.5	M50.4	M50.3	M50.2	M50.1	M50.0
453h	M51	M51.7	M51.6	M51.5	M51.4	M51.3	M51.2	M51.1	M51.0
454h	M52	M52.7	M52.6	M52.5	M52.4	M52.3	M52.2	M52.1	M52.0
455h	M53	M53.7	M53.6	M53.5	M53.4	M53.3	M53.2	M53.1	M53.0
456h	M54	M54.7	M54.6	M54.5	M54.4	M54.3	M54.2	M54.1	M54.0
457h	M55	M55.7	M55.6	M55.5	M55.4	M55.3	M55.2	M55.1	M55.0
458h	M56	M56.7	M56.6	M56.5	M56.4	M56.3	M56.2	M56.1	M56.0
459h	M57	M57.7	M57.6	M57.5	M57.4	M57.3	M57.2	M57.1	M57.0
45Ah	M58	M58.7	M58.6	M58.5	M58.4	M58.3	M58.2	M58.1	M58.0
45Bh	M59	M59.7	M59.6	M59.5	M59.4	M59.3	M59.2	M59.1	M59.0
45Ch	M60	M60.7	M60.6	M60.5	M60.4	M60.3	M60.2	M60.1	M60.0
45Dh	M61	M61.7	M61.6	M61.5	M61.4	M61.3	M61.2	M61.1	M61.0
45Eh	M62	M62.7	M62.6	M62.5	M62.4	M62.3	M62.2	M62.1	M62.0
45Fh	M63	M63.7	M63.6	M63.5	M63.4	M63.3	M63.2	M63.1	M63.0
460h	M64	M64.7	M64.6	M64.5	M64.4	M64.3	M64.2	M64.1	M64.0
461h	M65	M65.7	M65.6	M65.5	M65.4	M65.3	M65.2	M65.1	M65.0
462h	M66	M66.7	M66.6	M66.5	M66.4	M66.3	M66.2	M66.1	M66.0
463h	M67	M67.7	M67.6	M67.5	M67.4	M67.3	M67.2	M67.1	M67.0
464h	M68	M68.7	M68.6	M68.5	M68.4	M68.3	M68.2	M68.1	M68.0
465h	M69	M69.7	M69.6	M69.5	M69.4	M69.3	M69.2	M69.1	M69.0
466h	M70	M70.7	M70.6	M70.5	M70.4	M70.3	M70.2	M70.1	M70.0
467h	M71	M71.7	M71.6	M71.5	M71.4	M71.3	M71.2	M71.1	M71.0
468h	M72	M72.7	M72.6	M72.5	M72.4	M72.3	M72.2	M72.1	M72.0
469h	M73	M73.7	M73.6	M73.5	M73.4	M73.3	M73.2	M73.1	M73.0
46Ah	M74	M74.7	M74.6	M74.5	M74.4	M74.3	M74.2	M74.1	M74.0
46Bh	M75	M75.7	M75.6	M75.5	M75.4	M75.3	M75.2	M75.1	M75.0
46Ch	M76	M76.7	M76.6	M76.5	M76.4	M76.3	M76.2	M76.1	M76.0
46Dh	M77	M77.7	M77.6	M77.5	M77.4	M77.3	M77.2	M77.1	M77.0
46Eh	M78	M78.7	M78.6	M78.5	M78.4	M78.3	M78.2	M78.1	M78.0
46Fh	M79	M79.7	M79.6	M79.5	M79.4	M79.3	M79.2	M79.1	M79.0
4A0h	M80	M80.7	M80.6	M80.5	M80.4	M80.3	M80.2	M80.1	M80.0
4A1h	M81	M81.7	M81.6	M81.5	M81.4	M81.3	M81.2	M81.1	M81.0

(Continued)

TABLE 2.6 (CONTINUED)
Allocation of Individual Bits of 8-Bit SRAM Registers M0, M1, ..., M127

Address	Name	Bit 7	Bit 6	Bit 5	Bit 4	Bit 3	Bit 2	Bit 1	Bit 0
4A2h	M82	M82.7	M82.6	M82.5	M82.4	M82.3	M82.2	M82.1	M82.0
4A3h	M83	M83.7	M83.6	M83.5	M83.4	M83.3	M83.2	M83.1	M83.0
4A4h	M84	M84.7	M84.6	M84.5	M84.4	M84.3	M84.2	M84.1	M84.0
4A5h	M85	M85.7	M85.6	M85.5	M85.4	M85.3	M85.2	M85.1	M85.0
4A6h	M86	M86.7	M86.6	M86.5	M86.4	M86.3	M86.2	M86.1	M86.0
4A7h	M87	M87.7	M87.6	M87.5	M87.4	M87.3	M87.2	M87.1	M87.0
4A8h	M88	M88.7	M88.6	M88.5	M88.4	M88.3	M88.2	M88.1	M88.0
4A9h	M89	M89.7	M89.6	M89.5	M89.4	M89.3	M89.2	M89.1	M89.0
4AAh	M90	M90.7	M90.6	M90.5	M90.4	M90.3	M90.2	M90.1	M90.0
4ABh	M91	M91.7	M91.6	M91.5	M91.4	M91.3	M91.2	M91.1	M91.0
4ACh	M92	M92.7	M92.6	M92.5	M92.4	M92.3	M92.2	M92.1	M92.0
4ADh	M93	M93.7	M93.6	M93.5	M93.4	M93.3	M93.2	M93.1	M93.0
4AEh	M94	M94.7	M94.6	M94.5	M94.4	M94.3	M94.2	M94.1	M94.0
4AFh	M95	M95.7	M95.6	M95.5	M95.4	M95.3	M95.2	M95.1	M95.0
4B0h	M96	M96.7	M96.6	M96.5	M96.4	M96.3	M96.2	M96.1	M96.0
4B1h	M97	M97.7	M97.6	M97.5	M97.4	M97.3	M97.2	M97.1	M97.0
4B2h	M98	M98.7	M98.6	M98.5	M98.4	M98.3	M98.2	M98.1	M98.0
4B3h	M99	M99.7	M99.6	M99.5	M99.4	M99.3	M99.2	M99.1	M99.0
4B4h	M100	M100.7	M100.6	M100.5	M100.4	M100.3	M100.2	M100.1	M100.0
4B5h	M101	M101.7	M101.6	M101.5	M101.4	M101.3	M101.2	M101.1	M101.0
4B6h	M102	M102.7	M102.6	M102.5	M102.4	M102.3	M102.2	M102.1	M102.0
4B7h	M103	M103.7	M103.6	M103.5	M103.4	M103.3	M103.2	M103.1	M103.0
4B8h	M104	M104.7	M104.6	M104.5	M104.4	M104.3	M104.2	M104.1	M104.0
4B9h	M105	M105.7	M105.6	M105.5	M105.4	M105.3	M105.2	M105.1	M105.0
4BAh	M106	M106.7	M106.6	M106.5	M106.4	M106.3	M106.2	M106.1	M106.0
4BBh	M107	M107.7	M107.6	M107.5	M107.4	M107.3	M107.2	M107.1	M107.0
4BCh	M108	M108.7	M108.6	M108.5	M108.4	M108.3	M108.2	M108.1	M108.0
4BDh	M109	M109.7	M109.6	M109.5	M109.4	M109.3	M109.2	M109.1	M109.0
4BEh	M110	M110.7	M110.6	M110.5	M110.4	M110.3	M110.2	M110.1	M110.0
4BFh	M111	M111.7	M111.6	M111.5	M111.4	M111.3	M111.2	M111.1	M111.0
4C0h	M112	M112.7	M112.6	M112.5	M112.4	M112.3	M112.2	M112.1	M112.0
4C1h	M113	M113.7	M113.6	M113.5	M113.4	M113.3	M113.2	M113.1	M113.0
4C2h	M114	M114.7	M114.6	M114.5	M114.4	M114.3	M114.2	M114.1	M114.0
4C3h	M115	M115.7	M115.6	M115.5	M115.4	M115.3	M115.2	M115.1	M115.0
4C4h	M116	M116.7	M116.6	M116.5	M116.4	M116.3	M116.2	M116.1	M116.0
4C5h	M117	M117.7	M117.6	M117.5	M117.4	M117.3	M117.2	M117.1	M117.0
4C6h	M118	M118.7	M118.6	M118.5	M118.4	M118.3	M118.2	M118.1	M118.0
4C7h	M119	M119.7	M119.6	M119.5	M119.4	M119.3	M119.2	M119.1	M119.0
4C8h	M120	M120.7	M120.6	M120.5	M120.4	M120.3	M120.2	M120.1	M120.0
4C9h	M121	M121.7	M121.6	M121.5	M121.4	M121.3	M121.2	M121.1	M121.0
4CAh	M122	M122.7	M122.6	M122.5	M122.4	M122.3	M122.2	M122.1	M122.0

(Continued)

TABLE 2.6 (CONTINUED)
Allocation of Individual Bits of 8-Bit SRAM Registers M0, M1, …, M127

Address	Name	Bit 7	Bit 6	Bit 5	Bit 4	Bit 3	Bit 2	Bit 1	Bit 0
4CBh	M123	M123.7	M123.6	M123.5	M123.4	M123.3	M123.2	M123.1	M123.0
4CCh	M124	M124.7	M124.6	M124.5	M124.4	M124.3	M124.2	M124.1	M124.0
4CDh	M125	M125.7	M125.6	M125.5	M125.4	M125.3	M125.2	M125.1	M125.0
4CEh	M126	M126.7	M126.6	M126.5	M126.4	M126.3	M126.2	M126.1	M126.0
4CFh	M127	M127.7	M127.6	M127.5	M127.4	M127.3	M127.2	M127.1	M127.0

```
;----------------------------------------------------
;  Project name:   PIC16F1847-Based PLC
;  Date:           23 November 2016
;  Author:         Prof. Dr. Murat UZAM
;  E-mail:         murat_uzam@hotmail.com
;  Location:       Talas, Kayseri, Turkey
;----------------------------------------------------
;               Some specifications:
;               Up to 32 Digital Inputs, up to 32 Digital Outputs,
;               Up to 4 Analog Inputs, 1 Analog Output,
;               1 High Speed Counter,  2 PWM Outputs,
;               1024 Memory bits (Internal Relays)
;               80 Counters (8-bit or 16-bit counters):
;                   TON, TOF, RTO, TOS, TP, TEP
;               80 Timers (8-bit or 16-bit Timers):
;                   CTU, CTD, CTUD, GCTUD
;               Drum Sequencer Instruction
;                   with up to 16 steps and
;                   16 outputs on each step
;----------------------------------------------------
    #include <p16F1847.inc>                     ; Processor Specific Variable Definitions
    #include <PICPLC_PIC16F1847_macros_Bsc.inc>; PICPLC Definitions and Macros for "Basic
Concepts"
    #include <PICPLC_PIC16F1847_user_Bsc.inc>   ; User Program for "Basic Concepts"
    #include <PICPLC_PIC16F1847_subr.inc>       ; Subroutines
;----------------------------------- ----------------------------------------------------
; Flash Program Memory (FPM); Page 0 of FPM (size: 2K x 14 Bits)
;----------------------------------- ----------------------------------------------------
        org     0x00            ; Reset Vector
        goto    main            ;
        org     0x04            ; Interrupt Vector
        ISR                     ; Interrupt Service Routines (ISR) macro
main                            ;
        initialize              ; The "initialize" macro is for PLC initialization.
                                ;
                                ;
PLC_scan                        ; The first number (1, 2, 3, 4) defines the number
        get_inputs 4,4          ; of digital I/O extension boards.
                                ; The second number (0, 1, 2, 3, 4) defines the number
                                ; analog inputs to be used.
;-----------------------------------
        pagesel ML1             ; The "get_inputs" macro is for
        goto    ML1             ; obtaining bouncing digital input signals: bI0, bI1, bI2, bI3
        subroutines             ; and noisy analog input signals: nAI0, nAI1, nAI1, nAI3.
;-----------------------------------
```

FIGURE 2.11   (*1 of 2*) The file "PICPLC_PIC16F1847_main.asm".

```
              org      0x0800        ; Page 1 of FPM (size: 2K x 14 Bits)
ML1
              lpf_progs              ; Low-pass digital filter macros for analog inputs are
                                     ; allocated in the "lpf_progs" macro.
              pagesel ML2            ;
              goto     ML2           ;

              org      0x1000        ; Page 2 of FPM (size: 2K x 14 Bits)

                                     ; If the user program in the "user_program_1" macro
                                     ; can fit in Page 2 of FPM, Then
ML2                                  ; it is not necessary to use the "user_program_2" macro.
              user_program_1         ;

              pagesel ML3            ;
              goto     ML3           ;

              org      0x1800        ; Page 3 of FPM (size: 2K x 14 Bits)

                                     ; If the user program in the "user_program_1" macro
                                     ; cannot fit in Page 2 of the FPM,
                                     ; Then the remaining part of the user program can be
ML3                                  ; allocated here in the "user_program_2" macro.
              user_program_2         ;

              pagesel ML4            ;
              goto     ML4           ;

              org      0x1FCF        ; This is the end of 8K FPM (Page 3).

ML4                                  ; The "send_outputs" macro is for
              send_outputs           ; sending digital output signals Q0, Q1, Q2, Q3
                                     ; from SRAM registers of PIC16F1847
              pagesel PLC_scan       ; to physical output registers.
              goto     PLC_scan      ;

              end
```

FIGURE 2.11   Continued

to be used for analog inputs AI0, AI1, AI2, and AI3. Flash program memory (8K × 14 bits from 0x0000 to 0x1FFF) contains four 2K × 14 bits as follows: Page 0 of FPM (size: 2K × 14 bits), 0x0000 to 0x07FF; Page 1 of FPM (size: 2K × 14 bits), 0x0800 to 0x0FFF; Page 2 of FPM (size: 2K × 14 bits), 0x1000 to 0x17FF; and Page 3 of FPM (size: 2K × 14 bits), 0x1800 to 0x1FFF. In order to go from one FPM page to the next one starting from the beginning, labels ML1, ML2, ML3, and ML4 are defined. Then pagesel directives "pagesel ML1", "pagesel ML2", "pagesel ML3", and "pagesel ML4" and instructions "goto ML1", "goto ML2", "goto ML3", and "goto ML4" are used as shown in Figure 2.11. To go from the end of Page 3 to the beginning of Page 0, the "pagesel PLC_scan" directive and the "goto PLC_scan" instruction are used.

The PIC16F1847-Based PLC is fixed to run at 32 MHz with an internal oscillator (internal oscillator frequency = 8 MHz, PLL = 4). The watchdog timer is used to prevent user program lock-ups. As will be explained later, the hardware timer TMR6 is utilized to obtain free-running reference timing signals.

2.4   CONTENTS OF THE FILE "PICPLC_PIC16F1847_USER_BSC.INC"

The structure of the file "PICPLC_PIC16F1847_user_Bsc.inc" is shown in Figure 2.12. The "user_program_1" macro is defined to accommodate user programs. It is allocated on Page 2 of FPM from addresses 0x1000 to 0x17FF. Likewise the "user_program_2" macro is also defined to accommodate user programs. It is allocated on Page 3 of FPM from addresses 0x1800 to 0x1FCF. If the size of a user program, when compiled, is smaller than 2K × 14 bits, then it can be allocated either in the "user_program_1" macro or in the "user_program_2" macro. However, if the size of a user program, when compiled, is greater than 2K × 14 bits and smaller than 4K × 14 bits, then firstly it must be split into two parts. Then the first part of the user program must be allocated in the "user_program_1" macro, and the second part must be allocated in the "user_program_2" macro.

2.5   CONTENTS OF THE FILE "PICPLC_PIC16F1847_SUBR.INC"

The structure of the file "PICPLC_PIC16F1847_subr.inc" is shown in Figure 2.13. This file contains the "subroutines" macro and it is defined to obtain time delays at the expense of CPU clocks. The "subroutines" macro is allocated at the end of Page

```
; If the user program allocated in the "user_program_1" macro can fit in
; Page 2 of FPM, Then it is not necessary to use the "user_program_2" macro.
;
user_program_1        macro
;
;--- PLC codes to be allocated in the "user_program_1" macro start from here ---


;--- PLC codes to be allocated in the "user_program_1" macro end here ----------
                      endm
;
;
; If the user program allocated in the "user_program_1" macro cannot fit in
; Page 2 of Flash Program Memory, Then the remaining part of the user program
; can be allocated in the "user_program_2" macro.
;
user_program_2        macro
;
;--- PLC codes to be allocated in the "user_program_2" macro start from here ---


;--- PLC codes to be allocated in the "user_program_2" macro end here ----------
                      endm
;
```

FIGURE 2.12   The structure of the file "PICPLC_PIC16F1847_user_Bsc.inc".

```
subroutines     macro
;----------------------------------------------------------------------------
pause_1ms                               ; pause_1ms produces a one milisecond delay
                movlw  .96              ; 96*10,375us=1ms    ;125 ns
                movwf  Temp_2           ; Temp_2 <-96        ;125 ns
wait_for_1ms
                call    pause_10us      ; 10 us
                decfsz  Temp_2,f        ; 125 ns
                goto    wait_for_1ms    ; 2*125ns = 250 ns
                return                  ; 2*125ns = 250 ns
pause_10us      ;this subroutine together with call pause_10us consumes 10 us
                goto    $+1             ; 2*125 ns = 250 ns
                goto    $+1             ; 2*125 ns = 250 ns
                goto    $+1             ; 2*125 ns = 250 ns
                goto    $+1             ; 2*125 ns = 250 ns
;----------------------------------------------------; 9 us
pause_9us       goto    $+1             ; 2*125 ns = 250 ns
                goto    $+1             ; 2*125 ns = 250 ns
                goto    $+1             ; 2*125 ns = 250 ns
                goto    $+1             ; 2*125 ns = 250 ns
;----------------------------------------------------; 8 us
pause_8us       goto    $+1             ; 2*125 ns = 250 ns
                goto    $+1             ; 2*125 ns = 250 ns
                goto    $+1             ; 2*125 ns = 250 ns
                goto    $+1             ; 2*125 ns = 250 ns
;----------------------------------------------------; 7 us
pause_7us       goto    $+1             ; 2*125 ns = 250 ns
                goto    $+1             ; 2*125 ns = 250 ns
                goto    $+1             ; 2*125 ns = 250 ns
                goto    $+1             ; 2*125 ns = 250 ns
;----------------------------------------------------; 6 us
pause_6us       goto    $+1             ; 2*125 ns = 250 ns
                goto    $+1             ; 2*125 ns = 250 ns
                goto    $+1             ; 2*125 ns = 250 ns
                goto    $+1             ; 2*125 ns = 250 ns
;----------------------------------------------------; 5 us
pause_5us       goto    $+1             ; 2*125 ns = 250 ns
                goto    $+1             ; 2*125 ns = 250 ns
                goto    $+1             ; 2*125 ns = 250 ns
                goto    $+1             ; 2*125 ns = 250 ns
;----------------------------------------------------; 4 us
pause_4us       goto    $+1             ; 2*125 ns = 250 ns
                goto    $+1             ; 2*125 ns = 250 ns
                goto    $+1             ; 2*125 ns = 250 ns
                goto    $+1             ; 2*125 ns = 250 ns
;----------------------------------------------------; 3 us
pause_3us       goto    $+1             ; 2*125 ns = 250 ns
                goto    $+1             ; 2*125 ns = 250 ns
                goto    $+1             ; 2*125 ns = 250 ns
                goto    $+1             ; 2*125 ns = 250 ns
```

FIGURE 2.13 (*1 of 2*) The file "PICPLC_PIC16F1847_subr.inc".

```
;----------------------------------------------; 2 us
pause_2us    goto    $+1              ; 2*125 ns = 250 ns
             goto    $+1              ; 2*125 ns = 250 ns
             goto    $+1              ; 2*125 ns = 250 ns
             goto    $+1              ; 2*125 ns = 250 ns
;----------------------------------------------; 1 us
pause_1us    goto    $+1              ; 2*125 ns = 250 ns
             goto    $+1              ; 2*125 ns = 250 ns
;----------------------------------------------; 500 ns
             return                   ; 2*125 ns = 250 ns; 750 ns
;--------------------------------------------------------------------
             endm
```

FIGURE 2.13 Continued

0 of FPM. The "subroutines" macro contains two time-delay-related subroutines: "pause_1ms" and "pause_10us". Flowcharts of the subroutines "pause_1ms" and "pause_10us" are depicted in Figures 2.14 and 2.15, respectively. The "pause_1ms" subroutine calls the "pause_10us" subroutine 96 times in order to obtain a 1-millisecond time delay. Of course calculations carried out to obtain these time delays rely on the 32 MHz CPU clock frequency as explained before. Actually, the "pause_10us"

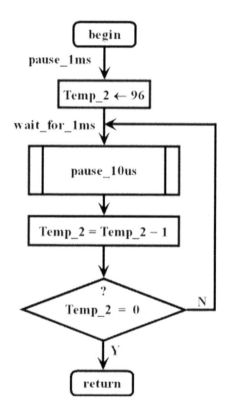

FIGURE 2.14 Flowchart of the "pause_1ms" subroutine.

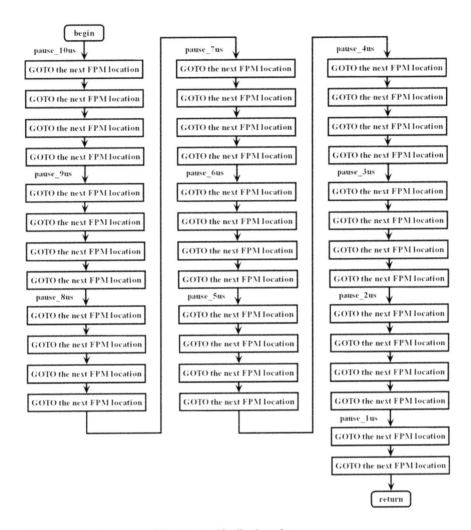

FIGURE 2.15   Flowchart of the "pause_10us" subroutine.

subroutine is a collection of ten subroutines, namely, "pause_10us", "pause_9us", …, "pause_1us". These subroutines can be used to obtain time delays of 10 μs, 9 μs, …, 1 μs, respectively. For example, in the "get_inputs" macro, the "pause_5us" subroutine is called to obtain necessary acquisition delays of 5 microseconds in the conversion of analog inputs AN0, AN1, AN2, and AN3 to digital values.

2.6   CONTENTS OF THE FILE "PICPLC _PIC16F1847_MACROS_BSC.INC"

The file "PICPLC_PIC16F1847_macros_Bsc.inc" contains not only the basic software macros but also all other macros explained in the following chapters. However, in this chapter we restrict our attention to the basic software macros shown below. The following subsections cover these macros in detail.

1. initialize (for PLC initialization)
2. ISR (interrupt service routines)
3. get_inputs (for handling the inputs)
4. lpf_progs (low-pass digital filter macros for analog inputs)
5. send_outputs (for sending the outputs)

The following text is from the beginning of the file "PICPLC_PIC16F1847_macros _Bsc.inc" and it is self-explanatory.

```
;
;------------------------------------------------------------------------
    #include <PICPLC_PIC16F1847_memory.inc>; 1024 memory bits
    are defined in this file.
;------------------------------------------------------------------------
;
;------------------------------------------------------------------------
    radix dec
;------------------------------------------------------------------------
    errorlevel   -302      ;Disables "Message[302]: Register in
                            operand not in Bank0.
                           ;Ensure that bank bits are correct."
;------------------------------------------------------------------------
;
;------------------------------------------------------------------------
; PIC16F1847 Configuration Bit Settings
;------------------------------------------------------------------------
__CONFIG _CONFIG1, _FOSC_INTOSC & _WDTE_ON & _PWRTE_ON &
_MCLRE_ON & _CP_OFF & _CPD_OFF & _BOREN_ON & _CLKOUTEN_OFF &
_IESO_ON & _FCMEN_ON
__CONFIG _CONFIG2, _WRT_OFF & _PLLEN_ON & _STVREN_OFF & _BORV_
LO & _LVP_OFF
;
;----- Port Assignments ----------------------------------------------
; RA0 = AN0 (Analog Input 0)—AI0
; RA1 = AN1 (Analog Input 1)—AI1
; RA2 = DACOUT—output of digital to analog converter (analog
   output)
; RA3 = VREF+ (ADC voltage reference input) (analog input)
; RA4 = PWM output (digital output)
; RA5 = MCLR input (digital input)
; RA6 = not used (digital input/output)
; RA7 = PWM output (digital output)
; RB0 = latch_out output for TPIC6B595 (digital output)
; RB1 = SDI1—SPI Data Input 1 (digital input)
; RB2 = SDO1—SPI Data Output 1 (digital output)
; RB3 = shft_ld output for 74HC165 (digital output)
; RB4 = SCK1—SPI Clock 1 (digital output)
```

```
; RB5 = AN7 (Analog Input 2)—AI2
; RB6 = T1CKI—high speed counter input
; RB7 = AN6 (Analog Input 3)—AI3
;-----------------------------------------------------------------------------
```

2.6.1 MACRO "INITIALIZE"

The macro "initialize" and its flowchart are shown in Figures 2.16 and 2.17, respectively. This macro is run only once when the PLC is switched on or when it is reset. There are mainly thirteen tasks carried out within this macro.

1. *Initialize the DAC*: DAC is enabled. DAC voltage output enable bit is set. Therefore, DAC voltage level is also an output on the DACOUT (RA2 pin). DAC positive reference source is selected as VREF+ (RA3 pin). DAC negative source VREF- is selected as Vss.

```
initialize          macro
          local     L1,L2
;------ Initialize the DAC ------------
          banksel   DACCON0   ;
          movlw     b'11100100'   ;DACEN=1, DACLPS=1, DACOE=1 (DACOUT is enabled),
          movwf     DACCON0       ;VREF+ is connected to RA3 pin,DACPSS=VREF+, DACNSS=Vss
;------ Initialize the ADC ------------
          banksel   ADCON1
          movlw     b'10100010'   ;Right justify, Fosc/32, A/D conversion will take about 1 us
          movwf     ADCON1        ;VREF+ is connected to RA3 pin, VREF- is connected to Vss.
;------ Initialize Port A ------------
          movlw     b'00101011'   ; TRISA and ADCON1 are both in BANK1.
          movwf     TRISA         ; Set RA0,RA1,RA3,RA5 to input, RA2,RA4,RA6,RA7 to output
          banksel   ANSELA        ;
          movlw     b'00001111'   ; Set RA0,RA1,RA2,RA3 to analog
          movwf     ANSELA        ; Set RA4,RA5,RA6,RA7 to digital
          banksel   PORTA         ;
          clrf      PORTA         ; Clear PORTA
;------ Clear TMR1 --------------------
          clrf      TMR1H         ; TMR1H, TMR1L, PORTA and PORTB are in BANK0.
          clrf      TMR1L         ;
;------ Initialize Port B ------------
          clrf      PORTB         ; Clear PORTB
          banksel   ANSELB        ;
          movlw     b'10100000'   ; Set RB5,RB7 to analog
          movwf     ANSELB        ; Set RB0,RB1,RB2,RB3,RB4,RB6 to digital
          banksel   TRISB         ;
          movlw     b'11100010'   ; Set RB1,RB5,RB6,RB7 as input
          movwf     TRISB         ; and set other bits of PORTB as outputs
;------ Initialize TMR6 interrupt Enable bit ------------
          bsf       PIE3,TMR6IE ; TRISB and PIE3 are in BANK1. Set TMR6 interrupt Enable bit
;------ Initialize SPI ------------
          banksel   SSP1STAT    ; SSP1STAT and SSP1CON1 are in BANK4.
          bsf       SSP1STAT,CKE;
          movlw     b'00100000'   ; SPI mode is selected. Enable serial port &
          movwf     SSP1CON1    ; configure SCK1, SDO1, SDI1 pins. Clock = Fosc/4
;------ Initialize Oscillator ------------
          banksel   OSCCON      ;
          movlw     b'01110000'   ; Internal Oscillator Frequency = 8 MHz
          movwf     OSCCON      ; PLL = 4. Therefore, operation frequency = 32 MHz
;------ Initialize TMR1IE ------------
          bsf       PIE1,TMR1IE ; OSCCON and PIE1 are in BANK1.
```

FIGURE 2.16 (*1 of 2*) The macro "initialize".

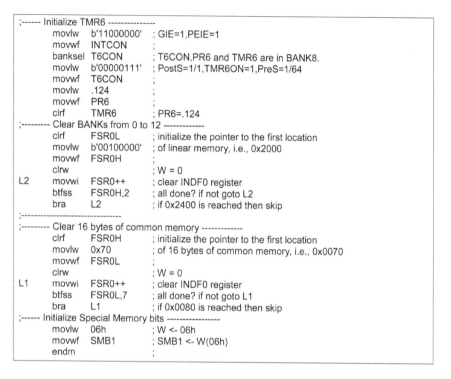

```
;------ Initialize TMR6 ---------------
        movlw   b'11000000'   ; GIE=1,PEIE=1
        movwf   INTCON        ;
        banksel T6CON         ; T6CON,PR6 and TMR6 are in BANK8.
        movlw   b'00000111'   ; PostS=1/1,TMR6ON=1,PreS=1/64
        movwf   T6CON         ;
        movlw   .124          ;
        movwf   PR6           ;
        clrf    TMR6          ; PR6=.124
;--------- Clear BANKs from 0 to 12 -------------
        clrf    FSR0L         ; initialize the pointer to the first location
        movlw   b'00100000'   ; of linear memory, i.e., 0x2000
        movwf   FSR0H         ;
        clrw                  ; W = 0
L2      movwi   FSR0++        ; clear INDF0 register
        btfss   FSR0H,2       ; all done? if not goto L2
        bra     L2            ; if 0x2400 is reached then skip
;-------------------------------
;--------- Clear 16 bytes of common memory -------------
        clrf    FSR0H         ; initialize the pointer to the first location
        movlw   0x70          ; of 16 bytes of common memory, i.e., 0x0070
        movwf   FSR0L         ;
        clrw                  ; W = 0
L1      movwi   FSR0++        ; clear INDF0 register
        btfss   FSR0L,7       ; all done? if not goto L1
        bra     L1            ; if 0x0080 is reached then skip
;------ Initialize Special Memory bits ------------------
        movlw   06h           ; W <- 06h
        movwf   SMB1          ; SMB1 <- W(06h)
        endm                  ;
```

FIGURE 2.16   Continued

2. *Initialize the ADC*: A/D result is right justified. A/D conversion clock is selected as Fosc/32 (A/D conversion will take about 1 μs). A/D negative voltage reference VREF- is selected as Vss. A/D positive voltage reference VREF+ is connected to external VREF+ pin (RA3 pin).

3. *Initialize PORTA*: Set RA0, RA1, RA3, RA5 as input, and RA2, RA4, RA6, RA7 as output. Set RA0, RA1, RA2, RA3 to analog, and set RA4, RA5, RA6, RA7 to digital. PORTA is organized as follows: RA0 (AN0, Analog Input 0—AI0), RA1 (AN1, Analog Input 1—AI1), RA2 (DACOUT—output of digital to analog converter), RA3 (VREF+—ADC voltage reference input), RA4 (PWM output), RA5 (MCLR input), RA6 (not used), RA7 (PWM output).

4. *Clear TMR1*.

5. *Initialize PORTB*: Set RB1, RB5, RB6, RB7 as input, and RB0, RB2, RB3, RB4 as output. Set RB5, RB7 to analog, and set RB0, RB1, RB2, RB3, RB4, RB6 to digital. PORTB is organized as follows: RB0 (latch out), RB1 (data in = SDI1—SPI Data Input 1), RB2 (data out = SDO1—SPI Data Output 1), RB3 (shift/load), RB4 (clock out = clock in = SCK1—SPI Clock 1), RB5 (AN7, Analog Input 2—AI2), RB6 (T1CKI—high speed counter input), RB7 (AN6, Analog Input 3—AI3).

6. *Initialize TMR6 interrupt enable bit*: PIE3,TMR6IE bit is set.

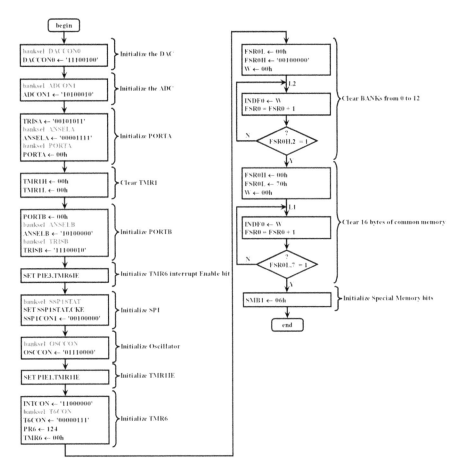

FIGURE 2.17   Flowchart of the macro "initialize".

7. *Initialize SPI*: SSP1STAT,CKE bit of the Serial Peripheral Interface (SPI1) is set. Therefore, transmit occurs on transition from active to idle clock state. SPI1 mode is selected. SPI1 Master mode is selected with clock = Fosc/4. Idle state for clock is a low level. Serial port is enabled and configured as follows: SCK1 = RB4, SDO1 = RB2, SDI1 = RB1.

8. *Initialize oscillator*: PLLEN = 1, according to the configuration bit setting. Therefore, 4xPLL is always enabled. 8 MHz internal oscillator frequency is selected. As a result, the CPU operation frequency is 32 MHz.

9. *Initialize TMR1IE*: PIE1,TMR1IE = 1. Therefore, TMR1 interrupt enable is activated.

10. *Initialize TMR6*: global interrupt enable bit and peripheral interrupt enable bit are both set (GIE=1, PEIE=1). TMR6 is set up as a free-running hardware timer with the ¼ of 32 MHz oscillator signal, i.e., 8 MHz, and with a prescaler arranged to divide the signal to 64 (PostS=1/1, TMR6ON=1, PreS=1/64). Timer6 module period register is loaded with the value of

124. Thus TMR6 generates an interrupt by setting the Timer6 interrupt flag TMR6IF every 1 millisecond starting from the beginning of the PLC operation. TMR6IF is handled in the ISR macro.

11. *Clear Banks from 0 to 12*: all 1,024 SRAM registers are loaded with initial "safe values". In other words, all utilized SRAM registers are cleared (loaded with 00h).

12. *Clear 16 bytes of common memory*: 16 bytes of common memory from 0x70 to 0x7F are cleared (loaded with 00h).

13. *Initialize special memory bits*: SMB1 register is loaded with 06h. As explained before, SMB1 holds special memory bits, therefore initial values of these special memory bits are put into SMB1 within this macro. As a result, these special memory bits are loaded with the following initial values: LOGIC0 (SMB1,0) = 0, LOGIC1 (SMB1,1) = 1, FRSTSCN (SMB1,2) = 1, SCNOSC (SMB1,3) = 0.

2.6.2 MACRO "ISR"

The "ISR" (interrupt service routines) macro and its flowchart are shown in Figures 2.18 and 2.19, respectively. The "ISR" macro is allocated at the address 0x04 of Page 0. This macro contains two interrupt service routines related to the timers/counters TMR6 and TMR1. The "ISR" is run when one of the interrupt flags TMR6IF or TMR1IF is set. Within the "ISR", these two interrupt flags are considered. If the related interrupt flag is set, then the instructions for that particular interrupt service routine are followed accordingly. Let us firstly consider the TMR6 "ISR". When TMR6IF is set, the following tasks are carried out (recall that TMR6IF will be set every 1 millisecond after the PLC is started.). First of all, TMR6IF is reset. Then, the T_2ms bit is toggled. This means that the T_2ms bit is toggled every 1 millisecond. Therefore, it has a period of 2 milliseconds. The CNT1 register is incremented. If CNT1 = 5, then the T_10ms bit is toggled. This means that the T_10ms bit is toggled every 5 milliseconds. Therefore, it has a period of 10 milliseconds. If CNT1 $\neq$ 5, then the CPU returns from interrupt. When the T_10ms bit is toggled, the CNT1 register is cleared and the CNT2 register is incremented. If CNT2 = 10, then the T_100ms bit is toggled. This means that the T_100ms bit is toggled every 50 milliseconds. Therefore, it has a period of 100 milliseconds. If CNT1 $\neq$ 10, then the CPU returns from interrupt. When the T_100ms bit is toggled, the CNT2 register is cleared and the CNT3 register is incremented. If CNT3 = 10, then the T_1s bit is toggled and then the CNT3 register is cleared. This means that the T_1s bit is toggled every 500 milliseconds. Therefore, it has a period of 1 second. If CNT3 $\neq$ 10, then the CPU returns from interrupt.

Let us secondly consider the TMR1 "ISR". When TMR1IF is set, the following tasks are carried out. First of all, TMR1IF is reset and the HSC_B2 register is incremented. If HSC_B2 = 0, then the HSC_B3 register is incremented. If HSC_B3 = 0 then, the TMR1 timer/counter is switched off (the T1CON,TMR1ON bit is reset), the status bit of the high speed counter (HSC_Q) bit is set, and the TMR1H and TMR1L registers are cleared. After that, the CPU returns from interrupt. In this project TMR1 is used to implement a 32-bit high speed up counter (HSC). When the

```
ISR     macro
        local    L0,L1,L2,L3,L4,L5
        pagesel  main                  ;This directive is very crutial!
        banksel  PIR3                  ;If TMR6IF is set
        btfss    PIR3,TMR6IF           ;Then skip
        goto     L1                    ;Else goto L1
;------- start of TMR6 ISR --------------------
        bcf      PIR3,TMR6IF           ;Clear TMR6IF
        banksel  CNT1                  ;
        btfss    T_2ms                 ;Toggle
        goto     L5                    ;T_2ms
        bcf      T_2ms                 ;bit
        btfsc    T_2ms                 ;after
L5      bsf      T_2ms                 ;1 ms
        incf     CNT1,F                ;CNT1 = CNT1 + 1

        ;----------------------------------------
        movlw    d'5'                  ;
        subwf    CNT1,W                ;If CNT1 = 5
        btfss    STATUS,Z              ;Then skip
        goto     L1                    ;Else goto L1.
        btfss    T_10ms                ;Toggle
        goto     L4                    ;T_10ms
        bcf      T_10ms                ;bit
        btfsc    T_10ms                ;after
L4      bsf      T_10ms                ;5 ms.
        clrf     CNT1                  ;Clear CNT1
        incf     CNT2,F                ;CNT2 = CNT2 + 1

        ;----------------------------------------
        movlw    d '10'                ;
        subwf    CNT2,W                ;If CNT2 = 10
        btfss    STATUS,Z              ;Then skip
        goto     L1                    ;Else goto L1.
        btfss    T_100ms               ;Toggle
        goto     L3                    ;T_100ms
        bcf      T_100ms               ;bit
        btfsc    T_100ms               ;after
L3      bsf      T_100ms               ;50 ms.
        clrf     CNT2                  ;Clear CNT2
        incf     CNT3,F                ;CNT3 = CNT3 + 1
```

FIGURE 2.18   (*1 of 2*) The macro "ISR" (interrupt service routines).

current count value of the HSC is equal to the preset value, the status bit of the HSC (HSC_Q) is set within the TMR1 "ISR" as explained.

2.6.3   ELIMINATION OF CONTACT BOUNCING PROBLEM
IN THE PIC16F1847-BASED PLC

2.6.3.1   Contact Bouncing Problem

When a mechanical contact such as a pushbutton, switch (examples of which are shown in Figure 2.20), user interface button, limit switch, relay, or contactor contact is opened or closed, the contact seldom demonstrates a clean transition from one state to another. There are two types of contacts: normally open (NO) and normally closed (NC). When a contact is closed or opened, it will close and open (technically speaking, make and break) many times due to mechanical vibration before finally settling in a stable state. As can be seen from Figure 2.21, this behavior of a contact is interpreted as multiple

```
          ;-----------------------------------------
          movlw   d'10'          ;
          subwf   CNT3,W         ;If CNT3 = 10
          btfss   STATUS,Z       ;Then skip
          goto    L1             ;Else goto L1.
          btfss   T_1s           ;Toggle
          goto    L2             ;T_1s
          bcf     T_1s           ;bit
          btfsc   T_1s           ;after
L2        bsf     T_1s           ;500 ms.
          clrf    CNT3           ;Clear CNT3
          ;------- end of TMR6 ISR --------------------
L1                               ;
          ;------- start of TMR1 ISR -------------------
          banksel PIR1           ;PIR1, PIR3, T1CON are in the same bank.
          btfss   PIR1,TMR1IF    ;If TMR1IF is set Then skip
          goto    L0             ;Else goto L0
          bcf     PIR1,TMR1IF    ;Clear TMR1IF
          banksel HSC_B2         ;HSC_B3 & HSC_B2 are in the same bank.
          incf    HSC_B2,f       ;HSC_B2 = HSC_B2 + 1
          ;-----------------------------------------
          btfss   STATUS,Z       ;If HSC_B2 = 0 Then skip
          goto    L0             ;Else goto L0
          incf    HSC_B3,f       ;HSC_B3 = HSC_B3 + 1
          ;-----------------------------------------
          btfss   STATUS,Z       ;If HSC_B3 = 0 Then skip
          goto    L0             ;Else goto L0
          ;-----------------------------------------
          banksel T1CON          ;PIR1, PIR3, T1CON are in the same bank.
          bcf     T1CON,TMR1ON   ;stop counting process
          banksel STP_bits       ;
          bsf     HSC_Q          ;Set the status bit of HSC
          banksel TMR1H          ;TMR1H & TMR1L are in the same bank.
          clrf    TMR1H          ;Clear TMR1H
          clrf    TMR1L          ;Clear TMR1L
          ;------- end of TMR1 ISR --------------------
L0        retfie                 ;PC=TOS, GIE=1
          endm                   ;end of ISR
```

FIGURE 2.18 Continued

false input signals and a digital circuit will respond to each of these on–off or off–on transitions. This problem is well-known as "contact bounce" and has always been a very important problem when interfacing switches, relays, etc. to a digital control system.

In some industrial applications, debouncing is required to eliminate both mechanical and electrical effects. Most switches seem to exhibit bounce duration under 10 milliseconds and therefore it is reasonable to pick a debounce period within the range of 20 to 50 milliseconds. On the other hand, when dealing with relay contacts, the debounce period should be large enough, i.e., within the range of 20 to 200 milliseconds. Nevertheless, a reasonable switch will not bounce longer than 500 milliseconds. Both closing and opening contacts suffer from the bouncing problem and therefore, in general, both the rising and falling edges of an input signal should be debounced, as seen from the timing diagram of Figure 2.22.

2.6.3.2 Understanding a Generic Single I/O Contact Debouncer

In order to understand how a debouncer works, let us now consider a generic single I/O debouncer. We can think of the generic single I/O debouncer as being

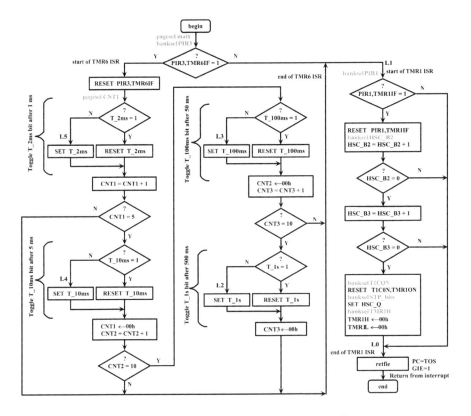

FIGURE 2.19   Flowchart of the macro "ISR" (interrupt service routines).

a single input/single output system, whose state-transition diagram is shown in Figure 2.23. In the state-transition diagram there are four states, namely s0, s1, s2, and s3, drawn as circles, and six transitions, namely, t1, t2, ..., t6, drawn as bars. States and transitions are connected by directed arcs. The following explains the behavior of the generic single I/O debouncer (also each channel of the 32-bit I/O contact debouncer, namely "dbncrN") based on the state-transition diagram shown in Figure 2.23:

1. Initially, it is assumed that the input signal IN and the output signal OUT are both LOW (state s0).
2. When the system is in s0 (the IN is LOW and the OUT is LOW), if the rising edge (↑) of IN is detected (transition t1), then the system moves from s0 to s1 and the debouncer starts a time delay, called debouncing time 1 (dt1).
3. While the system is in s1 (the IN is HIGH and the OUT is LOW), before the "dt1" time delay ends, if the falling edge (↓) of IN is detected (transition t5), then the system goes back to s0 from s1, and the time delay "dt1" is cancelled and the OUT remains LOW (no state change is issued).

FIGURE 2.20   Different type and make switches and buttons.

FIGURE 2.21   Contact bouncing problem, causing an input signal to bounce between 0 and 1.

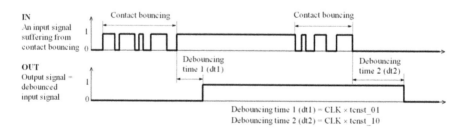

FIGURE 2.22   The timing diagram of a single I/O debouncer (also the timing diagram of each channel of the independent 32 bit I/O contact debouncers, dbncrN).

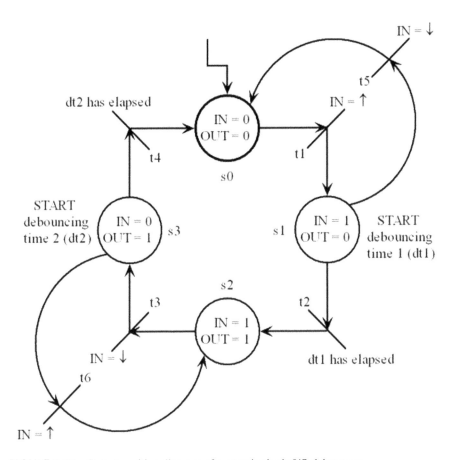

FIGURE 2.23 State-transition diagram of a generic single I/O debouncer.

4. When the system is in s1 (the IN is HIGH and the OUT is LOW), if the input signal is still HIGH and the time-delay "dt1" has elapsed (transition t2), then the system moves from s1 to s2. In this case, the state change is issued, i.e., the OUT is set to HIGH.

5. When the system is in s2 (the IN is HIGH and the OUT is HIGH), if the falling edge (↓) of IN is detected (transition t3), then the system moves from s2 to s3 and the debouncer starts a time delay, called debouncing time 2 (dt2).

6. While the system is in s3 (the IN is LOW and the OUT is HIGH), before the "dt2" time delay ends, if the rising edge (↑) of IN is detected (transition t6), then the system goes back to s2 from s3, and the time-delay "dt2" is cancelled and the OUT remains HIGH (no state change is issued).

7. When the system is in s3 (the IN is LOW and the OUT is HIGH), if the input signal is still LOW and the time delay "dt2" has elapsed (transition t4), then the system moves from s3 to s0. In this case, the state change is issued, i.e., the OUT is set to LOW.

2.6.3.3 Debouncer Macro "dbncrN"

The macro "dbncrN" and its flowchart are shown in Figures 2.24 and 2.25, respectively. Table 2.7 shows the schematic symbol of the macro "dbncrN". The detailed timing diagram of one channel of this debouncer is provided in Figure 2.26. It can be used for debouncing 32 independent buttons, switches, relay or contactor contacts, etc. It is seen that the output changes its state only after the input becomes stable and waits in the stable state for the predefined debouncing time "dt1" or "dt2". The debouncing is applied to both the rising and falling edges of the input signal. In this macro, each channel is intended for a "normally open contact" connected to the PIC

```
dbncrN  macro  num, regi,biti, re_Treg,re_Tbit, tcnst_01, tcnst_10, rego,bito
        local  L1,L2,L3,L4
;-----------------------------------------------------------------------
  if    num < 32      ;if num < 32 then carry on, else do not compile.
;-----------------------------------------------------------------------
        banksel DBNCR
        btfsc   rego,bito
        goto    L4
        btfsc   regi,biti
        goto    L2
        clrf    DBNCR+num
        goto    L1
L4      btfss   regi,biti
        goto    L3
        clrf    DBNCR+num
        goto    L1
L3      btfss   re_Treg,re_Tbit
        goto    L1
        incf    DBNCR+num,f
        movf    DBNCR+num,w
        xorlw   tcnst_10
        btfsc   STATUS,Z
        bcf     rego,bito
        goto    L1
L2      btfss   re_Treg,re_Tbit
        goto    L1
        incf    DBNCR+num,f
        movf    DBNCR+num,w
        xorlw   tcnst_01
        btfsc   STATUS,Z
        bsf     rego,bito
L1
;-----------------------------------------------------------------------
  else
  error "num must be one of 0, 1, ..., 31"
  endif
;-----------------------------------------------------------------------
        endm
```

FIGURE 2.24   The macro "dbncrN".

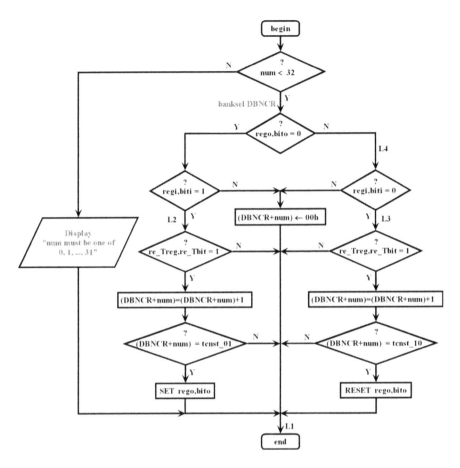

FIGURE 2.25 The flowchart of the macro "dbncrN".

by means of a pull-down resistor, as this is the case with the PIC16F1847-Based PLC. It can also be used without any problem for a "normally closed contact" connected to the PIC by means of a pull-up resistor. The "debouncing times" such as 20 millisec-onds, 50 milliseconds, or 100 milliseconds can be selected as required depending on the application. It is possible to pick up different debouncing times for each channel. It is also possible to choose different debouncing times for rising and falling edges of the same input signal if necessary. This gives a great deal of flexibility. This is simply done by changing the related time constant "tcnst_01" or "tcnst_10" defining the debouncing time delay for each channel and for both edges within the assembly program. Note that if the state change of the contact is shorter than the predefined "debouncing time", this will also be regarded as bouncing and it will not be taken into account. Therefore, no state change will be issued in this case. Each of the 32 input channels of the debouncer may be used independently from other channels. The activity of one channel does not affect the other channels.

Let us now briefly consider how the macro "dbncrN" works. First of all, one of the previously defined rising edges of a reference timing signal is chosen as

TABLE 2.7
The Schematic Symbol of the Macro "dbncrN"

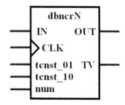

IN (regi,biti): a Boolean variable passed into the macro through regi,biti. It represents the input signal to be debounced, i.e., one of the bouncing input signals bI0.0, bI0.1, ..., bI3.7.

num: any number from 0 to 31. 32 independent debouncers are chosen by this number. It is used to define the 8-bit variable "DBNCR+num".

CLK (re_Treg,re_Tbit): one of the rising edge signals for reference timing signals, i.e., one of re_T2ms, re_T10ms, re_T100ms, re_T1s. It is used to define the timing period.

tcnst_01: an integer constant value from 1 to 255. Debouncing time 1 **(dt1)** is obtained by the formula: dt1 = the period of (re_Treg,re_Tbit) × tcnst_01.

tcnst _10: an integer constant value from 1 to 255. Debouncing time 2 **(dt2)** is obtained by the formula: dt2 = the period of (re_Treg,re_Tbit) × tcnst_10.

OUT(rego,bito): a Boolean variable passed out of the macro through rego,bito. It represents the output signal, which is the debounced version of the input signal, i.e., one of the debounced input signals I0.0, I0.1, ..., I3.7.

TV (8-bit timing count value hold in an 8-bit register) = DBNCR+num (num = 0, 1, ..., 31)

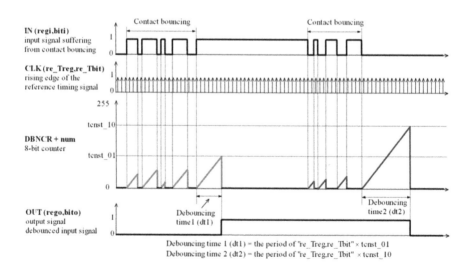

FIGURE 2.26   Detailed timing diagram of one of the channels of the macro "dbncrN".

"re_Treg,re_Tbit" to be used within this macro. Then, we can set up both debouncing times dt1 and dt2 by means of time constants "tcnst_01" and "tcnst_10", as dt1 = the period of (re_Treg,re_Tbit) × tcnst_01 and dt2 = the period of (re_Treg,re_Tbit) × tcnst_10 respectively. If the input signal (regi,biti) = 0 and the output signal (rego,bito) = 0 or the input signal (regi,biti) = 1 and the output signal (rego,bito) = 1, then the related counter "DBNCR+num" is loaded with "00h" and no state change is issued. If the output signal (rego,bito) = 0 and the input signal (regi,biti) = 1, then with each rising edge of the chosen reference timing signal "re_Treg,re_Tbit", the related counter "DBNCR+num" is incremented by one. In this case, when the count value of "DBNCR+num" is equal to the number "tcnst_01", this means that the input signal is debounced properly and then state change from 0 to 1 is issued for the output signal (rego,bito). Similarly, if the output signal (rego,bito) = 1 and the input signal (regi,biti) = 0, then with each rising edge of the chosen reference timing signal "re_Treg,re_Tbit", the related counter "DBNCR+num" is incremented by one. In this case, when the count value of "DBNCR+num" is equal to the number "tcnst_10", this means that the input signal is debounced properly and then state change from 1 to 0 is issued for the output signal (rego,bito). For this macro it is necessary to define 32 8-bit variables in a bank in consecutive SRAM locations, the first of which is to be defined as "DBNCR". As explained before, in this project 8-bit variable DBNCR is defined in the first GPR location of Bank12 (0x620) and the consecutive 31 SRAM locations from 0x621 to 0x63F are reserved to be used in the macro "dbncrN".

2.6.4   MACRO "GET_INPUTS"

The "get_inputs" macro contains the "re_RTS" and "dbncrN" macros. The macro "dbncrN" is considered in the previous section. Let us now consider the macro "re_RTS", before considering the "get_inputs" macro. The macro "re_RTS" and its flowchart are shown in Figures 2.27 and 2.28, respectively. It is used for generating rising edge signals from reference timing signals. In this macro the parameter "num" can be 0, 1, 2, and 3. When num = 0, the macro "re_RTS" takes the reference timing signal "T_2ms" (SMB2,0), generates the rising edge signal from "T_2ms", and stores this new signal as "re_T2ms" in SMB2,4. In order to detect the rising edge signal, the Boolean variable SMB1,4 is used. When num = 1, the macro "re_RTS" takes the reference timing signal "T_10ms" (SMB2,1), generates the rising edge signal from "T_10ms", and stores this new signal as "re_T10ms" in SMB2,5. In order to detect the rising edge signal, the Boolean variable SMB1,5 is used. When num = 2, the macro "re_RTS" takes the reference timing signal "T_100ms" (SMB2,2), generates the rising edge signal from "T_100ms", and stores this new signal as "re_T100ms" in SMB2,6. In order to detect the rising edge signal, the Boolean variable SMB1,6 is used. When num = 3, the macro "re_RTS" takes the reference timing signal "T_1s" (SMB2,3), generates the rising edge signal from "T_1s", and stores this new signal as "re_T1s" in SMB2,7. In order to detect the rising edge signal, the Boolean variable SMB1,7 is used.

The macro "get_inputs" and its flowchart are shown in Figures 2.29 and 2.30, respectively. The "get_inputs" macro is allocated on Page 0 of FPM after the label "PLC_scan". This macro has two parameters, namely "dIO_board_N" and "AI_N". The parameter "dIO_board_N" can be 1, 2, 3, or 4 and it defines the number of digital I/O extension

```
re_RTS  macro  num
        local  L1,L2
        btfss  SMB2,num
        bsf    SMB1,num+4
        btfss  SMB2,num
        goto   L2
        btfss  SMB1,num+4
        goto   L2
        bcf    SMB1,num+4
        bsf    SMB2,num+4
        goto   L1
L2      bcf    SMB2,num+4
L1
        endm
```

FIGURE 2.27   The macro "re_RTS".

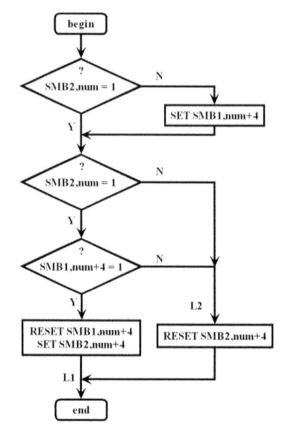

FIGURE 2.28   The flowchart of the macro "re_RTS".

```
get_inputs          macro  dIO_board_N,AI_N
                    local i
;------------------------------------------------------------------------
if (AI_N>=0)&&(AI_N<5)        ;if (AI_N>0)&&(AI_N<4) then carry on, else do not compile.
;------------------------------------------------------------------------
  if (dIO_board_N>0)&&(dIO_board_N<5)      ;if (dI/O_board_N>0)&&(dI/O_board_N< 5)
                                           ;then carry on, else do not compile.
;------------------------------------------------------------------------
   if (dIO_board_N==1)
#define dIO_board_1
   endif
   if (dIO_board_N==2)
#define dIO_board_2
   endif
   if (dIO_board_N==3)
#define dIO_board_3
   endif
   if (dIO_board_N==4)
#define dIO_board_4
   endif
;----------------------------
   if (AI_N>=1)
#define AI_num_is_at_least_1
   endif
   if (AI_N>=2)
#define AI_num_is_at_least_2
   endif
   if (AI_N>=3)
#define AI_num_is_at_least_3
   endif
   if (AI_N==4)
#define AI_num_is_4
   endif
;----------------------------
#define  Q0.0  Q0,0
#define  Q0.1  Q0,1
#define  Q0.2  Q0,2
#define  Q0.3  Q0,3
#define  Q0.4  Q0,4
#define  Q0.5  Q0,5
#define  Q0.6  Q0,6
#define  Q0.7  Q0,7
;----------------------------
;
```

FIGURE 2.29   *(1 of 7)* The macro "get_inputs".

boards to be connected to the CPU board. The parameter "AI_N" can be 0, 1, 2, 3, or 4 and it defines the number of analog inputs to be used. When AI_N = 0, no analog input is used. When AI_N = 1, analog input AI0 is used. When AI_N = 2, 3, and 4, respectively, analog inputs AI0 and AI1; AI0, AI1, and AI2; and AI0, AI1, AI2, and AI3 are used, respectively. The macro "get_inputs" is compiled based on these two numbers.

Let us now consider five tasks carried out within the macro "get_inputs".

1. *Definition of some labels*: based on the dIO_board_N, one of the following labels is defined. When dIO_board_N = 1 (or 2, 3, 4, respectively),
   label "dIO_board_1" (or "dIO_board_2", "dIO_board_3", "dIO_board_4",

```
;------------------------------
   ifdef dIO_board_2    ; then define the following.
#define  Q1.0  Q1,0
#define  Q1.1  Q1,1
#define  Q1.2  Q1,2
#define  Q1.3  Q1,3
#define  Q1.4  Q1,4
#define  Q1.5  Q1,5
#define  Q1.6  Q1,6
#define  Q1.7  Q1,7
   endif
;------------------------------
;
;
;------------------------------
   ifdef dIO_board_3    ; then define the following.
#define  Q1.0  Q1,0
#define  Q1.1  Q1,1
#define  Q1.2  Q1,2
#define  Q1.3  Q1,3
#define  Q1.4  Q1,4
#define  Q1.5  Q1,5
#define  Q1.6  Q1,6
#define  Q1.7  Q1,7

#define  Q2.0  Q2,0
#define  Q2.1  Q2,1
#define  Q2.2  Q2,2
#define  Q2.3  Q2,3
#define  Q2.4  Q2,4
#define  Q2.5  Q2,5
#define  Q2.6  Q2,6
#define  Q2.7  Q2,7
   endif
;------------------------------
;
;------------------------------
   ifdef dIO_board_4    ; then define the following.
#define  Q1.0  Q1,0
#define  Q1.1  Q1,1
#define  Q1.2  Q1,2
#define  Q1.3  Q1,3
#define  Q1.4  Q1,4
#define  Q1.5  Q1,5
#define  Q1.6  Q1,6
#define  Q1.7  Q1,7
```

FIGURE 2.29 Continued

respectively) is defined. Similarly, based on the AI_N, some of the following labels are defined. When $AI\_N \geq 1$, label "AI_num_is_at_least_1" is defined. When $AI\_N \geq 2$, label "AI_num_is_at_least_2" is defined. When $AI\_N \geq 3$, label "AI_num_is_at_least_3" is defined. When $AI\_N = 4$, label "AI_num_is_4" is defined.

2. *Definition of digital inputs and digital outputs*: bouncing inputs bI0.0, bI0.1, ..., bI0.7, debounced inputs I0.0, I0.1, ..., I0.7, and outputs Q0.0, Q0.1, ..., Q0.7 are defined. If the number of digital I/O boards ≥ 2 (dIO_board_N ≥ 2), then bouncing inputs bI1.0, bI1.1, ..., bI1.7, debounced

```
#define  Q2.0  Q2,0
#define  Q2.1  Q2,1
#define  Q2.2  Q2,2
#define  Q2.3  Q2,3
#define  Q2.4  Q2,4
#define  Q2.5  Q2,5
#define  Q2.6  Q2,6
#define  Q2.7  Q2,7

#define  Q3.0  Q3,0
#define  Q3.1  Q3,1
#define  Q3.2  Q3,2
#define  Q3.3  Q3,3
#define  Q3.4  Q3,4
#define  Q3.5  Q3,5
#define  Q3.6  Q3,6
#define  Q3.7  Q3,7
  endif
;----------------------------
;
;------------------------------------------
#define  bI0.0  bI0,0     ; b:bouncing
#define  bI0.1  bI0,1
#define  bI0.2  bI0,2
#define  bI0.3  bI0,3
#define  bI0.4  bI0,4
#define  bI0.5  bI0,5
#define  bI0.6  bI0,6
#define  bI0.7  bI0,7

#define  I0.0  I0,0        ; I0 = debounced bI0
#define  I0.1  I0,1
#define  I0.2  I0,2
#define  I0.3  I0,3
#define  I0.4  I0,4
#define  I0.5  I0,5
#define  I0.6  I0,6
#define  I0.7  I0,7
;------------------------------------------
;
;------------------------------------------
  if (dIO_board_N>=2) ;if # of digital I/O boards >=2 then define the following.
#define  bI1.0  bI1,0     ; b:bouncing
#define  bI1.1  bI1,1
#define  bI1.2  bI1,2
#define  bI1.3  bI1,3
#define  bI1.4  bI1,4
#define  bI1.5  bI1,5
#define  bI1.6  bI1,6
#define  bI1.7  bI1,7
```

FIGURE 2.29   Continued

inputs I1.0, I1.1, ..., I1.7, and outputs Q1.0, Q1.1, ..., Q1.7 are defined. If the number of digital I/O boards ≥ 3 (dIO_board_N ≥ 3), then bouncing inputs bI2.0, bI2.1, ..., bI2.7, debounced inputs I2.0, I2.1, ..., I2.7, and outputs Q2.0, Q2.1, ..., Q2.7 are defined. If the number of digital I/O boards = 4 (dIO_board_N = 4), then bouncing inputs bI3.0, bI3.1, ...,

```
#define I1.0 I1,0        ; I1 = debounced bI1
#define I1.1 I1,1
#define I1.2 I1,2
#define I1.3 I1,3
#define I1.4 I1,4
#define I1.5 I1,5
#define I1.6 I1,6
#define I1.7 I1,7
  endif
;-----------------------------------------
   if (dIO_board_N>=3) ;if # of digital I/O boards >=3 then define the following.
#define bI2.0 bI2,0     ; b:bouncing
#define bI2.1 bI2,1
#define bI2.2 bI2,2
#define bI2.3 bI2,3
#define bI2.4 bI2,4
#define bI2.5 bI2,5
#define bI2.6 bI2,6
#define bI2.7 bI2,7

#define I2.0 I2,0        ; I2 = debounced bI2
#define I2.1 I2,1
#define I2.2 I2,2
#define I2.3 I2,3
#define I2.4 I2,4
#define I2.5 I2,5
#define I2.6 I2,6
#define I2.7 I2,7
  endif
;-----------------------------------------
;
;-----------------------------------------
   if (dIO_board_N==4) ;if # of digital I/O boards ==4 then define the following.
#define bI3.0 bI3,0     ; b:bouncing
#define bI3.1 bI3,1
#define bI3.2 bI3,2
#define bI3.3 bI3,3
#define bI3.4 bI3,4
#define bI3.5 bI3,5
#define bI3.6 bI3,6
#define bI3.7 bI3,7

#define I3.0 I3,0        ; I3 = debounced bI3
#define I3.1 I3,1
#define I3.2 I3,2
#define I3.3 I3,3
#define I3.4 I3,4
#define I3.5 I3,5
#define I3.6 I3,6
#define I3.7 I3,7
  endif
```

FIGURE 2.29   Continued

bI3.7, debounced inputs I3.0, I3.1, …, I3.7, and outputs Q3.0, Q3.1, …, Q3.7 are defined.

3. *Generation of rising edge signals from reference timing signals*: the macro "re_RTS" is called four times with the parameters 0, 1, 2, and 3. This means that the rising edge signals re_T2ms, re_T10ms, re_T100ms, and re_T1s are

```
;
;          re_RTS   0
;          re_RTS   1
;          re_RTS   2
;          re_RTS   3
i=0
   while i<4
          re_RTS   i
i+=1
   endw
;------------------------------------------
;
;----------------- Handle Digital Inputs ----------------------------
;debounce all bouncing digital inputs and put them into related INPUT BITS
;debouncing time dt1 = dt2 = 2 ms x 10 = 20 ms
dbncrN 0,bI0.0,re_T2ms,.10,.10,I0.0    ; 4 input registers bI0,bI1,bI2 and bI3
dbncrN 1,bI0.1,re_T2ms,.10,.10,I0.1    ; taken from 74HC165 registers
dbncrN 2,bI0.2,re_T2ms,.10,.10,I0.2    ; are debounced here one by one.
dbncrN 3,bI0.3,re_T2ms,.10,.10,I0.3    ; Debounce all bits of
dbncrN 4,bI0.4,re_T2ms,.10,.10,I0.4    ; bI0.
dbncrN 5,bI0.5,re_T2ms,.10,.10,I0.5    ; The debounced input signals
dbncrN 6,bI0.6,re_T2ms,.10,.10,I0.6    ; are stored in the register
dbncrN 7,bI0.7,re_T2ms,.10,.10,I0.7    ; I0
   ;----------------------------------------------------------------------
   ;
   ;
   ;----------------------------------------------------------------------
if (dIO_board_N>=2)           ;if (dIO_board_N>=2) then compile the following code.
dbncrN 8,bI1.0,re_T2ms,.10,.10,I1.0    ; debounce all bits of
dbncrN 9,bI1.1,re_T2ms,.10,.10,I1.1    ;
dbncrN 10,bI1.2,re_T2ms,.10,.10,I1.2   ; bI1.
dbncrN 11,bI1.3,re_T2ms,.10,.10,I1.3   ;
dbncrN 12,bI1.4,re_T2ms,.10,.10,I1.4   ; The debounced input signals
dbncrN 13,bI1.5,re_T2ms,.10,.10,I1.5   ;
dbncrN 14,bI1.6,re_T2ms,.10,.10,I1.6   ; are stored in the register
dbncrN 15,bI1.7,re_T2ms,.10,.10,I1.7   ; I1
endif                                   ;
   ;----------------------------------------------------------------------
   ;
   ;
   ;----------------------------------------------------------------------
if (dIO_board_N>=3)           ;if (dIO_board_N>=3) then compile the following code.
dbncrN 16,bI2.0,re_T2ms,.10,.10,I2.0   ; debounce all bits of
dbncrN 17,bI2.1,re_T2ms,.10,.10,I2.1   ;
dbncrN 18,bI2.2,re_T2ms,.10,.10,I2.2   ; bI2.
dbncrN 19,bI2.3,re_T2ms,.10,.10,I2.3   ;
dbncrN 20,bI2.4,re_T2ms,.10,.10,I2.4   ; The debounced input signals
dbncrN 21,bI2.5,re_T2ms,.10,.10,I2.5   ;
dbncrN 22,bI2.6,re_T2ms,.10,.10,I2.6   ; are stored in the register
dbncrN 23,bI2.7,re_T2ms,.10,.10,I2.7   ; I2
endif                                   ;
   ;----------------------------------------------------------------------
```

FIGURE 2.29 Continued

generated from the reference timing signals T2ms, T10ms, T100ms, and T1s, respectively, as explained before.

4. *Debouncing of bouncing digital input signals*: bouncing digital input signals taken from four I/O extension boards are stored in the bI0, bI1, bI2, and bI3 registers of SRAM. Each register holds eight bouncing digital inputs:

```
if (dIO_board_N==4)              ;if (dIO_board_N==4) then compile the following code.
    dbncrN 24,bI3.0,re_T2ms,.10,.10,I3.0   ; debounce all bits of
    dbncrN 25,bI3.1,re_T2ms,.10,.10,I3.1   ;
    dbncrN 26,bI3.2,re_T2ms,.10,.10,I3.2   ; bI3.
    dbncrN 27,bI3.3,re_T2ms,.10,.10,I3.3   ;
    dbncrN 28,bI3.4,re_T2ms,.10,.10,I3.4   ; The debounced input signals
    dbncrN 29,bI3.5,re_T2ms,.10,.10,I3.5   ;
    dbncrN 30,bI3.6,re_T2ms,.10,.10,I3.6   ; are stored in the register
    dbncrN 31,bI3.7,re_T2ms,.10,.10,I3.7   ; I3
endif                                      ;
    ;-----------------------------------------------------------------
    ;
    ;----------------- Handle Analog Inputs ------------------------------
    ;
    ;------- convert AN0 to noisy Digital Value nAI0 = nAI0H&nAI0L ---------
if (AI_N>=1)                               ; if (AI_N>=1) then compile the following code.
    banksel ADCON0                         ;
    movlw   b'00000001'                    ; Select channel AN0
    movwf   ADCON0                         ; Turn ADC On
    call    pause_5us                      ; Acquisiton delay for 5 micro seconds
    bsf     ADCON0,ADGO                    ; Start conversion
    btfsc   ADCON0,ADGO                    ; Is conversion done?
    goto    $-1                            ; No, test again
    banksel ADRESH                         ;
    movf    ADRESH,W                       ; Read upper 2 bits
    banksel nAI0H                          ;
    movwf   nAI0H                          ; store in GPR space
    banksel ADRESL                         ;
    movf    ADRESL,W                       ; Read lower 8-bits
    banksel nAI0L                          ;
    movwf   nAI0L                          ; Store in GPR space
endif                                      ;
    ;-----------------------------------------------------------------
    ;
    ;------- convert AN1 to noisy Digital Value nAI1 = nAI1H&nAI1L ---------
if (AI_N>=2)                               ; if (AI_N>=2) then compile the following code.
    banksel ADCON0                         ;
    movlw   b'00000101'                    ; Select channel AN1
    movwf   ADCON0                         ; Turn ADC On
    call    pause_5us                      ; Acquisiton delay for 5 micro seconds
    bsf     ADCON0,ADGO                    ; Start conversion
    btfsc   ADCON0,ADGO                    ; Is conversion done?
    goto    $-1                            ; No, test again
    banksel ADRESH                         ;
    movf    ADRESH,W                       ; Read upper 2 bits
    banksel nAI1H                          ;
    movwf   nAI1H                          ; store in GPR space
    banksel ADRESL                         ;
    movf    ADRESL,W                       ; Read lower 8-bits
    banksel nAI1L                          ;
    movwf   nAI1L                          ; Store in GPR space
endif                                      ;
```

FIGURE 2.29 Continued

bI0,i (i = 0, 1, ..., 7), bI1,i (i = 0, 1, ..., 7), bI2,i (i = 0, 1, ..., 7), and bI3,i (i = 0, 1, ..., 7), respectively. By using the macro "dbncrN", each bit of bI0,i (i = 0, 1, ..., 7) is debounced and each debounced input signal is stored in the related bit I0,i (i=0, 1, ..., 7). If the number of digital I/O boards ≥ 2 (dIO_board_N ≥ 2), then each bit of bI1,i (i = 0, 1, ..., 7) is debounced by

```
      ;------- convert AN7 to noisy Digital Value nAI2 = nAI2H&nAI2L ---------
   if (AI_N>=3)                        ; if (AI_N>=3) then compile the following code.
      banksel  ADCON0                  ;
      movlw    b'00011101'             ; Select channel AN7
      movwf    ADCON0                  ; Turn ADC On
      call     pause_5us               ; Acquisiton delay for 5 micro seconds
      bsf      ADCON0,ADGO             ; Start conversion
      btfsc    ADCON0,ADGO             ; Is conversion done?
      goto     $-1                     ; No, test again
      banksel  ADRESH                  ;
      movf     ADRESH,W                ; Read upper 2 bits
      banksel  nAI2H                   ;
      movwf    nAI2H                   ; store in GPR space
      banksel  ADRESL                  ;
      movf     ADRESL,W                ; Read lower 8-bits
      banksel  nAI2L                   ;
      movwf    nAI2L                   ; Store in GPR space
   endif                               ;
      ;-----------------------------------------------------------------
      ;
      ;------- convert AN6 to noisy Digital Value nAI3 = nAI3H&nAI3L ---------
   if (AI_N==4)                        ; if (AI_N==4) then compile the following code.
      banksel  ADCON0                  ;
      movlw    b'00011001'             ; Select channel AN6
      movwf    ADCON0                  ; Turn ADC On
      call     pause_5us               ; Acquisiton delay for 5 micro seconds
      bsf      ADCON0,ADGO             ; Start conversion
      btfsc    ADCON0,ADGO             ; Is conversion done?
      goto     $-1                     ; No, test again
      banksel  ADRESH                  ;
      movf     ADRESH,W                ; Read upper 2 bits
      banksel  nAI3H                   ;
      movwf    nAI3H                   ; store in GPR space
      banksel  ADRESL                  ;
      movf     ADRESL,W                ; Read lower 8-bits
      banksel  nAI3L                   ;
      movwf    nAI3L                   ; Store in GPR space
   endif                               ;
   ;-----------------------------------------------------------------
   else                                ;
   error "The number of digital I/O extension boards must be one of the following: 1, 2, 3, 4"
   endif                               ;
   ;-----------------------------------------------------------------
   else                                ;
   error "The number of analog inputs must be one of the following: 0, 1, 2, 3, 4"
   endif                               ;
   ;-----------------------------------------------------------------
      endm
```

FIGURE 2.29 Continued

using the macro "dbncrN" and each debounced input signal is stored in the related bit I1,i (i = 0, 1, ..., 7). If the number of digital I/O boards ≥ 3 (dIO_board_N ≥ 3), then each bit of bI2,i (i = 0, 1, ..., 7) is debounced by using the macro "dbncrN" and each debounced input signal is stored in the related bit I2,i (i = 0, 1, ..., 7). If the number of digital I/O boards = 4 (dIO_board_N = 4), then each bit of bI3,i (i = 0, 1, ..., 7) is debounced by using the macro "dbncrN" and each debounced input signal is stored in the related

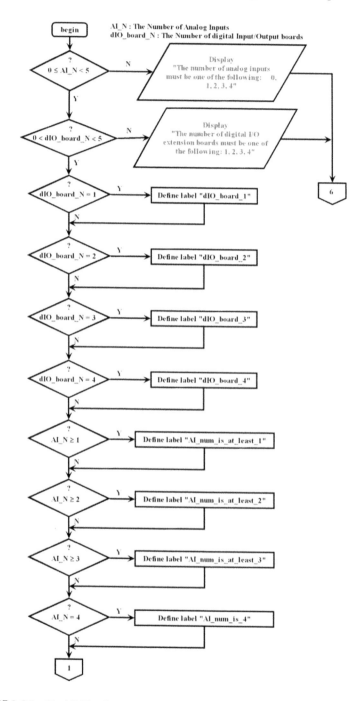

FIGURE 2.30   (*1 of 6*) The flowchart of the macro "get_inputs".

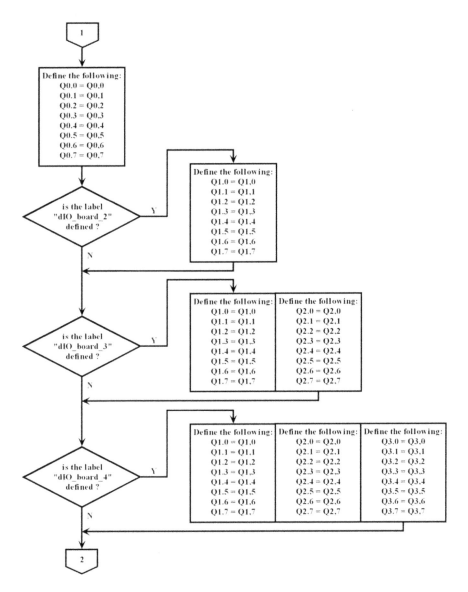

FIGURE 2.30   Continued

bit $I3,i$ (i = 0, 1, ..., 7). In general, a 20-millisecond time delay is enough for debouncing both the rising and falling edges of an input signal. Therefore, to achieve these time delays, the rising edge signal re_T2ms is chosen as "re_Treg,re_Tbit" (with a 2-millisecond period) and both "tcnst_01" and "tcnst_10" are chosen to be "10". Then we obtain the following: dt1 = the period of (re_Treg,re_Tbit) × tcnst_01 = 2 ms × 10 = 20 ms, dt2 = the period of (re_Treg,re_Tbit) × tcnst_10 = 2 ms × 10 = 20 ms.

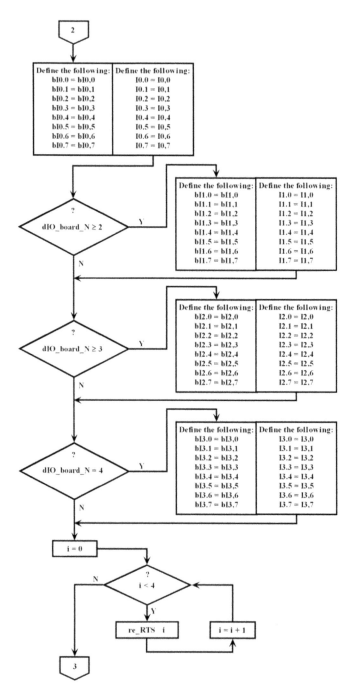

FIGURE 2.30 Continued

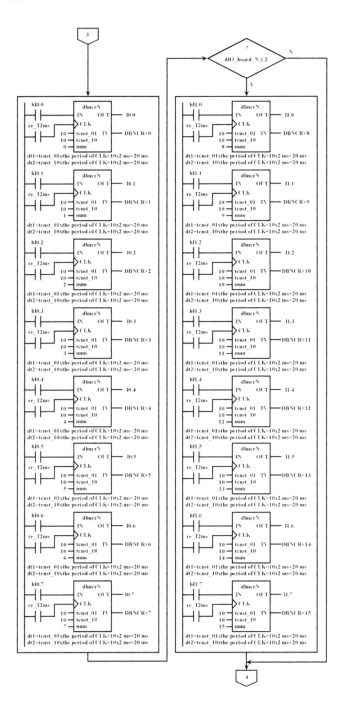

FIGURE 2.30 Continued

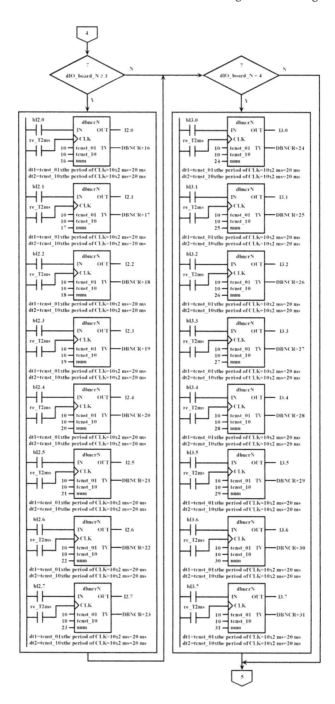

FIGURE 2.30 Continued

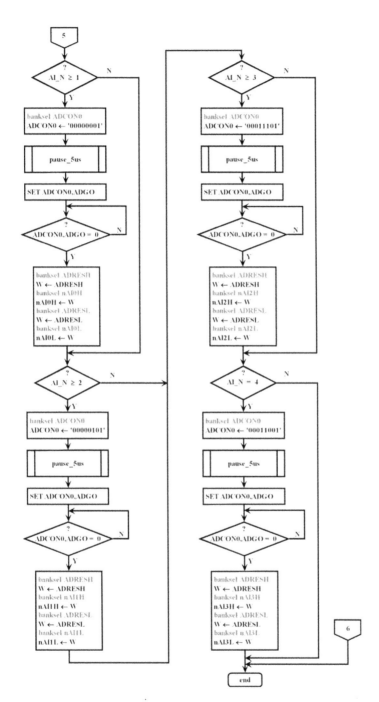

FIGURE 2.30 Continued

5. *Handling Analog Inputs*: the last task is about conversion of analog signals into 10-bit digital values and storing them in related registers. Note that the 10-bit digital values obtained suffer from high frequency noise. Therefore, they are called noisy inputs.

If the number of analog inputs ≥ 1 (AI_N ≥ 1), then the first analog input AN0 (RA0) is converted into the related 10-bit digital value as follows: AN0 channel of the ADC module is selected. The ADC is turned on. Wait for 5 microseconds as acquisition delay by calling a "pause_5us" subroutine from the "subroutines" macro. Start the conversion. When the conversion is done, read the upper 2 bits and store them in the nAI0H register. Likewise, read the lower 8 bits and store them in the nAI0L register.

   If the number of analog inputs ≥ 2 (AI_N ≥ 2), then the second analog input AN1 (RA1) is converted into the related 10-bit digital value as follows: AN1 channel of the ADC module is selected. The ADC is turned on. Wait for 5 microseconds as acquisition delay by calling a "pause_5us" subroutine from the "subroutines" macro. Start the conversion. When the conversion is done, read the upper 2 bits and store them in the nAI1H register. Likewise, read the lower 8 bits and store them in the nAI1L register.

   If the number of analog inputs ≥ 3 (AI_N ≥ 3), then the analog input AN7 (RB5) is converted into the related 10-bit digital value as follows: AN7 channel of the ADC module is selected. The ADC is turned on. Wait for 5 microseconds as acquisition delay by calling a "pause_5us" subroutine from the "subroutines" macro. Start the conversion. When the conversion is done, read the upper 2 bits and store them in the nAI2H register. Likewise, read the lower 8 bits and store them in the nAI2L register.

   If the number of analog inputs $= 4$ (AI_N $= 4$), then the analog input AN6 (RB7) is converted into the related 10-bit digital value as follows: AN6 channel of the ADC module is selected. The ADC is turned on. Wait for 5 microseconds as acquisition delay by calling a "pause_5us" subroutine from the "subroutines" macro. Start the conversion. When the conversion is done, read the upper 2 bits and store them in the nAI3H register. Likewise, read the lower 8 bits and store them in the nAI3L register.

   In these conversions in the PIC16F1847-Based PLC, we assume that there is a 5.00V reference voltage connected to the RA3 pin, since this pin is defined as the VREF+ (ADC voltage reference input). Thus Table 2.8 shows some examples 0V to 5V analog voltage input values vs. ideal digital values to be obtained from a 10-bit ADC of a 5V microcontroller.

2.6.5   Low-Pass Digital Filter Macro "Lpf_progs"

Filtering occurs frequently in the analog world. Unfortunately, in the digital world, engineers apply it mainly to the DSPs (digital signal processors) and not to the small 8-bit microcontrollers that designers commonly use. This situation occurs because the math for the filter design is more complicated than most engineers are willing to deal with. Moreover, digital filtering requires calculations on integers instead of on floating-point numbers. This scenario causes two problems. First, the rounding-off error from the limited number of bits can degrade the filter response or even make it unstable. Second, you must handle the fractional values with integer math.

TABLE 2.8

Some Example 0V–5V Analog Voltage Input Values vs. Ideal Digital Values to Be Obtained from a 10-Bit ADC of a 5V Microcontroller (VREF+ = 5.00V)

	AIH		AIL								
V_{IN}(V)	Bit 9 MSB	Bit 8	Bit 7	Bit 6	Bit 5	Bit 4	Bit 3	Bit 2	Bit 1	Bit 0 LSB	Decimal
5.00	1	1	1	1	1	1	1	1	1	1	1023
..	.	.	.	.	.	.	.	.	.	.	.
4.75	1	1	1	1	0	0	1	1	0	0	972
..	.	.	.	.	.	.	.	.	.	.	.
4.50	1	1	1	0	0	1	1	0	0	1	921
..	.	.	.	.	.	.	.	.	.	.	.
4.25	1	1	0	1	1	0	0	1	0	1	869
..	.	.	.	.	.	.	.	.	.	.	.
4.00	1	1	0	0	1	1	0	0	1	0	818
..	.	.	.	.	.	.	.	.	.	.	.
3.75	1	0	1	1	1	1	1	1	1	1	767
..	.	.	.	.	.	.	.	.	.	.	.
3.50	1	0	1	1	0	0	1	1	0	0	716
..	.	.	.	.	.	.	.	.	.	.	.
3.25	1	0	1	0	0	1	1	0	0	1	665
..	.	.	.	.	.	.	.	.	.	.	.
3.00	1	0	0	1	1	0	0	1	0	1	613
..	.	.	.	.	.	.	.	.	.	.	.
2.75	1	0	0	0	1	1	0	0	1	0	562
..	.	.	.	.	.	.	.	.	.	.	.
2.50	0	1	1	1	1	1	1	1	1	1	511
..	.	.	.	.	.	.	.	.	.	.	.
2.25	0	1	1	1	0	0	1	1	0	0	460
..	.	.	.	.	.	.	.	.	.	.	.
2.00	0	1	1	0	0	1	1	0	0	1	409
..	.	.	.	.	.	.	.	.	.	.	.
1.75	0	1	0	1	1	0	0	1	0	1	357
..	.	.	.	.	.	.	.	.	.	.	.
1.50	0	1	0	0	1	1	0	0	1	0	306
..	.	.	.	.	.	.	.	.	.	.	.
1.25	0	0	1	1	1	1	1	1	1	1	255
..	.	.	.	.	.	.	.	.	.	.	.
1.00	0	0	1	1	0	0	1	1	0	0	204
..	.	.	.	.	.	.	.	.	.	.	.
0.75	0	0	1	0	0	1	1	0	0	1	153
..	.	.	.	.	.	.	.	.	.	.	.
0.50	0	0	0	1	1	0	0	1	1	0	102
..	.	.	.	.	.	.	.	.	.	.	.
0.25	0	0	0	0	1	1	0	0	1	1	51
..	.	.	.	.	.	.	.	.	.	.	.
0.00	0	0	0	0	0	0	0	0	0	0	0

Several ways exist to solve these issues. For example, you can use operations with 16-, 32-, and 64-bit numbers, or you can scale for better accuracy. These and other methods usually require more memory, and, as a result, the program often does not fit into a small microcontroller. A literature search shows that published digital-filter firmware is written in C. Programs in C need more memory than those written in assembler. This situation often makes them unacceptable for small microcontrollers with limited memory resources [R2.1]. Leaving aside the sophisticated design methods based on Z transformation with its extensive math, the method proposed in this book uses a new approach based on a digital low-pass filter applied to each bit of the digital value of the noisy analog input signal. So for each 10-bit digital signal we use ten separate digital low-pass filters to obtain the filtered 10-bit digital signals. Then these filtered signals are stored in two 8-bit GPRs. A digital low-pass filter proposed here basically functions in the same manner as explained in the elimination of a contact bouncing problem. To solve a contact bouncing problem, debouncer macro "dbncrN" is used, as explained before. Here, to obtain a digital low-pass filter, we use the same method as used in the macro "dbncrN", but it is modified with new jargons and thus it is called the macro "lpf_b".

The macro "lpf_progs" is organized to accommodate 40 instances of the macro "lpf_b". The macro "lpf_b" and its flowchart are shown in Figures 2.31 and 2.32, respectively. Table 2.9 shows the schematic symbol of the macro "lpf_b". It can be used for filtering four analog signals with 10-bit resolution. This macro is intended for filtering noisy analog input signals nAI0, nAI1, nAI2, and nAI3 (note that these signals consist of 10-bit digital values stored in two 8-bit registers nAI0H, nAI0L, nAI1H, nAI1L, nAI2H, nAI2L, nAI3H, and nAI3L, respectively) and obtaining filtered analog input signals AI0, AI1, AI2, and AI3 (note that these signals consist of 10-bit digital values stored in two 8-bit registers AI0H, AI0L, AI1H, AI1L, AI2H, AI2L, AI3H, and AI3L, respectively).

The output signal changes its state only after the input becomes stable and waits in the stable state for the predefined filtering time "ft1" or "ft2". The filtering is applied to both the rising and falling edges of the input signal. The "filtering times", such as 20 milliseconds (the cut-off frequency of the low-pass filter is 50 Hz), 50 milliseconds (the cut-off frequency of the low-pass filter is 20 Hz), or 100 milliseconds (the cut-off frequency of the low-pass filter is 10 Hz) can be selected as required. It is possible to pick up different filtering times for each channel. It is also possible to choose different filtering times for rising and falling edges of the same input signal if necessary. This is simply done by changing the related time constant "tcnst_01" or "tcnst_10" defining the filtering time delay for each channel and for both edges within the assembly program. Note that if the state change of the signal is shorter than the predefined "filtering time", this will also be regarded as a noisy input signal and it will not be taken into account. Therefore, no state change will be issued in this case. Each of the 40 input channels of the filter may be used independently from other channels. The activity of one channel does not affect the other channels.

Let us now briefly consider how the macro "lpf_b" works. First of all, one of the previously defined rising edges of a reference timing signal is chosen as "re_Treg,re_Tbit" to be used within this macro. Then, we can set up both filtering times ft1 and ft2 by means of time constants "tcnst_01" and "tcnst_10", as ft1 = the period of

```
lpf_b    macro   num, regi,biti, re_Treg,re_Tbit, tcnst_01, tcnst_10, rego,bito
         local   L1,L2,L3,L4
;-----------------------------------------------------------------------------
   if     num < 40          ;if num < 40 then carry on, else do not compile.
;-----------------------------------------------------------------------------
         banksel rego
         btfsc   rego,bito
         goto    L4
         banksel regi
         btfsc   regi,biti
         goto    L2
         banksel LPF
         clrf    LPF+num
         goto    L1
L4
         banksel regi
         btfss   regi,biti
         goto    L3
         banksel LPF
         clrf    LPF+num
         goto    L1
L3       btfss   re_Treg,re_Tbit
         goto    L1
         banksel LPF
         incf    LPF+num,F
         movf    LPF+num,W
         xorlw   tcnst_10
         btfss   STATUS,Z
         goto    L1
         banksel rego
         bcf     rego,bito
         goto    L1
L2       btfss   re_Treg,re_Tbit
         goto    L1
         banksel LPF
         incf    LPF+num,F
         movf    LPF+num,W
         xorlw   tcnst_01
         btfss   STATUS,Z
         goto    L1
         banksel rego
         bsf     rego,bito
L1
;-----------------------------------------------------------------------------
   else
   error "num must be one of 0, 1, 2, ..., 39"
;-----------------------------------------------------------------------------
   endif
         endm
```

FIGURE 2.31   The macro "lpf_b".

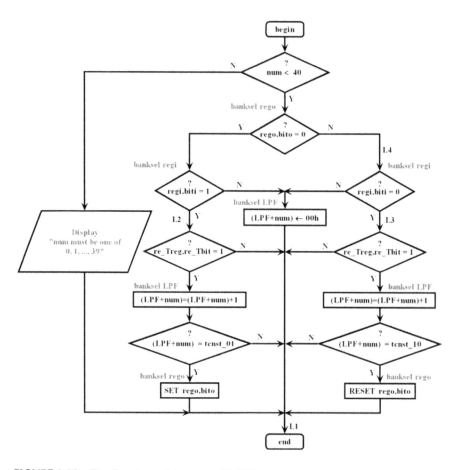

FIGURE 2.32 The flowchart of the macro "lpf_b".

(re_Treg,re_Tbit) × tcnst_01 and ft2 = the period of (re_Treg,re_Tbit) × tcnst_10, respectively. If the input signal (regi,biti) = 0 and the output signal (rego,bito) = 0 or the input signal (regi,biti) = 1 and the output signal (rego,bito) = 1, then the related counter "LPF+num" is loaded with "00h" and no state change is issued. If the output signal (rego,bito) = 0 and the input signal (regi,biti) = 1, then with each rising edge of the chosen reference timing signal "re_Treg,re_Tbit", the related counter "LPF+num" is incremented by one. In this case, when the count value of "LPF+num" is equal to the number "tcnst_01", this means that the input signal is filtered properly and then state change from 0 to 1 is issued for the output signal (rego,bito). Similarly, if the output signal (rego,bito) = 1 and the input signal (regi,biti) = 0, then with each rising edge of the chosen reference timing signal "re_Treg,re_Tbit", the related counter "LPF+num" is incremented by one. In this case, when the count value of "LPF+num" is equal to the number "tcnst_10", this means that the input signal is filtered properly and then state change from 1 to 0 is issued for the output signal (rego,bito). For this macro it is necessary to define 40 8-bit variables in a bank in consecutive SRAM locations, the first of which is to be defined as "LPF". As explained before, in this

TABLE 2.9
The Schematic Symbol of the Macro "lpf_b"

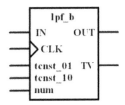

IN (regi,biti): a Boolean variable passed into the macro through regi,biti. It represents the noisy input
signal to be filtered, i.e., one of the noisy input signals nAI0L,0, nAI0L,1, ..., nAI0L,7, nAI0H,0
nAI0H,1; nAI1L,0, nAI1L,1, ..., nAI1L,7, nAI1H,0, nAI1H,1; nAI2L,0, nAI2L,1, ..., nAI2L,7,
nAI2H,0, nAI2H,1; nAI3L,0, nAI3L,1, ..., nAI3L,7, nAI3H,0, nAI3H,1.

num: any number from 0 to 39. 40 independent 1-bit filters are chosen by this number. It is used to
define the 8-bit variable "LPF+num".

CLK (re_Treg,re_Tbit): one of the rising edge signals for reference timing signals, i.e., one of
re_T2ms, re_T10ms, re_T100ms, re_T1s. It is used to define the timing period.

tcnst_01: an integer constant value from 1 to 255. Filtering time 1 (**ft1**) is obtained by the formula: ft1
= the period of (re_Treg,re_Tbit) × tcnst_01.

tcnst _10: an integer constant value from 1 to 255. Filtering time 2 (**ft2**) is obtained by the formula: ft2
= the period of (re_Treg,re_Tbit) × tcnst_10.

OUT(rego,bito): a Boolean variable passed out of the macro through rego,bito. It represents the output
signal, which is the low-pass filtered version of the noisy input signal, i.e., one of the filtered input
signals AI0L,0, AI0L,1, ..., AI0L,7, AI0H,0, AI0H,1; AI1L,0, AI1L,1, ..., AI1L,7, AI1H,0, AI1H,1;
AI2L,0, AI2L,1, ..., AI2L,7, AI2H,0, AI2H,1; nAI3L,0, AI3L,1, ..., AI3L,7, AI3H,0, AI3H,1.

TV (8-bit timing count value hold in an 8-bit register) = LPF+num (num = 0, 1, ..., 39)

project the 8-bit variable LPF is defined in the first GPR location of Bank11 (0x5A0)
and 39 consecutive SRAM locations from 0x5A1 to 0x5C7 are reserved to be used
in the macro "lpf_b".

The macro "lpf_progs" and its flowchart are shown in Figure 2.33 and 2.34,
respectively. This macro represents 40 instances of the macro "lpf_b" with different
noisy digital input signals and filtered digital output signals. Filtering of noisy digital
input signals is carried out as follows: noisy digital input signals obtained from the
four analog inputs AI0, AI1, AI2, and AI3 are stored in register pairs nAI0L, nAI0H,
nAI1L, nAI1H, nAI2L, nAI2H, nAI3L, and nAI3H, respectively. Each register pair
holds ten noisy digital inputs: nAI0L,0, nAI0L,1, ..., nAI0L,7, nAI0H,0, nAI0H,1;
nAI1L,0, nAI1L,1, ..., nAI1L,7, nAI1H,0, nAI1H,1; nAI2L,0, nAI2L,1, ..., nAI2L,7,
nAI2H,0, nAI2H,1; nAI3L,0, nAI3L,1, ..., nAI3L,7, nAI3H,0, nAI3H,1.

If the number of analog inputs is at least 1 (if the label "AI_num_is_at_least_1"
is defined—if AI_N = 1, then this label is defined in the macro "get_inputs"), then
bits of nAI0L,i (i = 0, 1, ..., 7) and nAI0H,0 nAI0H,1 are filtered by using the macro
"lpf_b" and filtered input signals are stored in the related bits AI0L,i (i = 0, 1, ..., 7)
and AI0H,0, AI0H,1, respectively.

If the number of analog inputs is at least 2 (if the label "AI_num_is_at_least_2"
is defined—if AI_N = 2, then this label is defined in the macro "get_inputs"), then

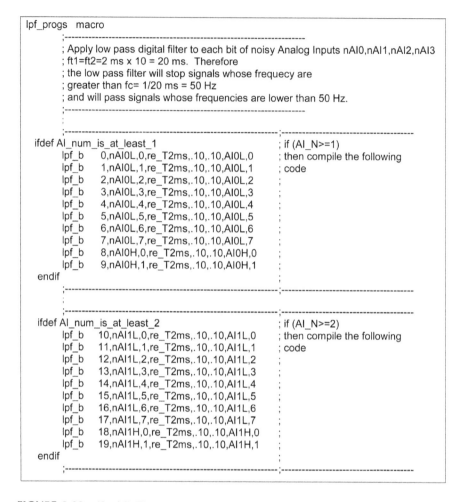

FIGURE 2.33 *(1 of 2)* The macro "lpf_progs" (low-pass digital filter macros for analog inputs).

bits of nAI1L,i (i = 0, 1, …, 7) and nAI1H,0 nAI1H,1 are filtered by using the macro "lpf_b" and filtered input signals are stored in the related bits AI1L,i (i = 0, 1, …, 7) and AI1H,0, AI1H,1, respectively.

If the number of analog inputs is at least 3 (if the label "AI_num_is_at_least_3" is defined—if AI_N = 3, then this label is defined in the macro "get_inputs"), then bits of nAI2L,i (i = 0, 1, …, 7) and nAI2H,0 nAI2H,1 are filtered by using the macro "lpf_b" and filtered input signals are stored in the related bits AI2L,i (i = 0, 1, …, 7) and AI2H,0, AI2H,1, respectively.

If the number of analog inputs is 4 (if the label "AI_num_is_4" is defined—if AI_N = 4, then this label is defined in the macro "get_inputs"), then bits of nAI3L,i (i = 0, 1, …, 7) and nAI3H,0 nAI3H,1 are filtered by using the macro "lpf_b" and filtered input signals are stored in the related bits AI3L,i (i = 0, 1, …, 7) and AI3H,0, AI3H,1, respectively.

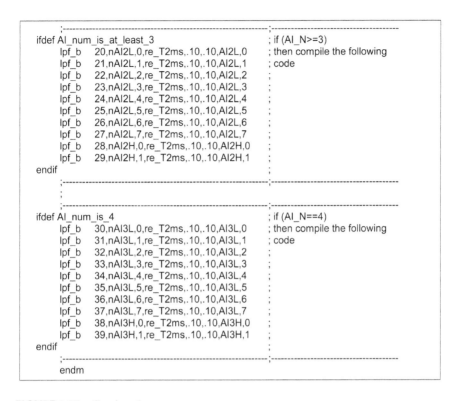

```
;------------------------------------------------;----------------------------------------
ifdef AI_num_is_at_least_3                        ; if (AI_N>=3)
    lpf_b   20,nAI2L,0,re_T2ms,.10,.10,AI2L,0     ; then compile the following
    lpf_b   21,nAI2L,1,re_T2ms,.10,.10,AI2L,1     ; code
    lpf_b   22,nAI2L,2,re_T2ms,.10,.10,AI2L,2     ;
    lpf_b   23,nAI2L,3,re_T2ms,.10,.10,AI2L,3     ;
    lpf_b   24,nAI2L,4,re_T2ms,.10,.10,AI2L,4     ;
    lpf_b   25,nAI2L,5,re_T2ms,.10,.10,AI2L,5     ;
    lpf_b   26,nAI2L,6,re_T2ms,.10,.10,AI2L,6     ;
    lpf_b   27,nAI2L,7,re_T2ms,.10,.10,AI2L,7     ;
    lpf_b   28,nAI2H,0,re_T2ms,.10,.10,AI2H,0     ;
    lpf_b   29,nAI2H,1,re_T2ms,.10,.10,AI2H,1     ;
endif                                             ;
;------------------------------------------------;----------------------------------------
;
;------------------------------------------------;----------------------------------------
ifdef AI_num_is_4                                 ; if (AI_N==4)
    lpf_b   30,nAI3L,0,re_T2ms,.10,.10,AI3L,0     ; then compile the following
    lpf_b   31,nAI3L,1,re_T2ms,.10,.10,AI3L,1     ; code
    lpf_b   32,nAI3L,2,re_T2ms,.10,.10,AI3L,2     ;
    lpf_b   33,nAI3L,3,re_T2ms,.10,.10,AI3L,3     ;
    lpf_b   34,nAI3L,4,re_T2ms,.10,.10,AI3L,4     ;
    lpf_b   35,nAI3L,5,re_T2ms,.10,.10,AI3L,5     ;
    lpf_b   36,nAI3L,6,re_T2ms,.10,.10,AI3L,6     ;
    lpf_b   37,nAI3L,7,re_T2ms,.10,.10,AI3L,7     ;
    lpf_b   38,nAI3H,0,re_T2ms,.10,.10,AI3H,0     ;
    lpf_b   39,nAI3H,1,re_T2ms,.10,.10,AI3H,1     ;
endif                                             ;
;------------------------------------------------;----------------------------------------
    endm
```

FIGURE 2.33 Continued

A 20-millisecond time delay is used for filtering both the rising and falling edges of an input signal. Therefore, to achieve these time delays, the rising edge signal re_T2ms is chosen as "re_Treg,re_Tbit" (with a 2-millisecond period) and both "tcnst_01" and "tcnst_10" are chosen to be "10". Then we obtain the following: ft1 = the period of (re_Treg,re_Tbit) × tcnst_01 = 2 ms × 10 = 20 ms, ft2 = the period of (re_Treg,re_Tbit) × tcnst_10 = 2 ms × 10 = 20 ms. Therefore, the cut-off frequency fc is 50 Hz.

It is important to understand that if the number of analog inputs AI_N = 0, then the above-mentioned code will not be compiled and stored in the flash program memory.

2.6.6 MACRO "SEND_OUTPUTS"

The macro "send_outputs" and its flowchart are shown in Figure 2.35 and 2.36, respectively. There are mainly two tasks carried out within this macro. *In the first task*, digital inputs are taken from the physical input registers 74HC/LS165 of I/O extension boards and they are stored in the bouncing input registers bI0, bI1, bI2, and bI3 of SRAM, and concurrently, states of digital outputs Q0, Q1, Q2, and Q3 are taken from the SRAM and they are sent to physical output registers (TPIC6B595 ICs) of I/O extension boards. In our previous PLC design based on PIC16F648A,

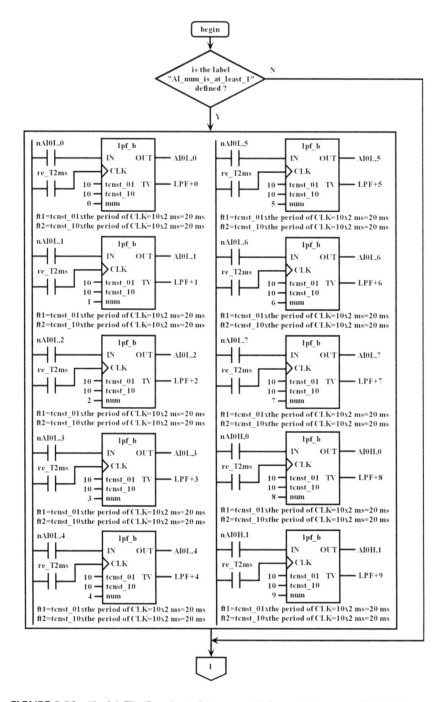

FIGURE 2.34 (*1 of 4*) The flowchart of the macro "lpf_progs" (low-pass digital filter macros for analog inputs).

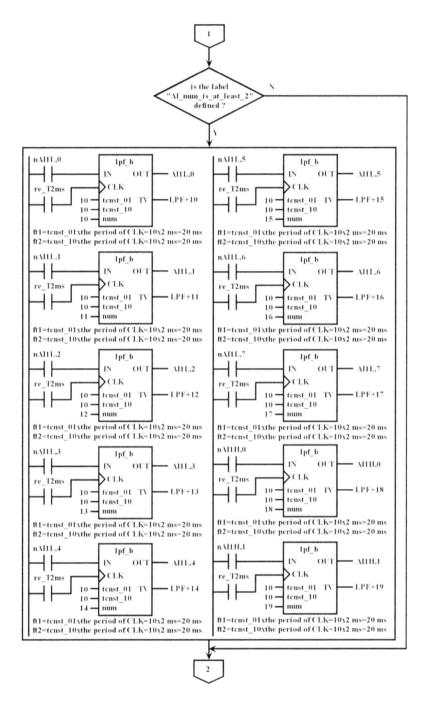

FIGURE 2.34 Continued

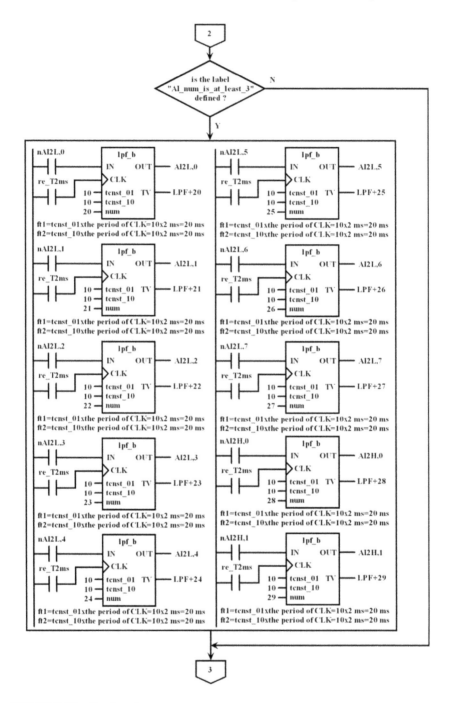

FIGURE 2.34 Continued

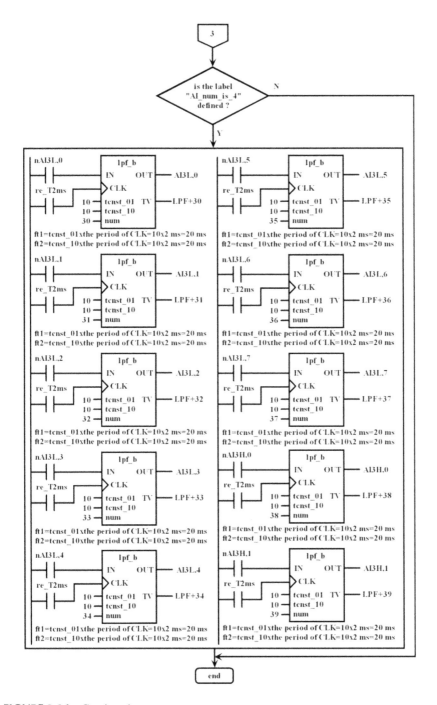

FIGURE 2.34 Continued

```
send_outputs   macro
         local     lp1,lp2,lp3,lp4,L1,L2
         banksel PORTB          ; "shft_ld" bit is in PORTB
         bcf       shft_ld       ; generate
         nop                     ; a shft_ld
         bsf       shft_ld       ; pulse
;-----------------------------------------------------
;
;
;--- If # of digital I/O boards is 4 then compile the following code ---------
    ifdef dIO_board_4            ; then compile the following code.
         banksel SSP1BUF        ; select bank for SSP1BUF
         movf      Q3,W          ;
         movwf     SSP1BUF       ;
lp1      btfss     SSP1STAT,BF; SSP1BUF and SSP1STAT are in BANK4.
         bra       lp1           ;
         movf      SSP1BUF,W    ;
         banksel bl0            ;
         movwf     bl0           ;
         banksel SSP1BUF        ; select bank for SSP1BUF
         movf      Q2,W          ;
         movwf     SSP1BUF       ;
lp2      btfss     SSP1STAT,BF;
         bra       lp2           ;
         movf      SSP1BUF,W    ;
         banksel bl1            ;
         movwf     bl1           ;
         banksel SSP1BUF        ; select bank for SSP1BUF
         movf      Q1,W          ;
         movwf     SSP1BUF       ;
lp3      btfss     SSP1STAT,BF;
         bra       lp3           ;
         movf      SSP1BUF,W    ;
         banksel bl2            ;
         movwf     bl2           ;
         banksel SSP1BUF        ; select bank for SSP1BUF
         movf      Q0,W          ;
         movwf     SSP1BUF       ;
lp4      btfss     SSP1STAT,BF;
         bra       lp4           ;
         movf      SSP1BUF,W    ;
         banksel bl3            ;
         movwf     bl3           ;
    endif                        ;
;-----------------------------------------------------
```

FIGURE 2.35   *(1 of 3)* The macro "send_outputs".

these two operations are carried out by using two separate macros in two different time intervals. However, in this project these two tasks are carried out concurrently by using the SPI1 (Serial Peripheral Interface) module. There are two Master
Synchronous Serial Port Modules (MSSP1 and MSSP2) in the PIC16F1847 microcontroller. These modules can be configured either as an SPI (Serial Peripheral

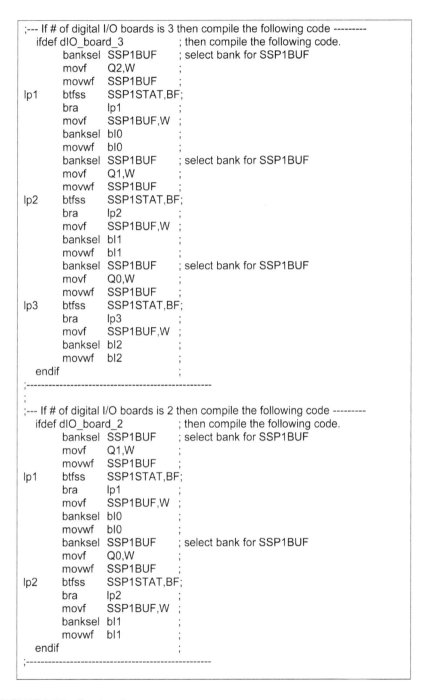

```
;--- If # of digital I/O boards is 3 then compile the following code ---------
     ifdef dIO_board_3           ; then compile the following code.
           banksel SSP1BUF       ; select bank for SSP1BUF
           movf    Q2,W          ;
           movwf   SSP1BUF       ;
lp1        btfss   SSP1STAT,BF;
           bra     lp1           ;
           movf    SSP1BUF,W     ;
           banksel bl0           ;
           movwf   bl0           ;
           banksel SSP1BUF       ; select bank for SSP1BUF
           movf    Q1,W          ;
           movwf   SSP1BUF       ;
lp2        btfss   SSP1STAT,BF;
           bra     lp2           ;
           movf    SSP1BUF,W     ;
           banksel bl1           ;
           movwf   bl1           ;
           banksel SSP1BUF       ; select bank for SSP1BUF
           movf    Q0,W          ;
           movwf   SSP1BUF       ;
lp3        btfss   SSP1STAT,BF;
           bra     lp3           ;
           movf    SSP1BUF,W     ;
           banksel bl2           ;
           movwf   bl2           ;
     endif                       ;
;----------------------------------------------
;
;
;--- If # of digital I/O boards is 2 then compile the following code ---------
     ifdef dIO_board_2           ; then compile the following code.
           banksel SSP1BUF       ; select bank for SSP1BUF
           movf    Q1,W          ;
           movwf   SSP1BUF       ;
lp1        btfss   SSP1STAT,BF;
           bra     lp1           ;
           movf    SSP1BUF,W     ;
           banksel bl0           ;
           movwf   bl0           ;
           banksel SSP1BUF       ; select bank for SSP1BUF
           movf    Q0,W          ;
           movwf   SSP1BUF       ;
lp2        btfss   SSP1STAT,BF;
           bra     lp2           ;
           movf    SSP1BUF,W     ;
           banksel bl1           ;
           movwf   bl1           ;
     endif                       ;
;----------------------------------------------
```

FIGURE 2.35 Continued

```
;--- If # of digital I/O boards is 1 then compile the following code ---------
     ifdef dIO_board_1            ; then compile the following code.
            banksel SSP1BUF       ; select bank for SSP1BUF
            movf    Q0,W          ;
            movwf   SSP1BUF       ;
lp1         btfss   SSP1STAT,BF;
            bra     lp1           ;
            movf    SSP1BUF,W     ;
            banksel bI0           ;
            movwf   bI0           ;
     endif                        ;
;----------------------------------------------------
;
;
;----------------------------------------------------
            banksel PORTB         ;" latch_out" bit is in PORTB
            bsf     latch_out     ; generate
            nop                   ; a latch_out
            bcf     latch_out     ; pulse
            clrwdt                ; clear the watchdog timer
            bcf     FRSTSCN       ; reset the FRSTSCN bit
            btfss   SCNOSC        ; toggle
            goto    L2            ; the SCNOSC bit
            bcf     SCNOSC        ; after a program
            goto    L1            ; scan
L2          bsf     SCNOSC        ;
L1                                ;
            endm                  ;
```

FIGURE 2.35 Continued

Interface) or as an Inter-Integrated Circuit (I²C). As explained in the initialization macro in this project, the MSSP1 is selected as an SPI. SPI1 Master mode is selected with clock = Fosc/4. Idle state for the clock is a low level. Serial port is enabled and configured as follows: SCK1 = RB4, SDO1 = RB2, SDI1 = RB1.

Let us now consider the contents of the macro "send_outputs". Firstly, a shift/load clock is generated by clearing the "shft_ld"(RB3 pin) signal and then after a time delay (1 cycle = 4T = 4 × 31.25 μs = 125 ns) obtained with a "nop" operation by setting the "shft_ld"(RB3 pin) signal again. Since this signal is connected to the parallel load input of the physical input registers 74HC/LS165 of I/O extension boards, data at the parallel data inputs of 74HC/LS165 ICs are asynchronously loaded into each of the eight internal stages.

If the number of digital I/O boards is 4 (if the label "dIO_board_4" is defined—if dIO_board_N = 4, then this label is defined in the macro "get_inputs"), then by using the SPI1 the following operations are carried out: the digital output byte Q3 (consisting of eight digital outputs Q3.0, Q3.1, ..., Q3.7) is sent out serially to be stored in the physical output register TPIC6B595 of the I/O extension board 3, and at the same time the bouncing digital input byte bI0 (consisting of eight digital inputs bI0.0, bI0.1, ..., bI0.7) is taken serially from the physical input register 74HC/LS165 of the

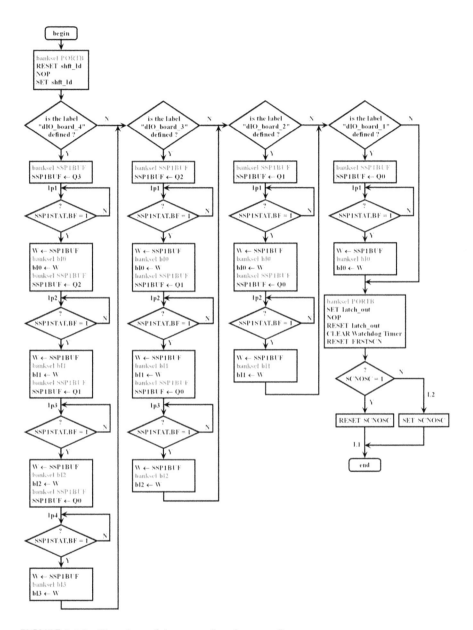

FIGURE 2.36 Flowchart of the macro "send_outputs".

I/O extension board 0. Next, the digital output byte Q2 (consisting of eight digital outputs Q2.0, Q2.1, …, Q2.7) is sent out serially to be stored in the physical output register TPIC6B595 of the I/O extension board 2, and at the same time the bouncing digital input byte bI1 (consisting of eight digital inputs bI1.0, bI1.1, …, bI1.7) is taken serially from the physical input register 74HC/LS165 of the I/O extension board 1. Then, the digital output byte Q1 (consisting of eight digital outputs Q1.0, Q1.1, …,

Q1.7) is sent out serially to be stored in the physical output register TPIC6B595 of the I/O extension board 1, and at the same time the bouncing digital input byte bI2 (consisting of eight digital inputs bI2.0, bI2.1, ..., bI2.7) is taken serially from the physical input register 74HC/LS165 of the I/O extension board 2. Finally, the digital output byte Q0 (consisting of eight digital outputs Q0.0, Q0.1, ..., Q0.7) is sent out serially to be stored in the physical output register TPIC6B595 of the I/O extension board 0, and at the same time the bouncing digital input byte bI3 (consisting of eight digital inputs bI3.0, bI3.1, ..., bI3.7) is taken serially from the physical input register 74HC/LS165 of the I/O extension board 3.

If the number of digital I/O boards is 3 (if the label "dIO_board_3" is defined—if dIO_board_N = 3, then this label is defined in the macro "get_inputs"), then by using the SPI1 the following operations are carried out: the digital output byte Q2 (consisting of eight digital outputs Q2.0, Q2.1, ..., Q2.7) is sent out serially to be stored in the physical output register TPIC6B595 of the I/O extension board 2, and at the same time the bouncing digital input byte bI0 (consisting of eight digital inputs bI0.0, bI0.1, ..., bI0.7) is taken serially from the physical input register 74HC/LS165 of the I/O extension board 0. Next, the digital output byte Q1 (consisting of eight digital outputs Q1.0, Q1.1, ..., Q1.7) is sent out serially to be stored in the physical output register TPIC6B595 of the I/O extension board 1, and at the same time the bouncing digital input byte bI1 (consisting of eight digital inputs bI1.0, bI1.1, ..., bI1.7) is taken serially from the physical input register 74HC/LS165 of the I/O extension board 1. Finally, the digital output byte Q0 (consisting of eight digital outputs Q0.0, Q0.1, ..., Q0.7) is sent out serially to be stored in the physical output register TPIC6B595 of the I/O extension board 0, and at the same time the bouncing digital input byte bI2 (consisting of eight digital inputs bI2.0, bI2.1, ..., bI2.7) is taken serially from the physical input register 74HC/LS165 of the I/O extension board 2.

If the number of digital I/O boards is 2 (if the label "dIO_board_2" is defined—if dIO_board_N = 2, then this label is defined in the macro "get_inputs"), then by using the SPI1 the following operations are carried out: the digital output byte Q1 (consisting of eight digital outputs Q1.0, Q1.1, ..., Q1.7) is sent out serially to be stored in the physical output register TPIC6B595 of the I/O extension board 1, and at the same time the bouncing digital input byte bI0 (consisting of eight digital inputs bI0.0, bI0.1, ..., bI0.7) is taken serially from the physical input register 74HC/LS165 of the I/O extension board 0. Next, the digital output byte Q0 (consisting of eight digital outputs Q0.0, Q0.1, ..., Q0.7) is sent out serially to be stored in the physical output register TPIC6B595 of the I/O extension board 0, and at the same time the bouncing digital input byte bI1 (consisting of eight digital inputs bI1.0, bI1.1, ..., bI1.7) is taken serially from the physical input register 74HC/LS165 of the I/O extension board 1.

If the number of digital I/O boards is 1 (if the label "dIO_board_1" is defined—if dIO_board_N = 1, then this label is defined in the macro "get_inputs"), then by using the SPI1 the following operations are carried out: the digital output byte Q0 (consisting of eight digital outputs Q0.0, Q0.1, ..., Q0.7) is sent out serially to be stored in the physical output register TPIC6B595 of the I/O extension board 0, and at the same time the bouncing digital input byte bI0 (consisting of eight digital inputs bI0.0, bI0.1, ..., bI0.7) is taken serially from the physical input register 74HC/LS165 of the I/O extension board 0.

Next, a latch_out clock is generated by setting the "latch_out" (RB0 pin) signal, and then after a time delay (1 cycle = 4T = 4 × 31.25 μs = 125 ns), obtained with a "nop" operation by clearing the "latch_out"(RB0 pin) signal again. Since this signal is connected to the register clock (RCK) input of the physical output registers TPIC6B595 of the I/O extension boards, digital output data (Q0, Q1, Q2, and Q3) are stored into each of the 8-bit D-type storage registers of TPIC6B595 ICs.

The watchdog timer is cleared within the macro "send_outputs".

In the second and last task, within this macro the special memory bit FRSTSCN is reset and the special memory bit SCNOSC is toggled after a program scan; i.e., when it is "1" it is reset, and when it is "0" it is set.

2.7   EXAMPLE PROGRAMS

Up to now we have seen the hardware and basic software necessary for the PIC16F1847-Based PLC. It is now time to consider a few examples. Before you can run the example programs considered here, you are expected to construct your own PIC16F1847-Based PLC hardware by using the necessary PCB files and by producing your PCBs with their components. For an effective use of examples, all example programs considered in this book (*Hardware and Basic Concepts*) are allocated within the file "PICPLC_PIC16F1847_user_Bsc.inc", which is downloadable from this book's webpage under the downloads section. Initially all example programs are commented out by putting a semicolon ";" in front of each line, as can be seen from the user program of Example 2.1 shown below:

```
;;__Example_Bsc_2.1_starts_from_here_____
;;_____
;
;       ld              LOGIC1
;       move_R          I0,Q0
;       move_R          I1,Q1
;       move_R          I2,Q2
;       move_R          I3,Q3
;;_____
;;__Example_Bsc_2.1_ends_here_____
```

When you would like to test one of the example programs you must uncomment each line of the example program by following the steps shown below:

1. Highlight the block of source lines you want to uncomment by dragging the mouse over these lines with the left mouse button held down. With default coloring in MPLAB X IDE you will now see green characters on a blue background.
2. Release the mouse button.
3. Press Ctrl/Shift/C or press the Alt, S, and M keys in succession, or select "Toggle Comment" from the toolbar Source menu. Now a semicolon will be removed from all selected source lines. With default coloring you will see red characters on a white background.

For example, when uncommented the user program of Example 2.1 will be seen as shown below:

```
;__Example_Bsc_2.1_starts_from_here_____
;_____
        ld              LOGIC1
        move_R          I0,Q0
        move_R          I1,Q1
        move_R          I2,Q2
        move_R          I3,Q3
;_____
;__Example_Bsc_2.1_ends_here_____
```

Then, you can run the project by pressing the symbol ▷ from the toolbar. Next, the MPLAB X IDE will produce the "PICPLC_PIC16F1847.X.production.hex" file for the project. Then the MPLAB X IDE will be connected to the PICkit 3 programmer and finally it will program the PIC16F1847 microcontroller within the CPU board of the PIC16F1847-Based PLC. During these steps, make sure that in the CPU board of the PIC16F1847-Based PLC, the 4PDT switch is in the "PROG" position and the power switch is in the "OFF" position. After loading the program file to the PIC16F1847 microcontroller, switch the 4PDT to "RUN" and the power switch to the "ON" position. Finally, you are ready to test the example program. **Warning**: When you finish your study with an example and try to take a look at another example, do not forget to comment the current example program before uncommenting another one. In other words, make sure that only one example program is uncommented and tested at the same time. Otherwise, if you somehow leave more than one example uncommented, the example you are trying to test will probably not function as expected since it may try to access the same resources that are being used and changed by other examples.

2.7.1 EXAMPLE 2.1

This example is aimed at testing the PIC16F1847-Based PLC with all digital inputs and all digital outputs to see if everything functions properly as expected. The ladder diagram of Example 2.1 and its user program are shown in Figures 2.37 and 2.38, respectively. When the project file of the PIC16F1847-Based PLC is open in the MPLAB X IDE, from the file "PICPLC_PIC16F1847_user_Bsc.inc", if you uncomment this example and run the project by pressing the symbol ▷ from the toolbar, then the PIC16F1847 microcontroller within the CPU board of the PIC16F1847-Based PLC will be programmed. Next you are ready to test the operation of this example. Although the macro "move_R" is explained in Chapter 4 of *Intermediate Concepts*, its function is straightforward. When enabled, i.e., when EN = 1, OUT := IN. So the macro "move_R" copies the contents of the 8-bit input register IN to the 8-bit output register OUT when it is enabled. Since all "move_R" functions are enabled with a LOGIC1, as soon as the PLC is switched on, Q0 := I0, Q1 := I1, Q2 := I2, and Q3 := I3. In other words, the logic states of the 8-bit digital inputs I0.0, I0.1, ..., I0.7 are transferred to the 8-bit digital outputs Q0.0, Q0.1, ..., Q0.7, respectively.

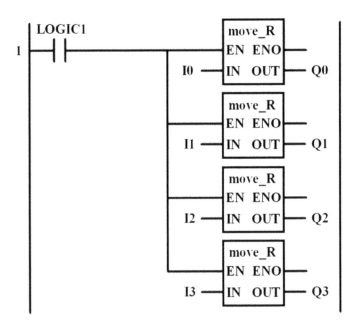

FIGURE 2.37   Ladder diagram of Example 2.1.

```
user_program_1   macro

;
;--- PLC codes to be allocated in the "user_program_1" macro start from here ---
;__Example_Bsc_2.1

        ld              LOGIC1
        move_R          I0,Q0
        move_R          I1,Q1
        move_R          I2,Q2
        move_R          I3,Q3
;
;--- PLC codes to be allocated in the "user_program_1" macro end here ----------
        endm
```

FIGURE 2.38   The user program for the ladder diagram shown in Figure 2.37.

The logic states of the 8-bit digital inputs I1.0, I1.1, …, I1.7 are transferred to the 8-bit digital outputs Q1.0, Q1.1, …, Q1.7, respectively. The logic states of the 8-bit digital inputs I2.0, I2.1, …, I2.7 are transferred to the 8-bit digital outputs Q2.0, Q2.1, …, Q2.7, respectively. Finally, the logic states of the 8-bit digital inputs I3.0, I3.1, …, I3.7 are transferred to the 8-bit digital outputs Q3.0, Q3.1, …, Q3.7, respectively.

2.7.2   EXAMPLE 2.2

This example is aimed at observing the contents of special memory bits allocated in the 8-bit registers SMB1 and SMB2. The ladder diagram of Example 2.2 and its user program are shown in Figures 2.39 and 2.40, respectively. When the project file of

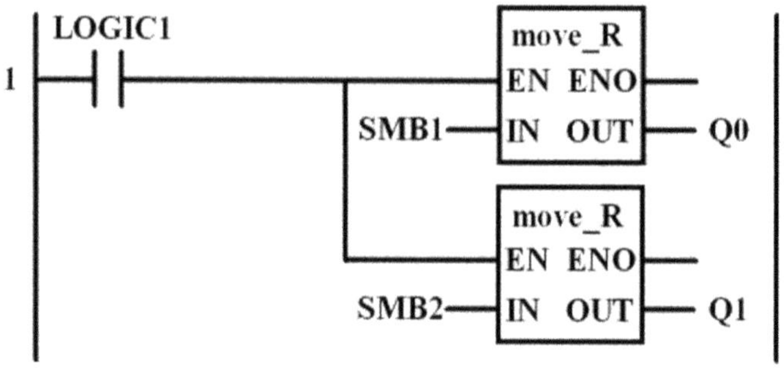

FIGURE 2.39   Ladder diagram of Example 2.2.

```
user_program_1   macro
;
;--- PLC codes to be allocated in the "user_program_1" macro start from here ---
;__Example_Bsc_2.2

        ld              LOGIC1
        move_R          SMB1,Q0
        move_R          SMB2,Q1

;--- PLC codes to be allocated in the "user_program_1" macro end here ----------
        endm
```

FIGURE 2.40   The user program for the ladder diagram shown in Figure 2.39.

the PIC16F1847-Based PLC is open in the MPLAB X IDE, from the file "PICPLC _PIC16F1847_user_Bsc.inc", if you uncomment this example and run the project by pressing the symbol ▷ from the toolbar, then the PIC16F1847 microcontroller within the CPU board of the PIC16F1847-Based PLC will be programmed. Next you are ready to test the operation of this example. The macro "move_R" copies the contents of the 8-bit input register IN to the 8-bit output register OUT when it is enabled, i.e., when EN = 1. Since two "move_R" functions are enabled with a LOGIC1, as soon as the PLC is switched on, Q0 := SMB1 and Q1 := SMB2. In other words, the logic states of the special memory bits LOGIC0, LOGIC1, FRSTSCN, and SCNOSC of SMB1 are transferred to the digital outputs Q0.0, Q0.1, Q0.2, and Q0.3, respectively. Bit 4, Bit 5, Bit 6, and Bit 7 of SMB1 are also transferred to the digital outputs Q0.4, Q0.5, Q0.6, and Q0.7, respectively, but since they are used in the "re_RTS" macro they have no particular meaning in this example. Logic states of the special memory bits T_2ms, T_10ms, T_100ms, T_1s, re_T2ms, re_T1s, re_T10ms, and re_T100ms of SMB2 are transferred to the digital outputs Q1.0, Q1.1, ..., Q1.7, respectively.

2.7.3   EXAMPLE 2.3

The purpose of this example is to show the effect of a contact bouncing problem. The ladder diagram of Example 2.3 and its user program are shown in Figures 2.41 and 2.42, respectively. When the project file of the PIC16F1847-Based PLC is open

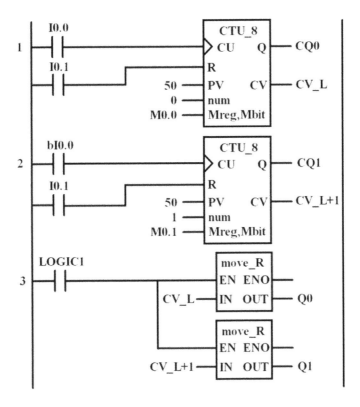

FIGURE 2.41   Ladder diagram of Example 2.3.

```
user_program_1   macro
;
;--- PLC codes to be allocated in the "user_program_1" macro start from here ---
;__Example_Bsc_2.3

                                           ;rung 1
                          ;num,Mreg,Mbit,CUreg,CUbit,Rreg,Rbit,PV
          CTU_8           0,M0.0,I0.0,I0.1,.50

                                           ;rung 2
                          ;num,Mreg,Mbit,CUreg,CUbit,Rreg,Rbit,PV
          CTU_8           1,M0.1,bI0.0,I0.1,.50

          ld              LOGIC1                    ;rung 3
          move_R          CV_L,Q0
          move_R          CV_L+1,Q1

;--- PLC codes to be allocated in the "user_program_1" macro end here ----------
          endm
```

FIGURE 2.42   The user program for the ladder diagram shown in Figure 2.41.

in the MPLAB X IDE, from the file "PICPLC_PIC16F1847_user_Bsc.inc", if you uncomment this example and run the project by pressing the symbol ▷ from the tool-bar, then the PIC16F1847 microcontroller within the CPU board of the PIC16F1847-Based PLC will be programmed. Next you are ready to test the operation of this example. Although the macro "CTU_8" is explained in Chapter 6 of this book, its function is as follows. This macro implements an 8-bit up counter. The up counter counts the number of rising edges (↑) detected at the Boolean input CU. PV defines the maximum value for the counter. For 8-bit resolution, PVmax = 255. Each time the counter is called with a new rising edge (↑) on CU, the count value CV is incre-mented by one. When the current count value CV is greater than or equal to the PV, the counter output Q is set true (ON – 1). The reset input R can be used to set the output Q false (OFF – 0), and clear the count value CV to zero. "num" is the counter number 0, 1, ..., 79. "Mreg,Mbit" is a unique memory bit and it is used to detect the rising edge of the input CU. The first rung in the ladder diagram defines the first up counter with CU = I0.0, R = I0.1, PV = 50, num = 0, Mreg,Mbit = M0.0, the counter output = CQ0 (in this example it is not used), and CV = CV_L. Rung 2 in the ladder diagram defines the second up counter with CU = bI0.0, R = I0.1, PV = 50, num = 1, Mreg,Mbit = M0.1, the counter output = CQ1 (in this example it is not used), and CV = CV_L+1. The macro "move_R" copies the contents of the 8-bit input register IN to the 8-bit output register OUT when it is enabled, i.e., when EN = 1. In the last rung of the ladder diagram, since two "move_R" functions are enabled with a LOGIC1, as soon as the PLC is switched on, Q0 = CV_L and Q1 = CV_L+1. It is interesting to see that the CU input of the first counter is I0.0, while the CU input of the second counter is bI0.0. This means that when R1 = R2 = I0.1 = 0, the first counter will count up with each rising edge (↑) of the debounced digital input I0.0, while the second counter will count up with each rising edge (↑) of the bouncing digital input bI0.0. This means that when you test this example, if you press the push button connected to the I0.0 input once, the first counter will definitely count up once but the second counter may count up twice or even three or four times. This can be observed from the outputs Q0 and Q1 where the current count values of the first counter and the second counter are stored, respectively.

2.7.4 EXAMPLE 2.4

The purpose of this example is to show the effect of low-pass digital filters applied on the analog inputs. The ladder diagram of Example 2.4 and its user program are shown in Figures 2.43 and 2.44, respectively. Figure 2.45 shows the test circuit for Example 2.4. In this test circuit, the output of the 5.00V voltage reference module is connected to the VREF+ input (RA3 pin). By means of the P1 potentiometer, an analog input value from 0V to 5V can be applied to the AI0 analog input. An RC low-pass filter with a cut-off frequency of 48 Hz is connected just before the AI0 pin. Note that this RC low-pass filter is not enough to filter noisy high-frequency signals from analog inputs. Therefore, a digital filter is used for each analog input within this project, as explained before. A voltmeter is used to measure the value of the input voltage. Table 2.8 shows some example 0V–5V analog voltage input values vs. ideal digital values to be obtained from a 10-bit ADC of a 5V microcontroller.

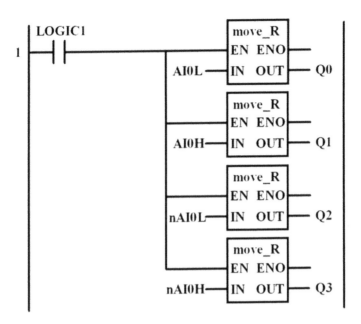

FIGURE 2.43 Ladder diagram of Example 2.4.

```
user_program_1   macro
;
;--- PLC codes to be allocated in the "user_program_1" macro start from here ---
;__Example_Bsc_2.4

        ld              LOGIC1                  ;rung 1
        move_R          AI0L,Q0
        move_R          AI0H,Q1
        move_R          nAI0L,Q2
        move_R          nAI0H,Q3

;--- PLC codes to be allocated in the "user_program_1" macro end here ----------
        endm
```

FIGURE 2.44 The user program for the ladder diagram shown in Figure 2.43.

When the project file of the PIC16F1847-Based PLC is open in the MPLAB X IDE, from the file "PICPLC_PIC16F1847_user_Bsc.inc", if you uncomment this example and run the project by pressing the symbol ▷ from the toolbar, then the PIC16F1847 microcontroller within the CPU board of the PIC16F1847-Based PLC will be programmed. Next you are ready to test the operation of this example. The macro "move_R" copies the contents of the 8-bit input register IN to the 8-bit output register OUT when it is enabled. Since all "move_R" functions are enabled with a LOGIC1, as soon as the PLC is switched on, Q0 := AI0L, Q1 := AI0H, Q2 := nAI0L, and Q3 := nAI0H. Register pair AI0H and AI0L hold the filtered 10-bit digital value of the analog input AI0. Likewise, register pair nAI0H and nAI0L hold the

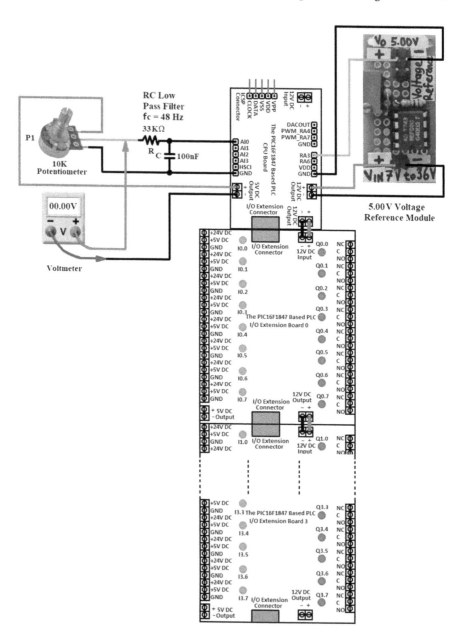

FIGURE 2.45   Test circuit for Example 2.4.

noisy 10-bit digital value of the analog input AI0. Table 2.10 depicts input test values applied to and digital values obtained from the test circuit shown in Figure 2.45. You can check these digital values by chancing the input voltage value. In addition, you can observe the difference between the 10-bit noisy digital value and the 10-bit filtered digital value for the analog input AI0.

TABLE 2.10
Input Test Values Applied to and Digital Values Obtained from the Test
Circuit Shown in Figure 2.45

| V_IN(V) | AI0H | | AI0L | | | | | | | | Decimal |
	Q1.1 MSB	Q1.0	Q0.7	Q0.6	Q0.5	Q0.4	Q0.3	Q0.2	Q0.1	Q0.0 LSB	
5.00	1	1	1	1	1	1	1	1	0	1	1021
..											
4.75	1	1	1	1	0	0	0	1	1	1	967
..											
4.50	1	1	1	0	0	1	0	1	0	1	917
..											
4.25	1	1	0	1	1	0	0	0	0	1	865
..											
4.00	1	1	0	0	1	0	1	1	1	1	815
..											
3.75	1	0	1	1	1	1	1	1	0	1	765
..											
3.50	1	0	1	1	0	0	1	0	0	1	713
..											
3.25	1	0	1	0	0	1	0	1	1	0	662
..											
3.00	1	0	0	1	1	0	0	1	0	0	612
..											
2.75	1	0	0	0	1	1	0	0	1	0	562
..											
2.50	1	0	0	0	0	0	0	0	0	0	512
..											
2.25	0	1	1	1	0	0	1	1	0	0	460
..											
2.00	0	1	1	0	0	1	1	0	0	1	409
..											
1.75	0	1	0	1	1	0	0	1	1	1	359
..											
1.50	0	1	0	0	1	1	0	1	0	1	309
..											
1.25	0	1	0	0	0	0	0	0	1	0	258
..											
1.00	0	0	1	1	0	0	1	1	1	1	207
..											
0.75	0	0	1	0	0	1	1	1	0	0	156
..											
0.50	0	0	0	1	1	0	1	0	0	1	105
..											
0.25	0	0	0	0	1	1	0	1	1	0	54
..											
0.00	0	0	0	0	0	0	0	0	1	0	2

2.7.5 EXAMPLE 2.5

The purpose of this example is to show not only the effect of low-pass digital filters applied on all analog inputs, but also the use of all analog inputs at the same time. The ladder diagram of Example 2.5 and its user program are shown in Figures 2.46 and 2.47, respectively. Figure 2.48 shows the test circuit for Example 2.5. In this test circuit, the output of the 5.00V voltage reference module is connected to the VREF+ input (RA3 pin). By means of the P0 (or P1, P2, and P3, respectively) potentiometer, an analog input value from 0V to 5V can be applied to the AI0 (or AI1, AI2, and AI3, respectively) analog input. An RC low-pass filter with a cut-off frequency of 48 Hz is connected just before the AI0 (or AI1, AI2, and AI3, respectively) pin. Note that these RC low-pass filters are not enough to filter noisy high-frequency signals from analog inputs. Therefore, a digital filter is used for each analog input within this project, as explained before. A voltmeter is used to measure the value of the input voltages. Table 2.8 shows some example 0V–5V analog voltage input values vs. ideal digital values to be obtained from a 10-bit ADC of a 5V microcontroller.

When the project file of the PIC16F1847-Based PLC is open in the MPLAB X IDE, from the file "PICPLC_PIC16F1847_user_Bsc.inc", if you uncomment this example and run the project by pressing the symbol ▷ from the toolbar, then the PIC16F1847 microcontroller within the CPU board of the PIC16F1847-Based PLC will be programmed. Next you are ready to test the operation of this example. Although the macro "B_mux_4_1_E" is explained in Chapter 4 of *Intermediate Concepts*, its function is as follows. When enabled (E = 1), the contents of the 8-bit destination variable Y will be taken from the 8-bit source variable R0 (or R1, R2, R3, respectively) if select inputs s1s0 = 00 (or 01, 10, 11, respectively). When not enabled (E = 0), the contents of the destination register Y remain unchanged. Register pair AI0H and AI0L (or AI1H and AI1L, AI2H and AI2L, AI3H and AI3L, respectively) holds the 10-bit digital value of the **filtered** analog input AI0 (or AI1, AI2, AI3, respectively). Register pair nAI0H and nAI0L (or nAI1H and nAI1L, nAI2H and nAI2L, nAI3H and nAI3L, respectively) holds the 10-bit digital value of the **noisy** analog input AI0 (or AI1, AI2, AI3, respectively).

Now let us take a look at the ladder diagram. In this ladder diagram, there are four 4x1 byte multiplexers. Enable inputs (E) are all connected to LOGIC1. Therefore, these 4x1 byte multiplexers are always enabled as soon as the PLC is switched on. Rung 1: the contents of the 8-bit destination variable Q0 will be taken from 8-bit source variable AI0L (or AI1L, AI2L, AI3L, respectively) if select inputs s1s0 = 00 (or 01, 10, 11, respectively). Rung 2: the contents of the 8-bit destination variable Q1 will be taken from 8-bit source variable AI0H (or AI1H, AI2H, AI3H, respectively) if select inputs s1s0 = 00 (or 01, 10, 11, respectively). Rung 3: the contents of the 8-bit destination variable Q2 will be taken from 8-bit source variable nAI0L (or nAI1L, nAI2L, nAI3L, respectively) if select inputs s1s0 = 00 (or 01, 10, 11, respectively). Rung 4: the contents of the 8-bit destination variable Q3 will be taken from 8-bit source variable nAI0H (or nAI1H, nAI2H, nAI3H, respectively) if select inputs s1s0 = 00 (or 01, 10, 11, respectively).

The following table summarizes the operation of this example program. Select inputs of these four 4x1 byte multiplexers are connected to the same inputs, i.e., s1 =

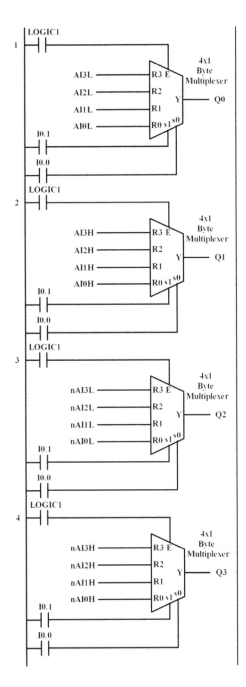

FIGURE 2.46   Ladder diagram of Example 2.5.

```
user_program_1   macro
;
;--- PLC codes to be allocated in the "user_program_1" macro start from here ---
;__Example_Bsc_2.5

        ld              LOGIC1              ;rung 1
                                            ;s1,s0,R3,R2,R1,R0,OUT
        B_mux_4_1_E  I0.1,I0.0,AI3L,AI2L,AI1L,AI0L,Q0

        ld              LOGIC1              ;rung 2
                                            ;s1,s0,R3,R2,R1,R0,OUT
        B_mux_4_1_E  I0.1,I0.0,AI3H,AI2H,AI1H,AI0H,Q1

        ld              LOGIC1              ;rung 3
                                            ;s1,s0,R3,R2,R1,R0,OUT
        B_mux_4_1_E  I0.1,I0.0,nAI3L,nAI2L,nAI1L,nAI0L,Q2

        ld              LOGIC1              ;rung 4
                                            ;s1,s0,R3,R2,R1,R0,OUT
        B_mux_4_1_E  I0.1,I0.0,nAI3H,nAI2H,nAI1H,nAI0H,Q3

;--- PLC codes to be allocated in the "user_program_1" macro end here ----------
        endm
```

FIGURE 2.47   The user program for the ladder diagram shown in Figure 2.46.

I0.1, s0 = I0.0. Thus when s1s0 = 00 (or 01, 10, 11, respectively), output pair Q1 and Q0 will hold the contents of the register pair AI0H and AI0L (or AI1H and AI1L, AI2H and AI2L, AI3H and AI3L, respectively) and output pair Q3 and Q2 will hold the contents of the register pair nAI0H and nAI0L (or nAI1H and nAI1L, nAI2H and nAI2L, nAI3H and nAI3L, respectively).

s1 I0.1	s0 I0.0	Q0	Q1	Q2	Q3
0	0	AI0L	AI0H	nAI0L	nAI0H
0	1	AI1L	AI1H	nAI1L	nAI1H
1	0	AI2L	AI2H	nAI2L	nAI2H
1	1	AI3L	AI3H	nAI3L	nAI3H

Table 2.10 depicts input test values applied to AI0 when s1s0 = 00 by using the potentiometer P0 and digital values obtained from the test circuit shown in Figure 2.48. Similar digital values can be obtained for AI1 (or AI2, AI3, respectively) when s1s0 = 01 (or 10, 11, respectively), by using the potentiometer P1 (or P2, P3, respectively). You can check these digital values by chancing the input voltage values. In addition, you can observe the difference between the 10-bit noisy digital values and the 10-bit filtered digital values for the analog inputs AI0, AI1, AI2, and AI3.

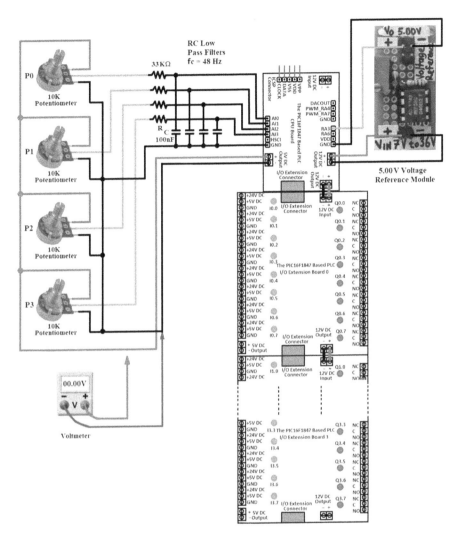

FIGURE 2.48   Test circuit for Example 2.5.

2.7.6 EXAMPLE 2.6

The purpose of this example is to test the operation of the digital-to-analog converter (DAC) module of the PIC16F1847-Based PLC. The ladder diagram of Example 2.6 and its user program are shown in Figures 2.49 and 2.50, respectively. There are two test circuits for this example, as shown in Figures 2.51 and 2.52. In the former, the output of the 5.00V voltage reference module is connected to the VREF+ input (RA3 pin). By means of the lower five bits of I0, namely I0.4 (MSB), I0.3, I0.2, I0.1, and I0.0 (LSB), DAC voltage output select bits DACR<4>, DACR<3>, DACR<2>, DACR<1>, and DACR<0> are applied. A voltmeter is used to measure the value of the input voltages. In the latter, the 5.17V output of the LM2596 step-down voltage regulator module is connected to the VREF+ input (RA3 pin). By means of the

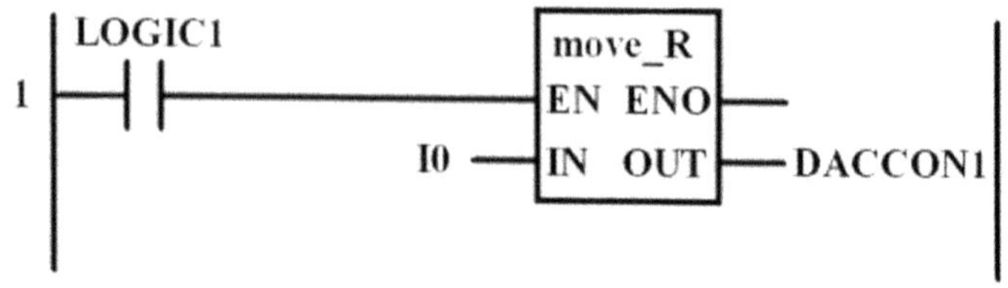

FIGURE 2.49 Ladder diagram of Example 2.6.

```
user_program_1  macro
;
;--- PLC codes to be allocated in the "user_program_1" macro start from here ---
;__Example_Bsc_2.6

        ld              LOGIC1                  ;rung 1
        move_R          I0,DACCON1

;--- PLC codes to be allocated in the "user_program_1" macro end here ----------
        endm
```

FIGURE 2.50 The user program for the ladder diagram shown in Figure 2.49.

lower five bits of I0, namely I0.4 (MSB), I0.3, I0.2, I0.1, and I0.0 (LSB), DAC voltage output select bits DACR<4>, DACR<3>, DACR<2>, DACR<1>, and DACR<0> are applied. A voltmeter is used to measure the value of the input voltages. It can be seen that the only difference between these two circuits is that of the reference voltage value applied to the VREF+ input (RA3 pin). The same example program is tested with these two different circuits.

When the project file of the PIC16F1847-Based PLC is open in the MPLAB X IDE, from the file "PICPLC_PIC16F1847_user_Bsc.inc", if you uncomment this example and run the project by pressing the symbol ▷ from the toolbar, then the PIC16F1847 microcontroller within the CPU board of the PIC16F1847-Based PLC will be programmed. Next you are ready to test the operation of this example. Since the "move_R" function is enabled with a LOGIC1, as soon as the PLC is switched on, DACCON1 = I0. In other words, the logic states of the 8-bit digital inputs I0.0, I0.1, ..., I0.7 are transferred to the 8-bit DACCON1 register. Recall that the DAC module has a 5-bit resolution and supplies a variable voltage reference, ratiometric with the input source, with 32 selectable output voltage levels, which are set with the DACR<4:0> bits of the DACCON1 register. The lower five bits of DACCON1 register, namely DACR<4>, DACR<3>, DACR<2>, DACR<1>, and DACR<0>, are used as DAC voltage output select bits, while the upper three bits are unimplemented. Therefore, the "move_R" function will transfer the state of the lower five bits of I0 to the lower five bits of the DACCON1 register, i.e., DACR<4> = I0.4, DACR<3> = I0.3, DACR<2> = I0.2, DACR<1> = I0.1, and DACR<0> = I0.0. Table 2.11 shows 5-bit digital input test values applied to and analog voltage output values obtained from the test circuit 1 shown in Figure 2.51. Similarly, Table 2.12 depicts 5-bit digital input test values applied to and analog voltage output values obtained from the test

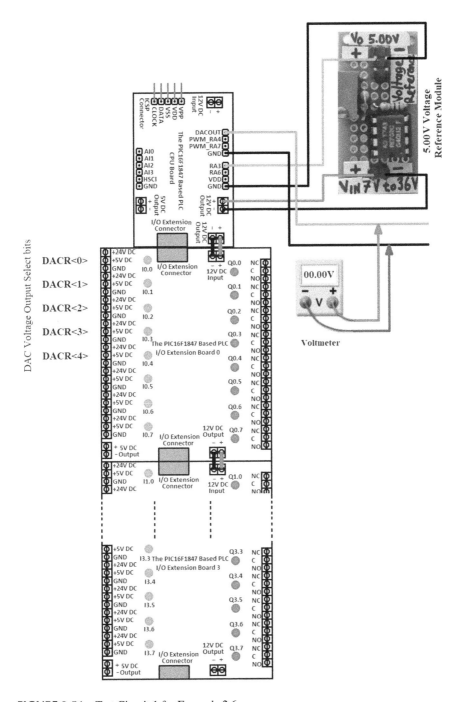

FIGURE 2.51   Test Circuit 1 for Example 2.6.

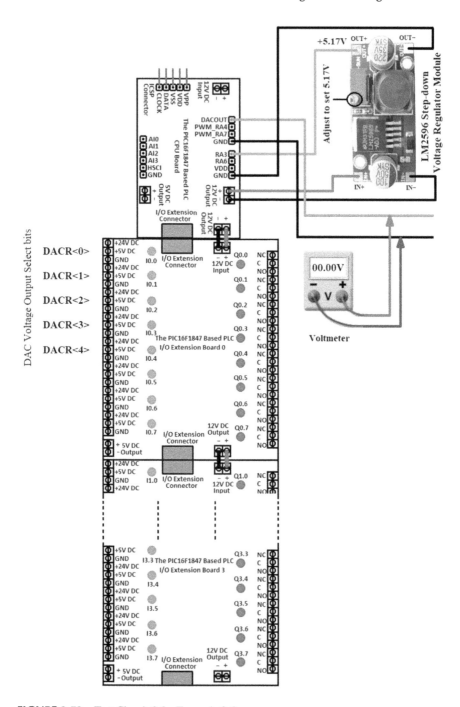

FIGURE 2.52   Test Circuit 2 for Example 2.6.

TABLE 2.11
5-Bit Digital Input Test Values Applied to and Analog Voltage Output Values Obtained from the Test Circuit 1 Shown in Figure 2.51

Decimal	I0.4 MSB	I0.3	I0.2	I0.1	I0.0 LSB	$V_{OUT}(V)$
31	1	1	1	1	1	4.87
30	1	1	1	1	0	4.71
29	1	1	1	0	1	4.55
28	1	1	1	0	0	4.40
27	1	1	0	1	1	4.24
26	1	1	0	1	0	4.08
25	1	1	0	0	1	3.92
24	1	1	0	0	0	3.76
23	1	0	1	1	1	3.61
22	1	0	1	1	0	3.45
21	1	0	1	0	1	3.29
20	1	0	1	0	0	3.14
19	1	0	0	1	1	2.98
18	1	0	0	1	0	2.80
17	1	0	0	0	1	2.66
16	1	0	0	0	0	2.51
15	0	1	1	1	1	2.35
14	0	1	1	1	0	2.19
13	0	1	1	0	1	2.03
12	0	1	1	0	0	1.88
11	0	1	0	1	1	1.72
10	0	1	0	1	0	1.56
9	0	1	0	0	1	1.41
8	0	1	0	0	0	1.25
7	0	0	1	1	1	1.10
6	0	0	1	1	0	0.95
5	0	0	1	0	1	0.79
4	0	0	1	0	0	0.63
3	0	0	0	1	1	0.47
2	0	0	0	1	0	0.31
1	0	0	0	0	1	0.16
0	0	0	0	0	0	0.00

circuit 2 shown in Figure 2.52. To conclude, you can check these two circuits by chancing 5-bit digital inputs and by measuring the output voltage values and compare your results with those provided in Tables 2.11 and 2.12. It is important to see that the highest output voltage value obtained from the circuit shown in Figure 2.51 is 4.87V, while it is 5.00V from the circuit shown in Figure 2.52.

TABLE 2.12

5-Bit Digital Input Test Values Applied to and Analog Voltage Output Values Obtained from the Test Circuit 2 Shown in Figure 2.52

Decimal	I0.4 MSB	I0.3	I0.2	I0.1	I0.0 LSB	$V_{OUT}(V)$
31	1	1	1	1	1	5.00
30	1	1	1	1	0	4.84
29	1	1	1	0	1	4.68
28	1	1	1	0	0	4.52
27	1	1	0	1	1	4.36
26	1	1	0	1	0	4.19
25	1	1	0	0	1	4.03
24	1	1	0	0	0	3.87
23	1	0	1	1	1	3.71
22	1	0	1	1	0	3.55
21	1	0	1	0	1	3.38
20	1	0	1	0	0	3.23
19	1	0	0	1	1	3.06
18	1	0	0	1	0	2.90
17	1	0	0	0	1	2.74
16	1	0	0	0	0	2.58
15	0	1	1	1	1	2.42
14	0	1	1	1	0	2.26
13	0	1	1	0	1	2.10
12	0	1	1	0	0	1.94
11	0	1	0	1	1	1.78
10	0	1	0	1	0	1.61
9	0	1	0	0	1	1.45
8	0	1	0	0	0	1.29
7	0	0	1	1	1	1.13
6	0	0	1	1	0	0.97
5	0	0	1	0	1	0.81
4	0	0	1	0	0	0.65
3	0	0	0	1	1	0.48
2	0	0	0	1	0	0.32
1	0	0	0	0	1	0.16
0	0	0	0	0	0	0.00

REFERENCE

R2.1. Abel Raynus, Armatron International, Malden, M.A.; Edited by Charles H. Small and Fran Granville, "8-bit microcontroller implements digital lowpass filter", January 24, 2008. Accessed on 26.04.2018 (https://www.edn.com/design/integrated-circuit-design/4323639/8-bit-microcontroller-implements-digital-lowpass-filter).

3 Contact and Relay-Based Macros

INTRODUCTION

In this chapter, the following contact and relay-based macros are described:

ld (load),
ld_not (load_not),
not,
or,
or_not,
nor,
and,
and_not,
nand,
xor,
xor_not,
xnor,
out,
out_not,
mid_out (midline output),
mid_out_not (inverted midline output),
in_out,
inv_out,
_set,
_reset,
SR (set–reset),
RS (reset–set),
r_edge (rising edge detector),
f_edge (falling edge detector),
r_toggle (output toggle with rising edge detector),
f_toggle (output toggle with falling edge detector),
adrs_re (Address rising edge detector),
adrs_fe (Address falling edge detector),
setBF (set bit field),
resetBF (reset bit field).

The file "PICPLC_PIC16F1847_macros_Bsc.inc", which is downloadable from this book's webpage under the downloads section, contains macros defined for the PIC16F1847-Based PLC explained in this book (*Hardware and Basic Concepts*).

The contact and relay-based macros are defined to operate on Boolean (1-bit) variables. Let us now consider these macros.

3.1 MACRO "LD" (LOAD)

The truth table and symbols of the macro "ld" are depicted in Table 3.1. Figure 3.1 shows the macro "ld" and its flowchart. This macro has a Boolean input variable IN passed into it as "reg,bit" and a Boolean output variable OUT passed out through "W". In ladder logic, this macro is represented by a normally open (NO) contact. When IN is 0 (or 1, respectively), OUT (W,0) is forced to be 0 (or 1, respectively). Assumption: the operand IN (reg,bit) can be in any bank. Operands for the instruction "ld" are shown in Table 3.2.

TABLE 3.1

Truth Table and Symbols of the Macro "ld"

Truth Table		Ladder Diagram Symbol	Schematic Symbol
IN	OUT	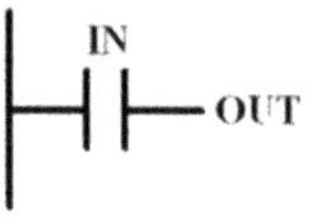	
reg,bit	W		
0	0		
1	1		

input & output

IN (Boolean input variable, reg,bit) = 0 or 1

OUT (Boolean output variable, through W) = 0 or 1

```
ld    macro   reg,bit
      banksel reg
      movlw   0
      btfsc   reg,bit
      movlw   1
      endm
```

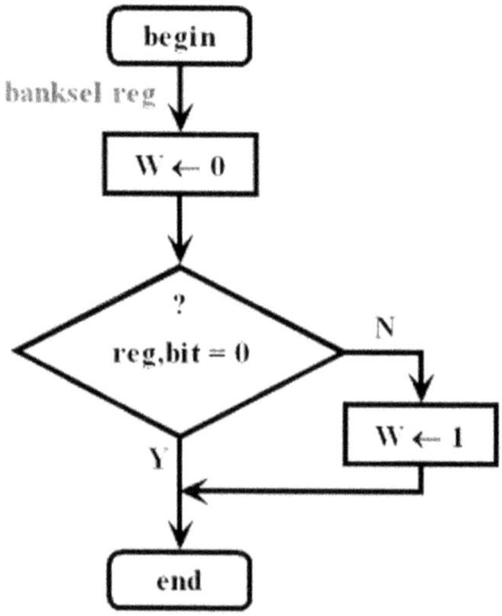

FIGURE 3.1 The macro "ld" and its flowchart.

TABLE 3.2
Operands for the Instruction "ld"

IN (reg,bit)	Data type	Operands
Bit	BOOL	I, Q, M, T_Q, C_Q, C_QD, C_QU, drum_Q, drum_TQ, drum_steps, drum_events, SF, MB, nAI, AI, LOGIC1, LOGIC0, FRSTSCN, SCNOSC, any Boolean variable from SRAM.

3.2 MACRO "LD_NOT" (LOAD_NOT)

The truth table and symbols of the macro "ld_not" are depicted in Table 3.3. Figure 3.2 shows the macro "ld_not" and its flowchart. This macro has a Boolean input variable IN passed into it as "reg,bit" and a Boolean output variable OUT passed out through "W". In ladder logic, this macro is represented by a normally closed (NC) contact. When IN is 0 (or 1, respectively), OUT (W,0) is forced to be 1 (or 0, respectively). Assumption: the operand IN (reg,bit) can be in any bank. Operands for the instruction "ld_not" are shown in Table 3.4.

TABLE 3.3
Truth Table and Symbols of the Macro "ld_not"

Truth Table		Ladder Diagram Symbol	Schematic Symbol

IN	OUT
reg,bit	W
0	1
1	0

input & output

IN (Boolean input variable, reg,bit) = 0 or 1
OUT (Boolean output variable, through W) = 0 or 1

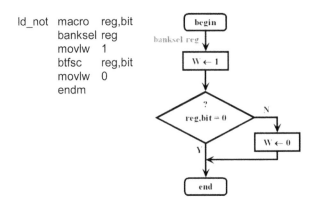

```
ld_not   macro   reg,bit
         banksel reg
         movlw   1
         btfsc   reg,bit
         movlw   0
         endm
```

FIGURE 3.2 The macro "ld_not" and its flowchart.

TABLE 3.4

Operands for the Instruction "ld_not"

IN (regi,bit)	Data type	Operands
Bit	BOOL	I, Q, M, T_Q, C_Q, C_QD, C_QU, drum_Q, drum_TQ, drum_steps, drum_events, SF, MB, nAI, AI, LOGIC1, LOGIC0, FRSTSCN, SCNOSC, any Boolean variable from SRAM.

TABLE 3.5

Truth Table and Symbols of the Macro "not"

Truth Table		Ladder Diagram Symbol	Schematic Symbol
IN	OUT	IN —|NOT|— OUT	IN —▷o— OUT
W	W		
0	1		
1	0		

input & output

IN (Boolean input variable, through W) = 0 or 1

OUT (Boolean output variable, through W) = 0 or 1

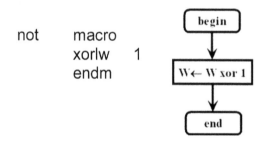

FIGURE 3.3   The macro "not" and its flowchart.

3.3   MACRO "NOT"

The truth table and symbols of the macro "not" are depicted in Table 3.5. Figure 3.3 shows the macro "not" and its flowchart. This macro is used as a logical NOT gate. The Boolean input variable IN is taken from "W" and the Boolean output variable OUT is send out by "W". When IN is 0 (or 1, respectively), OUT (W,0) is forced to be 1 (or 0, respectively).

3.4   MACRO "OR"

The truth table and symbols of the macro "or" are depicted in Table 3.6. Figure 3.4 shows the macro "or" and its flowchart. This macro has two Boolean input variables IN1, passed in through "W", and IN2, passed in through "reg,bit", and a Boolean

TABLE 3.6
Truth Table and Symbols of the Macro "or"

Truth Table			Ladder Diagram Symbol	Schematic Symbol

IN1	IN2	OUT
W	reg,bit	W
0	0	0
0	1	1
1	0	1
1	1	1

inputs & output

IN1 (Boolean input variable 1, through W) = 0 or 1
IN2 (Boolean input variable 2, reg,bit) = 0 or 1
OUT (Boolean output variable, through W) = 0 or 1

```
or    macro    reg,bit
      banksel  reg
      movwf    Temp_1
      movlw    0
      btfsc    reg,bit
      movlw    1
      iorwf    Temp_1,w
      endm
```

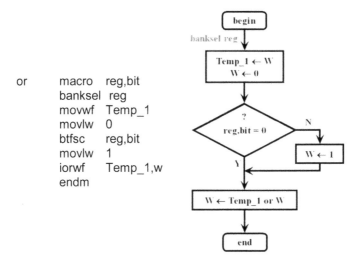

FIGURE 3.4 The macro "or" and its flowchart.

output variable OUT, passed out through "W". This macro is used as a two-input logical OR gate. IN1 should be loaded into (W,0) before this macro is invoked. The operand IN2 (reg,bit) can be in any bank. Operands for the instruction "or" are shown in Table 3.7.

3.5 MACRO "OR_NOT"

The truth table and symbols of the macro "or_not" are depicted in Table 3.8. Figure 3.5 shows the macro "or_not" and its flowchart. This macro has two Boolean input variables IN1, passed in through "W", and IN2, passed in through "reg,bit", and a Boolean output variable OUT, passed out through "W". This macro is also used as a

TABLE 3.7

Operands for the Instruction "or"

IN2 (reg,bit)	Data type	Operands
Bit	BOOL	I, Q, M, T_Q, C_Q, C_QD, C_QU, drum_Q, drum_TQ, drum_steps, drum_events, SF, MB, nAI, AI, LOGIC1, LOGIC0, FRSTSCN, SCNOSC, any Boolean variable from SRAM.

TABLE 3.8

Truth Table and Symbols of the Macro "or_not"

Truth Table			Ladder Diagram Symbol	Schematic Symbol
IN1	IN2	OUT		
W	reg,bit	W		
0	0	1		
0	1	0		
1	0	1		
1	1	1		

inputs & output

IN1 (Boolean input variable 1, through W) = 0 or 1
IN2 (Boolean input variable 2, reg,bit) = 0 or 1
OUT (Boolean output variable, through W) = 0 or 1

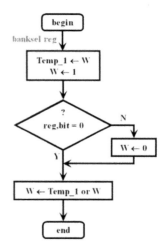

```
or_not  macro   reg,bit
        banksel reg
        movwf   Temp_1
        movlw   1
        btfsc   reg,bit
        movlw   0
        iorwf   Temp_1,w
        endm
```

FIGURE 3.5   The macro "or_not" and its flowchart.

TABLE 3.9
Operands for the Instruction "or_not"

IN2 (reg,bit)	Data type	Operands
Bit	BOOL	I, Q, M, T_Q, C_Q, C_QD, C_QU, drum_Q, drum_TQ, drum_steps, drum_events, SF, MB, nAI, AI, LOGIC1, LOGIC0, FRSTSCN, SCNOSC, any Boolean variable from SRAM.

TABLE 3.10
Truth Table and Symbols of the Macro "nor"

Truth Table		Ladder Diagram Symbol	Schematic Symbol
IN1	IN2		
W	reg,bit		
0	0		
0	1		
1	0		
1	1		

inputs & output

IN1 (Boolean input variable 1, through W) = 0 or 1
IN2 (Boolean input variable 2, reg,bit) = 0 or 1
OUT (Boolean output variable, through W) = 0 or 1

two-input logical OR gate, but this time IN2 is inverted. IN1 should be loaded into (W,0) before this macro is invoked. The operand IN2 (reg,bit) can be in any bank. Operands for the instruction "or_not" are shown in Table 3.9.

3.6 MACRO "NOR"

The truth table and symbols of the macro "nor" are depicted in Table 3.10. Figure 3.6 shows the macro "nor" and its flowchart. This macro has two Boolean input variables IN1, passed in through "W", and IN2, passed in through "reg,bit", and a Boolean output variable OUT, passed out through "W". This macro is used as a two-input logical NOR gate. IN1 should be loaded into (W,0) before this macro is invoked. The operand IN2 (reg,bit) can be in any bank. Operands for the instruction "nor" are shown in Table 3.11.

3.7 MACRO "AND"

The truth table and symbols of the macro "and" are depicted in Table 3.12. Figure 3.7 shows the macro "and" and its flowchart. This macro has two Boolean input

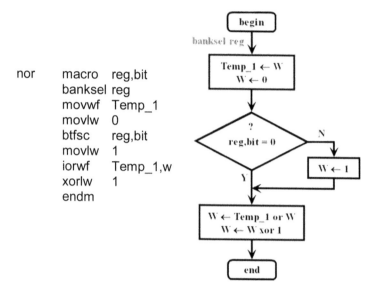

```
nor     macro   reg,bit
        banksel reg
        movwf   Temp_1
        movlw   0
        btfsc   reg,bit
        movlw   1
        iorwf   Temp_1,w
        xorlw   1
        endm
```

FIGURE 3.6   The macro "nor" and its flowchart.

TABLE 3.11

Operands for the Instruction "nor"

IN2 (reg,bit)	Data type	Operands
Bit	BOOL	I, Q, M, T_Q, C_Q, C_QD, C_QU, drum_Q, drum_TQ, drum_steps, drum_events, SF, MB, nAI, AI, LOGIC1, LOGIC0, FRSTSCN, SCNOSC, any Boolean variable from SRAM.

TABLE 3.12

Truth Table and Symbols of the Macro "and"

Truth Table			Ladder Diagram Symbol	Schematic Symbol
IN1	IN2	OUT	IN1 IN2	IN1
W	reg,bit	W	⊢⊢⊢OUT	IN2 OUT
0	0	0		
0	1	0		
1	0	0		
1	1	1		

inputs & output

IN1 (Boolean input variable 1, through W) = 0 or 1
IN2 (Boolean input variable 2, reg,bit) = 0 or 1
OUT (Boolean output variable, through W) = 0 or 1

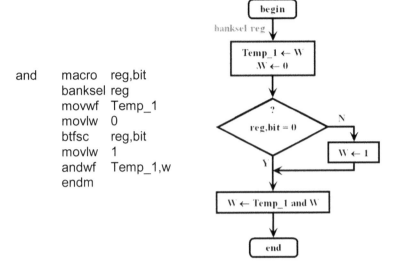

```
and     macro   reg,bit
        banksel reg
        movwf   Temp_1
        movlw   0
        btfsc   reg,bit
        movlw   1
        andwf   Temp_1,w
        endm
```

FIGURE 3.7 The macro "and" and its flowchart.

TABLE 3.13
Operands for the Instruction "and"

IN2 (reg,bit)	Data type	Operands
Bit	BOOL	I, Q, M, T_Q, C_Q, C_QD, C_QU, drum_Q, drum_TQ, drum_steps, drum_events, SF, MB, nAI, AI, LOGIC1, LOGIC0, FRSTSCN, SCNOSC, any Boolean variable from SRAM.

variables IN1, passed in through "W", and IN2, passed in through "reg,bit", and a Boolean output variable OUT, passed out through "W". This macro is used as a two-input logical AND gate. IN1 should be loaded into (W,0) before this macro is invoked. The operand IN2 (reg,bit) can be in any bank. Operands for the instruction "and" are shown in Table 3.13.

3.8 MACRO "AND_NOT"

The truth table and symbols of the macro "and_not" are depicted in Table 3.14. Figure 3.8 shows the macro "and_not" and its flowchart. This macro has two Boolean input variables IN1, passed in through "W", and IN2, passed in through "reg,bit", and a Boolean output variable OUT, passed out through "W". This macro is also used as a two-input logical AND gate, but this time one of the inputs is inverted. IN1 should be loaded into (W,0) before this macro is invoked. The operand IN2 (reg,bit) can be in any bank. Operands for the instruction "and_not" are shown in Table 3.15.

TABLE 3.14
Truth Table and Symbols of the Macro "and_not"

Truth Table			Ladder Diagram Symbol	Schematic Symbol
IN1	IN2	OUT		
W	reg,bit	W		
0	0	0		
0	1	0		
1	0	1		
1	1	0		

inputs & output

IN1 (Boolean input variable 1, through W) = 0 or 1
IN2 (Boolean input variable 2, reg,bit) = 0 or 1
OUT (Boolean output variable, through W) = 0 or 1

```
and_not macro    reg,bit
        banksel  reg
        movwf    Temp_1
        movlw    1
        btfsc    reg,bit
        movlw    0
        andwf    Temp_1,w
        endm
```

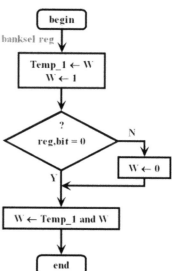

FIGURE 3.8   The macro "and_not" and its flowchart.

TABLE 3.15
Operands for the Instruction "and_not"

IN2 (reg,bit)	Data type	Operands
Bit	BOOL	I, Q, M, T_Q, C_Q, C_QD, C_QU, drum_Q, drum_TQ, drum_steps, drum_events, SF, MB, nAI, AI, LOGIC1, LOGIC0, FRSTSCN, SCNOSC, any Boolean variable from SRAM.

3.9   MACRO "NAND"

The truth table and symbols of the macro "nand" are depicted in Table 3.16. Figure 3.9 shows the macro "nand" and its flowchart. This macro has two Boolean input variables IN1, passed in through "W", and IN2, passed in through "reg,bit", and a Boolean output variable OUT, passed out through "W". This macro is used as a two-input logical NAND gate. IN1 should be loaded into (W,0) before this macro is invoked. The operand IN2 (reg,bit) can be in any bank. Operands for the instruction "nand" are shown in Table 3.17.

TABLE 3.16
Truth Table and Symbols of the Macro "nand"

Truth Table			Ladder Diagram Symbol	Schematic Symbol
IN1	IN2	OUT		
W	reg,bit	W		
0	0	1		
0	1	1		
1	0	1		
1	1	0		

inputs & output

IN1 (Boolean input variable 1, through W) = 0 or 1
IN2 (Boolean input variable 2, reg,bit) = 0 or 1
OUT (Boolean output variable, through W) = 0 or 1

```
nand    macro    reg,bit
        banksel  reg
        movwf    Temp_1
        movlw    0
        btfsc    reg,bit
        movlw    1
        andwf    Temp_1,w
        xorlw    1
        endm
```

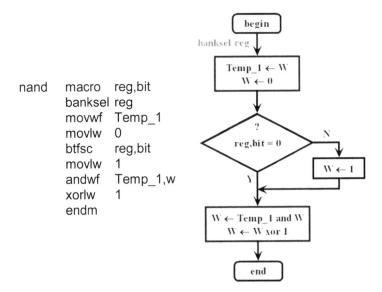

FIGURE 3.9   The macro "nand" and its flowchart.

TABLE 3.17

Operands for the Instruction "nand"

IN2 (reg,bit)	Data type	Operands
Bit	BOOL	I, Q, M, T_Q, C_Q, C_QD, C_QU, drum_Q, drum_TQ, drum_steps, drum_events, SF, MB, nAI, AI, LOGIC1, LOGIC0, FRSTSCN, SCNOSC, any Boolean variable from SRAM.

3.10   MACRO "XOR"

The truth table and symbols of the macro "xor" are depicted in Table 3.18. Figure 3.10 shows the macro "xor" and its flowchart. This macro has two Boolean input variables IN1, passed in through "W", and IN2, passed in through "reg,bit", and a Boolean output variable OUT, passed out through "W". This macro is used as a two-input logical EXOR gate. IN1 should be loaded into (W,0) before this macro is invoked. The operand IN2 (reg,bit) can be in any bank. Operands for the instruction "xor" are shown in Table 3.19.

3.11   MACRO "XOR_NOT"

The truth table and symbols of the macro "xor_not" are depicted in Table 3.20. Figure 3.11 shows the macro "xor_not" and its flowchart. This macro has two Boolean input variables IN1, passed in through "W", and IN2, passed in through "reg,bit", and a Boolean output variable OUT, passed out through "W". This macro is also used as a two-input logical EXOR gate, but this time one of the inputs is inverted. IN1 should be loaded into (W,0) before this macro is invoked. The operand

TABLE 3.18

Truth Table and Symbols of the Macro "xor"

Truth Table			Ladder Diagram Symbol	Schematic Symbol
IN1	IN2	OUT		
W	reg,bit	W		
0	0	0		
0	1	1		
1	0	1		
1	1	0		

inputs & output

IN1 (Boolean input variable 1, through W) = 0 or 1

IN2 (Boolean input variable 2, reg,bit) = 0 or 1

OUT (Boolean output variable, through W) = 0 or 1

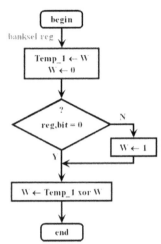

```
xor      macro    reg,bit
         banksel  reg
         movwf    Temp_1
         movlw    0
         btfsc    reg,bit
         movlw    1
         xorwf    Temp_1,w
         endm
```

FIGURE 3.10   The macro "xor" and its flowchart.

TABLE 3.19
Operands for the Instruction "xor"

IN2 (reg,bit)	Data type	Operands
Bit	BOOL	I, Q, M, T_Q, C_Q, C_QD, C_QU, drum_Q, drum_TQ, drum_steps, drum_events, SF, MB, nAI, AI, LOGIC1, LOGIC0, FRSTSCN, SCNOSC, any Boolean variable from SRAM.

TABLE 3.20
Truth Table and Symbols of the Macro "xor_not"

Truth Table			Ladder Diagram Symbol	Schematic Symbol
IN1	IN2	OUT		
W	reg,bit	W		
0	0	1		
0	1	0		
1	0	0		
1	1	1		

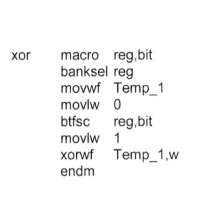

inputs & output

IN1 (Boolean input variable 1, through W) = 0 or 1
IN2 (Boolean input variable 2, reg,bit) = 0 or 1
OUT (Boolean output variable, through W) = 0 or 1

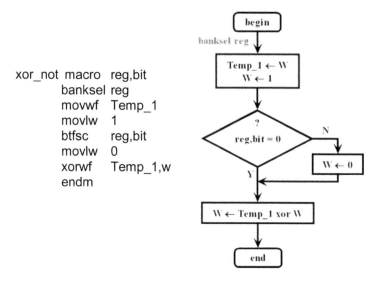

```
xor_not macro   reg,bit
        banksel reg
        movwf   Temp_1
        movlw   1
        btfsc   reg,bit
        movlw   0
        xorwf   Temp_1,w
        endm
```

FIGURE 3.11   The macro "xor_not" and its flowchart.

TABLE 3.21
Operands for the Instruction "xor_not"

IN2 (reg,bit)	Data type	Operands
Bit	BOOL	I, Q, M, T_Q, C_Q, C_QD, C_QU, drum_Q, drum_TQ, drum_steps, drum_events, SF, MB, nAI, AI, LOGIC1, LOGIC0, FRSTSCN, SCNOSC, any Boolean variable from SRAM.

IN2 (reg,bit) can be in any bank. Operands for the instruction "xor_not" are shown in Table 3.21.

3.12   MACRO "XNOR"

The truth table and symbols of the macro "xnor" are depicted in Table 3.22. Figure 3.12 shows the macro "xnor" and its flowchart. This macro has two Boolean input variables IN1, passed in through "W", and IN2, passed in through "reg,bit", and a Boolean output variable OUT, passed out through "W". This macro is used as a two-input logical EXNOR gate. IN1 should be loaded into (W,0) before this macro is invoked. The operand IN2 (reg,bit) can be in any bank. Operands for the instruction "xnor" are shown in Table 3.23.

3.13   MACRO "OUT"

The truth table and symbols of the macro "out" are depicted in Table 3.24. Figure 3.13 shows the macro "out" and its flowchart. This macro has a Boolean input

TABLE 3.22
Truth Table and Symbols of the Macro "xnor"

Truth Table			Ladder Diagram Symbol	Schematic Symbol
IN1	IN2	OUT	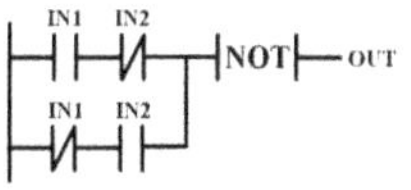	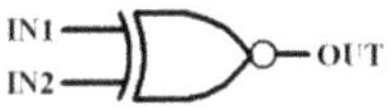
W	reg,bit	W		
0	0	1		
0	1	0		
1	0	0		
1	1	1		

inputs & output

IN1 (Boolean input variable 1, through W) = 0 or 1
IN2 (Boolean input variable 2, reg,bit) = 0 or 1
OUT (Boolean output variable, through W) = 0 or 1

```
xnor    macro   reg,bit
        banksel reg
        movwf   Temp_1
        movlw   0
        btfsc   reg,bit
        movlw   1
        xorwf   Temp_1,w
        xorlw   1
        endm
```

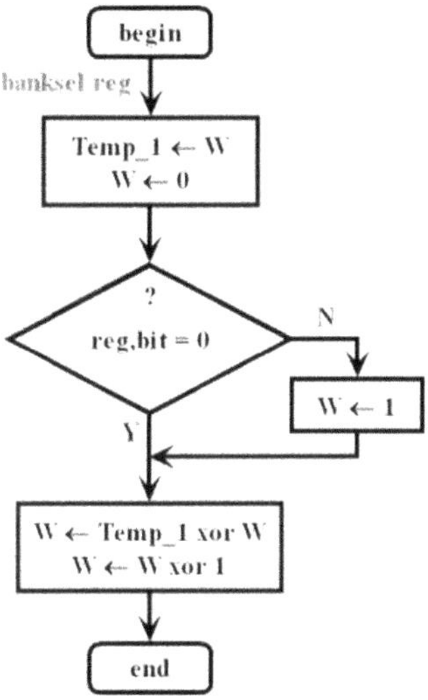

FIGURE 3.12 The macro "xnor" and its flowchart.

variable IN passed in through "W" and a Boolean output variable OUT passed out through "reg,bit". In ladder logic, this macro is represented by an output relay (internal or external relay). When IN is 0 (or 1, respectively), OUT is forced to be 0 (or 1, respectively). Operands for the instruction "out" are shown in Table 3.25.

3.14 MACRO "OUT_NOT"

The truth table and symbols of the macro "out_not" are depicted in Table 3.26. Figure 3.14 shows the macro "out_not" and its flowchart. This macro has a Boolean input variable IN passed in through "W" and a Boolean output variable OUT passed out

TABLE 3.23
Operands for the Instruction "xnor"

IN2

(reg,bit)	Data type	Operands
Bit	BOOL	I, Q, M, T_Q, C_Q, C_QD, C_QU, drum_Q, drum_TQ, drum_steps, drum_events, SF, MB, nAI, AI, LOGIC1, LOGIC0, FRSTSCN, SCNOSC, any Boolean variable from SRAM.

TABLE 3.24
Truth Table and Symbols of the Macro "out"

	Ladder Diagram	
Truth Table	Symbol	Schematic Symbol

IN	OUT
W	reg,bit
0	0
1	1

input & output

IN (Boolean input variable, through W) = 0 or 1
OUT (Boolean output variable, reg,bit) = 0 or 1

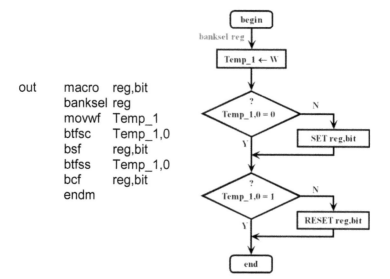

```
out    macro    reg,bit
       banksel  reg
       movwf    Temp_1
       btfsc    Temp_1,0
       bsf      reg,bit
       btfss    Temp_1,0
       bcf      reg,bit
       endm
```

FIGURE 3.13   The macro "out" and its flowchart.

TABLE 3.25
Operands for the Instruction "out"

OUT

(reg,bit)	Data type	Operands
Bit	BOOL	Q, M, T_Q, C_Q, C_QD, C_QU, drum_Q, drum_TQ, drum_steps, drum_events, SF, MB, any Boolean variable from SRAM.

TABLE 3.26
Truth Table and Symbols of the Macro "out_not"

Truth Table		Ladder Diagram Symbol	Schematic Symbol
IN	OUT		
W	reg,bit		
0	1		
1	0		

input & output

IN (Boolean input variable, through W) = 0 or 1

OUT (Boolean output variable, reg,bit) = 0 or 1

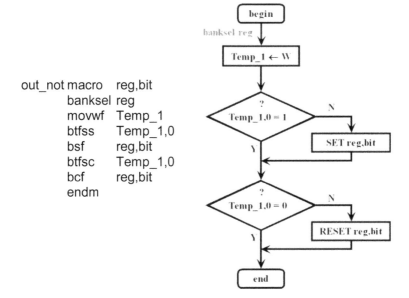

```
out_not macro   reg,bit
        banksel reg
        movwf   Temp_1
        btfss   Temp_1,0
        bsf     reg,bit
        btfsc   Temp_1,0
        bcf     reg,bit
        endm
```

FIGURE 3.14   The macro "out_not" and its flowchart.

TABLE 3.27
Operands for the Instruction "out_not"

OUT (reg,bit)	Data type	Operands
Bit	BOOL	Q, M, T_Q, C_Q, C_QD, C_QU, drum_Q, drum_TQ, drum_steps, drum_events, SF, MB, any Boolean variable from SRAM.

TABLE 3.28
The Symbols of the Macro "mid_out"

Truth Table		Ladder Diagram Symbol	Schematic Symbol
IN	OUT	OUT	OUT
W	reg,bit	IN —(#)—	#
0	0		IN
1	1		

input & output

IN (Boolean input variable, through W) = 0 or 1
OUT (Boolean midline output variable, Mreg,Mbit, a unique
memory bit, i.e. one of M0.0, M0.1, ..., M127.7) = 0 or 1

through "reg,bit". In ladder logic, this macro is represented by an inverted output relay (internal or external relay). When IN is 0 (or 1, respectively), OUT is forced to be 1 (or 0, respectively). Operands for the instruction "out_not" are shown in Table 3.27.

3.15   MACRO "MID_OUT" (MIDLINE OUTPUT)

The truth table and symbols of the macro "mid_out" are depicted in Table 3.28. Figure 3.15 shows the macro "mid_out" and its flowchart. ---(#)--- (midline output) is an intermediate assigning element which saves the logical result of the preceding branch elements of a rung taken from the Boolean input variable IN (W,0) to a specified Boolean midline output variable OUT (Mreg,Mbit). In series with other contacts, ---(#)--- is inserted like a contact. A ---(#)--- element may never be connected to the power rail or directly after a branch connection or at the end of a branch. IN should be loaded into (W,0) before this macro is invoked. Assumption: the operand "Mreg,Mbit" can be in any bank.

3.16   MACRO "MID_OUT_NOT" (INVERTED MIDLINE OUTPUT)

The symbols of the macro "mid_out_not" are depicted in Table 3.29. Figure 3.16 shows the macro "mid_out_not" and its flowchart. ---(I#)--- (inverted midline output) is an intermediate assigning element which saves the complement of the logical

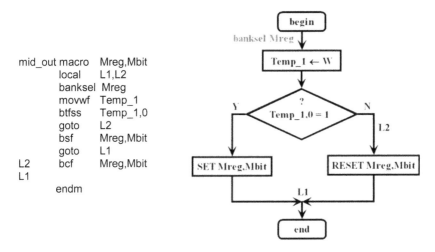

```
mid_out macro  Mreg,Mbit
         local  L1,L2
         banksel Mreg
         movwf  Temp_1
         btfss  Temp_1,0
         goto   L2
         bsf    Mreg,Mbit
         goto   L1
L2       bcf    Mreg,Mbit
L1
         endm
```

FIGURE 3.15 The macro "mid_out" and its flowchart.

TABLE 3.29
The Symbols of the Macro "mid_out_not"

Truth Table		Ladder Diagram Symbol	Schematic Symbol
IN	OUT	OUT IN —(I#)—	OUT I# IN
W	reg,bit		
0	1		
1	0		

input & output

IN (Boolean input variable, through W) = 0 or 1

OUT (Boolean midline output variable, Mreg,Mbit, a unique
memory bit, i.e. one of M0.0, M0.1, ..., M127.7) = 0 or 1

result of the preceding branch elements of a rung taken from the Boolean input variable IN (W,0) to a specified Boolean midline output signal OUT (Mreg,Mbit). In series with other contacts, ---(I#)--- is inserted like a contact. A ---(I#)--- element may never be connected to the power rail or directly after a branch connection or at the end of a branch. IN should be loaded into (W,0) before this macro is invoked. Assumption: the operand "Mreg,Mbit" can be in any bank.

3.17 MACRO "IN_OUT"

The truth table and symbols of the macro "in_out" are depicted in Table 3.30. Figure 3.17 shows the macro "in_out" and its flowchart. This macro has a Boolean input variable IN passed in through "regi,biti" and a Boolean output variable OUT passed out through "rego,bito". When IN is 0 (or 1, respectively), OUT is forced to be 0 (or

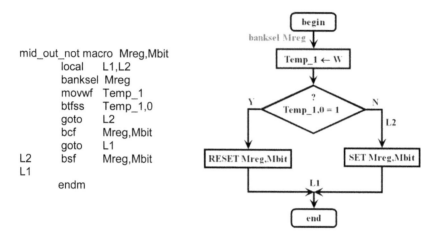

```
mid_out_not macro  Mreg,Mbit
          local    L1,L2
          banksel  Mreg
          movwf    Temp_1
          btfss    Temp_1,0
          goto     L2
          bcf      Mreg,Mbit
          goto     L1
L2        bsf      Mreg,Mbit
L1
          endm
```

FIGURE 3.16   The macro "mid_out_not" and its flowchart.

TABLE 3.30
Truth Table and Symbols of the Macro "in_out"

Truth Table		Ladder Diagram Symbol	Schematic Symbol
IN	OUT		
regi,biti	rego,bito		
0	0		
1	1		

input & output

IN (Boolean input variable, regi,biti) = 0 or 1

OUT (Boolean output variable, rego,bito) = 0 or 1

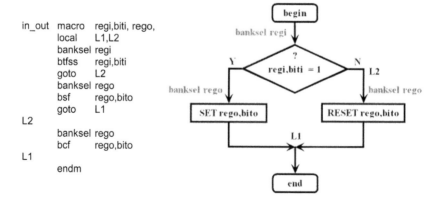

```
in_out  macro  regi,biti, rego,
        local  L1,L2
        banksel regi
        btfss  regi,biti
        goto   L2
        banksel rego
        bsf    rego,bito
        goto   L1
L2
        banksel rego
        bcf    rego,bito
L1
        endm
```

FIGURE 3.17   The macro "in_out" and its flowchart.

1, respectively). Assumption: the operands "regi,biti" and "rego,bito" can be in any bank. Operands for the instruction "in_out" are shown in Table 3.31.

3.18   MACRO "INV_OUT"

The truth table and symbols of the macro "inv_out" are depicted in Table 3.32. Figure 3.18 shows the macro "inv_out" and its flowchart. This macro has a Boolean

TABLE 3.31
Operands for the Instruction "in_out"

Input/Output	Data type	Operands
IN (regi,biti)	BOOL	I, Q, M, T_Q, C_Q, C_QD, C_QU, drum_Q, drum_TQ, drum_steps, drum_events, SF, MB, nAI, AI, LOGIC1, LOGIC0, FRSTSCN, SCNOSC, any Boolean variable from SRAM.
OUT (rego,bito)	BOOL	Q, M, T_Q, C_Q, C_QD, C_QU, drum_Q, drum_TQ, drum_steps, drum_events, SF, MB, any Boolean variable from SRAM.

TABLE 3.32
Truth Table and Symbols of the Macro "inv_out"

	Ladder Diagram	
Truth Table	Symbol	Schematic Symbol

IN	OUT
regi,biti	rego,bito
0	1
1	0

input & output

IN (Boolean input variable, regi,biti) = 0 or 1
OUT (Boolean output variable, rego,bito) = 0 or 1

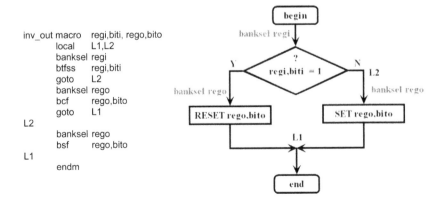

```
inv_out macro   regi,biti, rego,bito
        local   L1,L2
        banksel regi
        btfss   regi,biti
        goto    L2
        banksel rego
        bcf     rego,bito
        goto    L1
L2
        banksel rego
        bsf     rego,bito
L1
        endm
```

FIGURE 3.18   The macro "inv_out" and its flowchart.

TABLE 3.33
Operands for the Instruction "inv_out"

Input/Output	Data type	Operands
IN (regi,biti) Bit	BOOL	I, Q, M, T_Q, C_Q, C_QD, C_QU, drum_Q, drum_TQ, drum_steps, drum_events, SF, MB, nAI, AI, LOGIC1, LOGIC0, FRSTSCN, SCNOSC, any Boolean variable from SRAM.
OUT (rego,bito) Bit	BOOL	Q, M, T_Q, C_Q, C_QD, C_QU, drum_Q, drum_TQ, drum_steps, drum_events, SF, MB, any Boolean variable from SRAM.

TABLE 3.34
Truth Table and Symbols of the Macro "_set"

Truth Table		Ladder Diagram Symbol	Schematic Symbol
IN	OUT	OUT	OUT
W	reg,bit	IN ─(s)─	S
0	No change		IN
1	Set		

input & output

IN (Boolean input variable, through W) = 0 or 1
OUT (Boolean output variable to be set, reg,bit) = 0 or 1

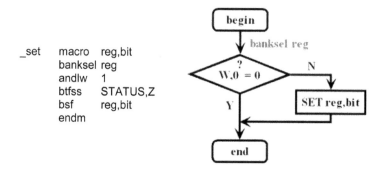

```
_set    macro   reg,bit
        banksel reg
        andlw   1
        btfss   STATUS,Z
        bsf     reg,bit
        endm
```

FIGURE 3.19 The macro "_set" and its flowchart.

input variable passed in through "regi,biti" and a Boolean output variable passed out through "rego,bito". When IN is 0 (or 1, respectively), OUT is forced to 1 (or 0, respectively). Assumption: the operands "regi,biti" and "rego,bito" can be in any bank. Operands for the instruction "inv_out" are shown in Table 3.33.

3.19 MACRO "_SET"

The truth table and symbols of the macro "_set" are depicted in Table 3.34. Figure 3.19 shows the macro "_set" and its flowchart. This macro has a Boolean input variable IN passed in through "W" and a Boolean output OUT variable passed out through "reg,bit". When IN is 0, no action is taken, but when IN is 1, OUT is set to 1. Operands for the instruction "_set" are shown in Table 3.35. This instruction can be placed anywhere in the program provided that IN (W,0) is loaded with a valid value of "0" or "1". Assumption: the operand "reg,bit" can be in any bank.

3.20   MACRO "_RESET"

The truth table and symbols of the macro "_reset" are depicted in Table 3.36. Figure 3.20 shows the macro "_reset" and its flowchart. This macro has a Boolean input variable IN passed in through "W" and a Boolean output variable OUT passed out

---

TABLE 3.35
Operands for the Instruction "_set"

OUT (reg,bit)	Data type	Operands
Bit	BOOL	Q, M, T_Q, C_Q, C_QD, C_QU, drum_Q, drum_TQ, drum_steps, drum_events, SF, MB, any Boolean variable from SRAM.

TABLE 3.36
Truth Table and Symbols of the Macro "_reset"

Truth Table		Ladder Diagram Symbol	Schematic Symbol
IN	OUT		
W	reg,bit	IN ──(R)──	
0	No change		
1	Reset		

input & output

IN (Boolean input variable, through W) = 0 or 1
OUT (Boolean output variable to be reset, reg,bit) = 0 or 1

```
_reset  macro   reg,bit
        banksel reg
        andlw   1
        btfss   STATUS,Z
        bcf     reg,bit
        endm
```

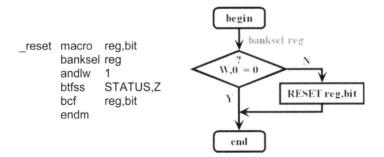

FIGURE 3.20   The macro "_reset" and its flowchart.

through "reg,bit". When IN is 0, no action is taken, but when IN is 1, OUT is reset. Operands for the instruction "_reset" are shown in Table 3.37. This instruction can be placed anywhere in the program provided that the IN (W,0) is loaded with a valid value of "0" or "1" - Assumption: the operand "reg,bit" can be in any bank.

3.21 MACRO "SR" (SET–RESET)

The truth table and symbols of the macro "SR" are depicted in Table 3.38. Figure 3.21 shows the macro "SR" and its flowchart. This macro has the two Boolean input variables S, passed in through "Sreg,Sbit", and R1, passed in through "R1reg,R1bit", and a Boolean output variable Q, passed out through "Qreg,Qbit". The macro "SR" defines a reset-dominant latch. If the set (S) and reset (R1) signals are both true, the output Q (Qreg,Qbit) will be 0. This instruction must be located at the end of a PLC rung. Operands for the instruction "SR" are shown in Table 3.39. Assumption: the operands "Sreg,Sbit", "R1reg,R1bit", and "Qreg,Qbit" can be in any bank.

3.22 MACRO "RS" (RESET–SET)

The truth table and symbols of the macro "RS" are depicted in Table 3.40. Figure 3.22 shows the macro "RS" and its flowchart. This macro has the two Boolean input variables S1, passed in through "S1reg,S1bit", and R, passed in through "Rreg,Rbit", and a Boolean output variable Q, passed out through "Qreg,Qbit". The macro "RS" defines a set-dominant latch. If the reset (R) and set (S1) signals are both true, the

---

TABLE 3.37
Operands for the Instruction "_reset"

OUT (reg,bit)	Data type	Operands
Bit	BOOL	Q, M, T_Q, C_Q, C_QD, C_QU, drum_Q, drum_TQ, drum_steps, drum_events, SF, MB, any Boolean variable from SRAM.

TABLE 3.38
The Truth Table and Symbol of the Macro "SR"

	Truth Table			Symbol
S	R1	Q_{t+1}	Comment	
0	0	Q_t	No change	
1	0	1	Set	
×	1	0	Reset	

×: don't care.

inputs & outputs

S (Boolean input variable S (set), Sreg,Sbit) = 0 or 1

R1 (Dominant Boolean input variable R1 (reset), R1reg,R1bit) = 0 or 1

Q (Boolean input variable Q, Qreg,Qbit) = 0 or 1

```
SR      macro    Sreg,Sbit, R1reg,R1bit, Qreg,Qbit
        local    L1,L2
        banksel  R1reg
        btfsc    R1reg,R1bit
        goto     L2
        banksel  Sreg
        btfss    Sreg,Sbit
        goto     L1
        banksel  Qreg
        bsf      Qreg,Qbit
        goto     L1
L2
        banksel  Qreg
        bcf      Qreg,Qbit
L1
        endm
```

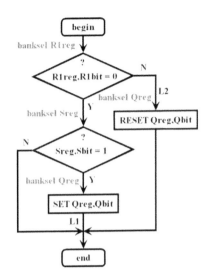

FIGURE 3.21 The macro "SR" and its flowchart.

TABLE 3.39
Operands for the Instruction "SR"

Input/Output	Data type	Operands
S (Sreg,Sbit)	BOOL	I, Q, M, T_Q, C_Q, C_QD, C_QU, drum_Q, drum_TQ, drum_steps, drum_events, SF, MB, nAI, AI, LOGIC1, LOGIC0, FRSTSCN, SCNOSC, any Boolean variable from SRAM.
R1 (R1reg,R1bit)	BOOL	I, Q, M, T_Q, C_Q, C_QD, C_QU, drum_Q, drum_TQ, drum_steps, drum_events, SF, MB, nAI, AI, LOGIC1, LOGIC0, FRSTSCN, SCNOSC, any Boolean variable from SRAM.
Q (Qreg,Qbit)	BOOL	Q, M, T_Q, C_Q, C_QD, C_QU, drum_Q, drum_TQ, drum_steps, drum_events, SF, MB, any Boolean variable from SRAM.

TABLE 3.40
Truth Table and Symbol- of the Macro "RS"

	Truth Table			Symbol
R	S1	Q_{t+1}	Comment	
0	0	Q_t	No change	
1	0	0	Reset	
×	1	1	Set	

× : don't care.

inputs & outputs

R (Boolean input variable R (reset), Rreg,Rbit) = 0 or 1
S1 (Dominant Boolean input variable S1 (set), S1reg,S1bit) = 0 or 1
Q (Boolean input variable Q, Qreg,Qbit) = 0 or 1

```
RS      macro  Rreg,Rbit, S1reg,S1bit, Qreg,Qbit
        local  L1,L2
        banksel S1reg
        btfsc  S1reg,S1bit
        goto   L2
        banksel Rreg
        btfss  Rreg,Rbit
        goto   L1
        banksel Qreg
        bcf    Qreg,Qbit
        goto   L1
L2
        banksel Qreg
        bsf    Qreg,Qbit
L1
        endm
```

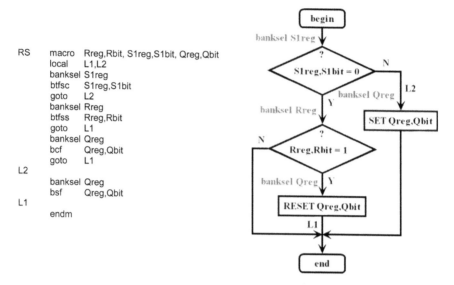

FIGURE 3.22 The macro "RS" and its flowchart.

TABLE 3.41

Operands for the Instruction "RS"

Input/Output	Data type	Operands
S1 (S1reg,S1bit)	BOOL	I, Q, M, T_Q, C_Q, C_QD, C_QU, drum_Q, drum_TQ, drum_steps, drum_events, SF, MB, nAI, AI, LOGIC1, LOGIC0, FRSTSCN, SCNOSC, any Boolean variable from SRAM.
R (Rreg,Rbit)	BOOL	I, Q, M, T_Q, C_Q, C_QD, C_QU, drum_Q, drum_TQ, drum_steps, drum_events, SF, MB, nAI, AI, LOGIC1, LOGIC0, FRSTSCN, SCNOSC, any Boolean variable from SRAM.
Q (Qreg,Qbit)	BOOL	Q, M, T_Q, C_Q, C_QD, C_QU, drum_Q, drum_TQ, drum_steps, drum_events, SF, MB, any Boolean variable from SRAM.

output Q (Qreg,Qbit) will be 1. This instruction must be located at the end of a PLC rung. Operands for the instruction "RS" are shown in Table 3.41. Assumption: the operands "S1reg,S1bit", "Rreg,Rbit", and "Qreg,Qbit" can be in any bank.

3.23   MACRO "R_EDGE" (RISING EDGE DETECTOR)

The symbols and the timing diagram of the macro "r_edge" are depicted in Table 3.42. Figure 3.23 shows the macro "r_edge" and its flowchart. This macro has a Boolean input variable IN, passed in through "W", a Boolean output variable OUT, passed out through "W", and a unique memory bit "Mreg,Mbit", used to detect the rising edge of the input signal IN. The input signal IN should be loaded into (W,0)

TABLE 3.42
The Symbols and the Timing Diagram of the Macro "r_edge"

Ladder Diagram Symbol   **Schematic Symbol**

inputs & outputs

IN (Boolean input variable, through W) = 0 or 1

Mreg,Mbit (A unique memory bit) = M0.0, M0.1, ..., M127.7

OUT (One shot output signal, through W) = 0 or 1

Timing diagram

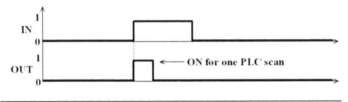

```
r_edge  macro   Mreg,Mbit
        local   L1,L2
        banksel Mreg
        movwf   Temp_1
        btfss   Temp_1,0
        bsf     Mreg,Mbit
        btfss   Temp_1,0
        goto    L1
        btfss   Mreg,Mbit
        goto    L2
        bcf     Mreg,Mbit
        movlw   1
        goto    L1
L2      movlw   0
L1
        endm
```

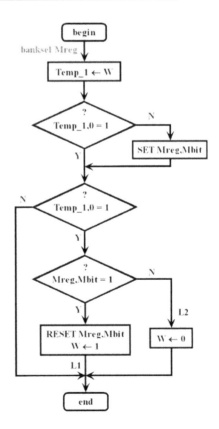

FIGURE 3.23   The macro "r_edge" and its flowchart.

before this macro is run, and the output signal OUT will be provided within "W" at the end of the macro. The identifier "P" represents positive (OFF [0] to ON [1]) transition detection. As can be seen from the timing diagram, OUT is ON for only one scan time when IN changes its state from OFF to ON. In the other instances OUT remains OFF. Assumption: the operand "Mreg,Mbit" can be in any bank.

3.24   MACRO "F_EDGE" (FALLING EDGE DETECTOR)

The symbols and the timing diagram of the macro "f_edge" are depicted in Table 3.43. Figure 3.24 shows the macro "f_edge" and its flowchart. This macro has a Boolean input variable IN, passed in through "W", a Boolean output variable OUT, passed out through "W", and a unique memory bit "Mreg,Mbit", used to detect the falling edge of the input signal IN. The input signal IN should be loaded into (W,0) before this macro is run, and the output signal OUT will be provided within "W" at the end of the macro. The identifier "N" represents negative (ON [1] to OFF [0]) transition detection. As can be seen from the timing diagram, OUT is ON for only one scan time when IN changes its state from ON to OFF. In the other instances OUT remains OFF. Assumption: the operand "Mreg,Mbit" can be in any bank.

3.25   MACRO "R_TOGGLE" (OUTPUT TOGGLE WITH RISING EDGE DETECTOR)

The symbols and the timing diagram of the macro "r_toggle" are depicted in Table 3.44. Figure 3.25 shows the macro "r_toggle" and its flowchart. This macro has a Boolean input variable IN, passed in through "W", a Boolean output variable OUT, passed out through "reg,bit", and a unique memory bit "Mreg,Mbit", used to detect

---

TABLE 3.43

The Symbols and the Timing Diagram of the Macro "f_edge"

Ladder Diagram Symbol   Schematic Symbol

inputs & outputs

IN (Boolean input variable, through W) = 0 or 1

Mreg,Mbit (A unique memory bit) = M0.0, M0.1, ..., M127.7

OUT (One shot output signal, through W) = 0 or 1

Timing diagram

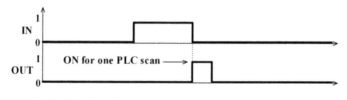

```
f_edge  macro   Mreg,Mbit
        local   L1,L2
        banksel Mreg
        movwf   Temp_1
        btfsc   Temp_1,0
        bsf     Mreg,Mbit
        btfsc   Temp_1,0
        goto    L2
        btfss   Mreg,Mbit
        goto    L2
        bcf     Mreg,Mbit
        movlw   1
        goto    L1
L2      movlw   0
L1
        endm
```

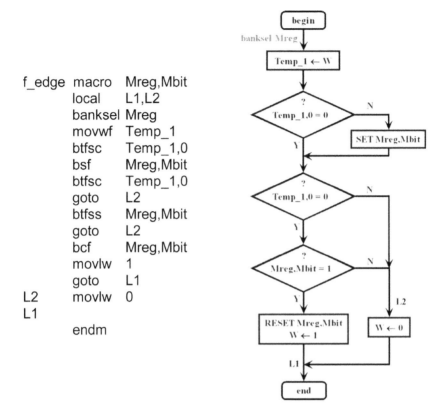

FIGURE 3.24   The macro "f_edge" and its flowchart.

the rising edge of the input signal IN. The input signal IN should be loaded into (W,0) before this macro is invoked. OUT is toggled when IN changes its state from OFF to ON. The identifier "PT" represents output toggle with positive (OFF [0] to ON [1]) transition detection. As can be seen from the timing diagram, OUT is toggled when IN changes its state from OFF to ON. In the other instances OUT remains unchanged. This instruction must be located at the end of a PLC rung. Operands for the instruction "r_toggle" are shown in Table 3.45. Assumption: the operands "Mreg,Mbit" and "reg,bit" can be in any bank.

3.26   MACRO "F_TOGGLE" (OUTPUT TOGGLE WITH FALLING EDGE DETECTOR)

The symbols and the timing diagram of the macro "f_toggle" are depicted in Table 3.46. Figure 3.26 shows the macro "f_toggle" and its flowchart. This macro has a Boolean input variable IN, passed in through "W", a Boolean output variable OUT, passed out through "reg,bit", and a unique memory bit "Mreg,Mbit", used to detect the falling edge of the input signal IN. The input signal IN should be loaded into (W,0) before this macro is invoked. OUT is toggled when IN changes its state from

TABLE 3.44

The Symbols and the Timing Diagram of the Macro "r_toggle"

Ladder Diagram Symbol   Schematic Symbol

inputs & outputs

IN (Boolean input variable, through W) = 0 or 1

Mreg,Mbit (A unique memory bit) = M0.0, M0.1, ..., M127.7

OUT (Boolean output variable, reg,bit) = 0 or 1

Timing diagram

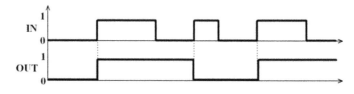

ON to OFF. The identifier "NT" represents output toggle with negative (ON [1] to OFF [0]) transition detection. As can be seen from the timing diagram, OUT is toggled when IN changes its state from ON to OFF. In the other instances OUT remains unchanged. This instruction must be located at the end of a PLC rung. Operands for the instruction "f_toggle" are shown in Table 3.47. Assumption: the operands "Mreg,Mbit" and "reg,bit" can be in any bank.

3.27 MACRO "ADRS_RE" (ADDRESS RISING EDGE DETECTOR)

The symbols and the timing diagram of the macro "adrs_re" are depicted in Table 3.48. Figure 3.27 shows the macro "adrs_re" and its flowchart. This macro has a Boolean enable input variable IN, passed in through "W", a Boolean input variable Address, passed in through "regi,biti", a Boolean output variable OUT, passed out through "W", and a unique memory bit "Mreg,Mbit", used to detect the rising edge of the input signal Address. The input signal IN should be loaded into (W,0) before this macro is invoked. When IN = 1, the macro "adrs_re" compares the signal state of Address (regi,biti) with the signal state from the previous scan, which is stored in "Mreg,Mbit". In this case, if the current signal state of Address (regi,biti) is "1" and the previous state was "0" (detection of rising edge), then the signal state of OUT (through "W") will be "1" after this instruction for one PLC scan time. The identifier "AP" represents positive (OFF [0] to ON [1]) transition detection of the Address. Operands for the instruction "adrs_re" are shown in Table 3.49. Assumption: the operands "Mreg,Mbit" and "regi,biti" can be in any bank.

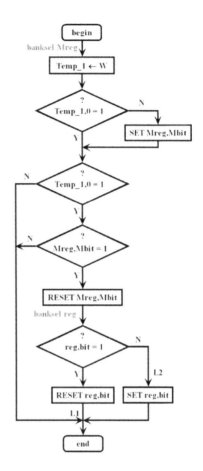

```
r_toggle macro  Mreg,Mbit, reg,bit
         local  L1,L2
         banksel Mreg
         movwf  Temp_1
         btfss  Temp_1,0
         bsf    Mreg,Mbit
         btfss  Temp_1,0
         goto   L1
         btfss  Mreg,Mbit
         goto   L1
         bcf    Mreg,Mbit
         banksel reg
         btfss  reg,bit
         goto   L2
         bcf    reg,bit
         goto   L1
L2       bsf    reg,bit
L1
         endm
```

FIGURE 3.25   The macro "r_toggle" and its flowchart.

TABLE 3.45
Operands for the Instruction "r_toggle"

OUT (reg,bit)	Data type	Operands
Bit	BOOL	Q, M, T_Q, C_Q, C_QD, C_QU, drum_Q, drum_TQ, drum_steps, drum_events, SF, MB, any Boolean variable from SRAM.

3.28   MACRO "ADRS_FE" (ADDRESS FALLING EDGE DETECTOR)

The symbols and the timing diagram of the macro "adrs_fe" are depicted in Table 3.50. Figure 3.28 shows the macro "adrs_fe" and its flowchart. This macro has a Boolean enable input variable IN, passed in through "W", a Boolean input variable Address, passed in through "regi,biti", a Boolean output variable OUT, passed out through "W", and a unique memory bit "Mreg,Mbit", used to detect

TABLE 3.46

The Symbols and the Timing Diagram of the Macro "f_toggle"

Ladder Diagram Symbol   Schematic Symbol

inputs & outputs

IN (Boolean input variable, through W) = 0 or 1

Mreg,Mbit (A unique memory bit) = M0.0, M0.1, ..., M127.7

OUT (Boolean output variable, reg,bit) = 0 or 1

Timing diagram

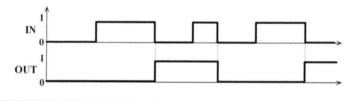

the falling edge of the input signal Address. The input signal IN should be loaded into (W,0) before this macro is invoked. When IN = 1, the macro "adrs_fe" compares the signal state of Address (regi,biti) with the signal state from the previous scan, which is stored in "Mreg,Mbit". In this case if the current signal state of Address (regi,biti) is "0" and the previous state was "1" (detection of falling edge), then the signal state of OUT (through W) will be "1" after this instruction for one PLC scan time. The identifier "AN" represents negative (ON [1] to OFF [0]) transition detection of the Address. Operands for the instruction "adrs_fe" are shown in Table 3.51. Assumption: the operands "Mreg,Mbit" and "regi,biti" can be in any bank.

3.29   MACRO "SETBF" (SET BIT FIELD)

The symbols of the macro "setBF" are depicted in Table 3.52. Figures 3.29 and 3.30 show the macro "setBF" and its flowchart, respectively. This macro has a Boolean enable input variable EN, passed in through "W", the parameter N (the number of consecutive bits to be set), and a Boolean output variable OUT (reg,bit), showing the first bit to set. The Boolean enable input variable EN should be loaded into (W,0) before this macro is invoked. When EN = 1, the macro "setBF" writes a data value of 1 to "N" consecutive bits starting from the address OUT. When EN = 0, the "setBF" macro is not activated, and the states of "N" consecutive bits starting from address OUT are not changed. This instruction must be located at the end of a PLC rung. Operands for the instruction "setBF" are shown in Table 3.53. Assumption 1: the operand "reg,bit" can be in any bank. Assumption 2: all consecutive bits are in the same bank.

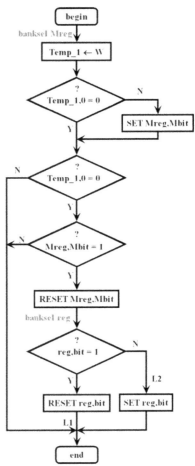

```
f_toggle macro   Mreg,Mbit, reg,bit
         local   L1,L2
         banksel Mreg
         movwf   Temp_1
         btfsc   Temp_1,0
         bsf     Mreg,Mbit
         btfsc   Temp_1,0
         goto    L1
         btfss   Mreg,Mbit
         goto    L1
         bcf     Mreg,Mbit
         banksel reg
         btfss   reg,bit
         goto    L2
         bcf     reg,bit
         goto    L1
L2       bsf     reg,bit
L1
         endm
```

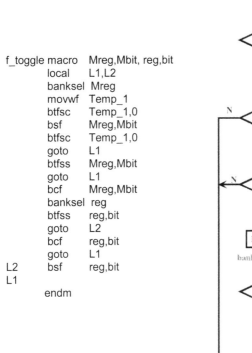

FIGURE 3.26   The macro "f_toggle" and its flowchart.

The macro "setBF" contains four "while" directives with three different conditions. The directive "while" is not an instruction, but used to control how code is assembled, not how it behaves at run-time. This directive is used for conditional assembly. It is not executed at run-time, but produces assembly code based on a condition. "while" is used to perform a loop while a certain condition is true. This directive

TABLE 3.47
Operands for the Instruction "f_toggle"

OUT (reg,bit)	Data type	Operands
Bit	BOOL	Q, M, T_Q, C_Q, C_QD, C_QU, drum_Q, drum_TQ, drum_steps, drum_events, SF, MB, any Boolean variable from SRAM.

TABLE 3.48

The Symbols and the Timing Diagram of the Macro "adrs_re"

Ladder Diagram Symbol   Schematic Symbol

inputs & output

IN (Boolean enable input variable, through W) = 0 or 1

Address (Boolean input variable, Address, regi,biti) = 0 or 1

Mreg,Mbit (A unique memory bit) = M0.0, M0.1, ..., M127.7

OUT (One shot output signal, through W) = 0 or 1

Timing Diagram

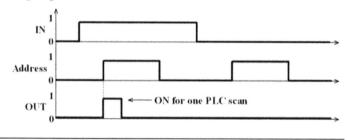

is used with the "endw" directive. The lines between the "while" and the "endw" are assembled as long as the condition evaluates to TRUE. An expression that evaluates to 0 is considered logically FALSE. An expression that evaluates to any other value is considered logically TRUE. A relational TRUE expression is guaranteed to return a non-0 value; FALSE a value of 0. A "while" loop can contain at most 100 lines and be repeated a maximum of 256 times. while loops can be nested up to 8 deep.

In the macro "setBF", the first "while" directive is based on the condition "if bit = 0", the second "while" directive is based on the condition "if bit > 0 and (bit+N) ≤ 8", and finally the last two "while" directives are based on the condition "if bit > 0 and (bit+N) > 8". At any time only one of these conditions is held according to "bit" and "N". Therefore, the code will be assembled based on the condition which evaluates to TRUE. Note that for the last condition two "while" directives are used. In order to understand the use of these directives in the macro "setBF", let us consider three different scenarios.

In the first scenario, consider the following program.

```
ld        I0.0
          ;reg,bit, N
setBF     Q0.0,11
```

Basically, according to this program, when I0.0 = 1, the bit field consisting of 11 bits starting from Q0.0 will be set.

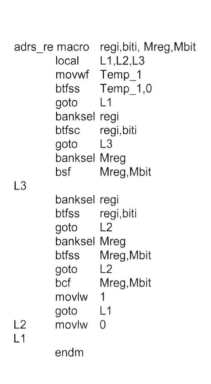

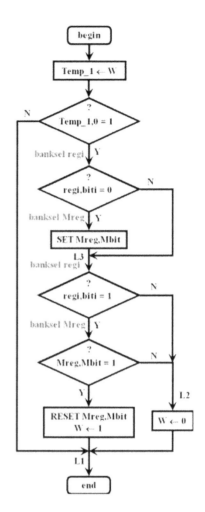

FIGURE 3.27 The macro "adrs_re" and its flowchart.

Now let us see how the code for this example is assembled by the macro "setBF". It is obvious that we have the following:

```
reg = Q0 = 74h
bit = 0
N = 11
```

TABLE 3.49
Operands for the Instruction "adrs_re"

Address (regi,biti)	Data type	Operands
Bit	BOOL	I, Q, M, T_Q, C_Q, C_QD, C_QU, drum_Q, drum_TQ, drum_steps, drum_events, SF, MB, nAI, AI, any Boolean variable from SRAM.

TABLE 3.50

The Symbols and the Timing Diagram of the Macro "adrs_fe"

Ladder Diagram Symbol   Schematic Symbol

inputs & output

IN (Boolean enable input variable, through W) = 0 or 1

Address (Boolean input variable, Address, regi,biti) = 0 or 1

Mreg,Mbit (A unique memory bit) = M0.0, M0.1, ..., M127.7

OUT (One shot output signal, through W) = 0 or 1

Timing Diagram

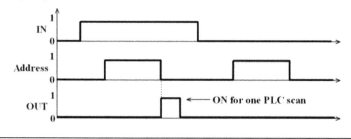

If we look at the macro "setBF" we can see that the condition of following part of the macro evaluates to TRUE, since bit = 0.

```
;-------------------------------------
   if        bit = 0
i = 0
   while i < N
      bsf        reg+i/8,i-(8*(i/8))
i += 1
   endw
   endif
;-------------------------------------
```

The following shows how the related part of the code is assembled for this example.

Note that the division $\dfrac{i}{8}$ is an integer divison. Thus we have following results:

$$\dfrac{i}{8} = 0 \, [i = 0,1, \ldots,7], \quad \dfrac{i}{8} = 1 \, [i = 8,9, \ldots,15], \quad \dfrac{i}{8} = 2 \, [i = 16,17, \ldots,23], \ldots$$

```
if        bit=0
   i = 0
```

```
adrs_fe macro    regi,biti, Mreg,Mbit
         local    L1,L2,L3
         movwf    Temp_1
         btfss    Temp_1,0
         goto     L1
         banksel regi
         btfss    regi,biti
         goto     L3
         banksel Mreg
         bsf      Mreg,Mbit
L3
         banksel regi
         btfsc    regi,biti
         goto     L2
         banksel Mreg
         btfss    Mreg,Mbit
         goto     L2
         bcf      Mreg,Mbit
         movlw    1
         goto     L1
L2       movlw    0
L1
         endm
```

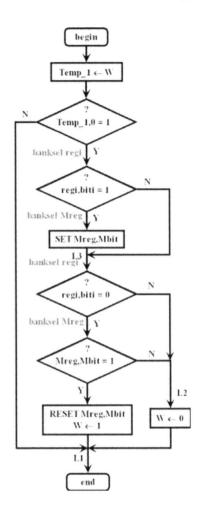

FIGURE 3.28 The macro "adrs_fe" and its flowchart.

```
while i < N [i = 0, N = 11]
```

$$\text{bsf} \quad reg + \frac{i}{8}, \; i - \left(8 * \left(\frac{i}{8}\right)\right) = \text{bsf } 74h + \frac{0}{8}, \; 0 - \left(8 * \left(\frac{0}{8}\right)\right) =$$

```
bsf   74h,0
i = i + 1 = 0 + 1 = 1
```

TABLE 3.51
Operands for the Instruction "adrs_fe"

Address (regi,biti)	Data type	Operands
Bit	BOOL	I, Q, M, T_Q, C_Q, C_QD, C_QU, drum_Q, drum_TQ, drum_steps, drum_events, SF, MB, nAI, AI, any Boolean variable from SRAM.

TABLE 3.52

The Symbols of the Macro "setBF"

Ladder Diagram Symbol	Schematic Symbol

inputs & output

EN (Boolean enable input variable, through W) = 0 or 1

N (the number of consecutive bits to be set) = 1, 2, ..., 256

OUT (the first bit to set, reg,bit) = 0 or 1

```
while i < N [i = 1, N = 11]
```

$$\text{bsf} \quad reg + \frac{i}{8}, \quad i - \left(8 * \left(\frac{i}{8}\right)\right) = \text{bsf} \quad 74h + \frac{1}{8}, \quad 1 - \left(8 * \left(\frac{1}{8}\right)\right) =$$

```
bsf   74h,1
i = i + 1 = 1 + 1 = 2
while i < N [i = 2, N = 11]
```

$$\text{bsf} \quad reg + \frac{i}{8}, \quad i - \left(8 * \left(\frac{i}{8}\right)\right) = \text{bsf} \ 74h + \frac{2}{8}, \quad 2 - \left(8 * \left(\frac{2}{8}\right)\right) =$$

```
bsf   74h,2
i = i + 1 = 2 + 1 = 3
while i < N [i = 3, N = 11]
```

$$\text{bsf} \quad reg + \frac{i}{8}, \quad i - \left(8 * \left(\frac{i}{8}\right)\right) = \text{bsf} \ 74h + \frac{3}{8}, \quad 3 - \left(8 * \left(\frac{3}{8}\right)\right) =$$

```
bsf   74h,3
i = i + 1 = 3 + 1 = 4
while i < N [i = 4, N = 11]
```

$$\text{bsf} \quad reg + \frac{i}{8}, \quad i - \left(8 * \left(\frac{i}{8}\right)\right) = \text{bsf} \ 74h + \frac{4}{8}, \quad 4 - \left(8 * \left(\frac{4}{8}\right)\right) =$$

```
bsf   74h,4
i = i + 1 = 4 + 1 = 5
while i < N [i = 5, N = 11]
```

$$\text{bsf} \quad reg + \frac{i}{8}, \quad i - \left(8 * \left(\frac{i}{8}\right)\right) = \text{bsf} \ 74h + \frac{5}{8}, \quad 5 - \left(8 * \left(\frac{5}{8}\right)\right) =$$

```
bsf   74h,5
i = i + 1 = 5 + 1 = 6
while i < N [i = 6, N = 11]
```

$$\text{bsf} \quad reg + \frac{i}{8}, \quad i - \left(8 * \left(\frac{i}{8}\right)\right) = \text{bsf} \ 74h + \frac{6}{8}, \quad 6 - \left(8 * \left(\frac{6}{8}\right)\right) =$$

```
bsf   74h,6
i = i + 1 = 6 + 1 = 7
while i < N [i = 7, N = 11]
```

```
setBF   macro  reg,bit, N
        local  i,L0
;-------------------------------------------------------------------------
    if    ((N>0)&&(N<257)) ;if (N>0)&&(N<257) then carry on, else do not compile.
;-------------------------------------------------------------------------
        movwf   Temp_1
        btfss   Temp_1,0
        goto    L0
        banksel reg
;---------------------------------
    if    bit==0
i=0
   while i < N
        bsf     reg+i/8,i-(8*(i/8))
i += 1
   endw
   endif
;---------------------------------
    if    bit>0 && ((bit+N)<=8)
i=0
   while i < N
        bsf     reg,bit+i
i += 1
   endw
   endif
;---------------------------------
    if    bit>0 && ((bit+N)>8)
i=0
   while i < 8–bit
        bsf     reg,bit+i
i += 1
   endw
i=0
   while i < N–(8–bit)
        bsf     reg+1+i/8,i–(8*(i/8))
i += 1
   endw
   endif
L0
;-------------------------------------------------------------------------
   else
   error "N must be a number from 1 to 256"
   endif
;-------------------------------------------------------------------------
        endm
```

FIGURE 3.29   The macro "setBF".

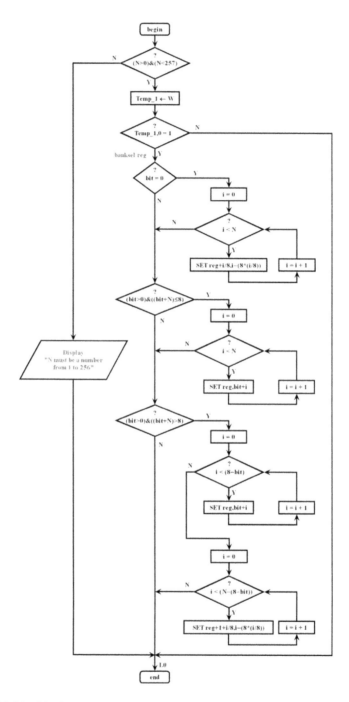

FIGURE 3.30   The flowchart of the macro "setBF".

TABLE 3.53

Operands for the Instruction "setBF"

OUT

(reg,bit)	Data type	Operands
Bit	BOOL	Q, M, T_Q, C_Q, C_QD, C_QU, drum_Q, drum_TQ, drum_steps, drum_events, SF, MB, any Boolean variable from SRAM.

$$\text{bsf} \quad reg + \frac{i}{8}, \; i - \left(8 * \left(\frac{i}{8}\right)\right) = \text{bsf } 74h + \frac{7}{8}, \; 7 - \left(8 * \left(\frac{7}{8}\right)\right) =$$

```
bsf  74h,7
i = i + 1 = 7 + 1 = 8
while i < N [i = 8, N = 11]
```

$$\text{bsf} \quad reg + \frac{i}{8}, \; i - \left(8 * \left(\frac{i}{8}\right)\right) = \text{bsf } 74h + \frac{8}{8}, \; 8 - \left(8 * \left(\frac{8}{8}\right)\right) =$$

```
bsf  75h,0
i = i + 1 = 8 + 1 = 9
while i < N [i = 9, N = 11]
```

$$\text{bsf} \quad reg + \frac{i}{8}, \; i - \left(8 * \left(\frac{i}{8}\right)\right) = \text{bsf } 74h + \frac{9}{8}, \; 9 - \left(8 * \left(\frac{9}{8}\right)\right) =$$

```
bsf  75h,1
i = i + 1 = 9 + 1 = 10
while i < N [i = 10, N = 11]
```

$$\text{bsf} \quad reg + \frac{i}{8}, \; i - \left(8 * \left(\frac{i}{8}\right)\right) = \text{bsf } 74h + \frac{10}{8}, \; 10 - \left(8 * \left(\frac{10}{8}\right)\right) =$$

```
bsf  75h,2
endwhile
endif
```

The assembled code has 11-bit set f "bsf" assembly instructions for consecutive bits starting from 74h,0 to 75h,2. When I0.0 = 1, the shaded cells in the following table shows the set 11 bits starting from Q0.0 due to the instruction "setBF Q0.0,11":

Address	Name	Bit 7	Bit 6	Bit 5	Bit 4	Bit 3	Bit 2	Bit 1	Bit 0
074h	Q0	Q0.7	Q0.6	Q0.5	Q0.4	Q0.3	Q0.2	Q0.1	Q0.0
075h	Q1	Q1.7	Q1.6	Q1.5	Q1.4	Q1.3	Q1.2	Q1.1	Q1.0

In the second scenario, consider the following program.

```
ld      I0.0
        ;reg,bit, N
setBF   Q0.1,7
```

Basically, according to this program, when I0.0 = 1, the bit field consisting of 7 bits starting from Q0.1 will be set.

Now let us see how the code for this example is assembled by the macro "setBF". It is obvious that we have the following:

```
reg = Q0 = 74h
bit = 1
N = 7
```

If we look at the macro "setBF" we can see that the condition of the following part of the macro evaluates to TRUE, since bit > 0 and bit+N = 8.

```
;------------------------------------
  if      bit > 0 && ((bit+N) <= 8)
i = 0
  while i < N
    bsf        reg,bit+i
i += 1
    endw
    endif
;------------------------------------
```

The following shows how the related part of the code is assembled for this example.

```
if      bit > 0 and (bit+N) ≤ 8
  i = 0
    while i < N [i = 0, N = 7]
    bsf    reg,bit+i = bsf    74h,1+0 = bsf   74h,1
  i = i + 1 = 0 + 1 = 1
    while i < N [i = 1, N = 7]
    bsf    reg,bit+i = bsf    74h,1+1 = bsf   74h,2
  i = i + 1 = 1 + 1 = 2
    while i < N [i = 2, N = 7]
    bsf    reg,bit+i = bsf    74h,1+2 = bsf   74h,3
  i = i + 1 = 2 + 1 = 3
    while i < N [i = 3, N = 7]
    bsf    reg,bit+i = bsf    74h,1+3 = bsf   74h,4
  i = i + 1 = 3 + 1 = 4
    while i < N [i = 4, N = 7]
    bsf    reg,bit+i = bsf    74h,1+4 = bsf   74h,5
  i = i + 1 = 4 + 1 = 5
    while i < N [i = 5, N = 7]
    bsf    reg,bit+i = bsf    74h,1+5 = bsf   74h,6
  i = i + 1 = 5 + 1 = 6
    while i < N [i = 6, N = 7]
    bsf    reg,bit+i = bsf    74h,1+6 = bsf   74h,7
    endwhile
endif
```

The assembled code has 7-bit set f "bsf" assembly instructions for consecutive bits starting from 74h,1 to 74h,7. When I0.0 = 1, the shaded cells in the following table shows the set 7 bits starting from Q0.1 due to the instruction "setBF Q0.1,7":

Address	Name	Bit 7	Bit 6	Bit 5	Bit 4	Bit 3	Bit 2	Bit 1	Bit 0
074h	Q0	Q0.7	Q0.6	Q0.5	Q0.4	Q0.3	Q0.2	Q0.1	Q0.0

In the third scenario, consider the following program.

```
ld        I0.0
          ;reg,bit, N
setBF     Q0.6,6
```

Basically, according to this program when I0.0 = 1, the bit field consisting of 6 bits starting from Q0.6 will be set.

Now let us see how the code for this example is assembled by the macro "setBF". It is obvious that we have the following:

```
reg = Q0 = 74h
bit = 6
N = 6
```

If we look at the macro "setBF" we can see that the condition of following part of the macro evaluates to TRUE, since bit > 0 and bit+N > 8.

```
;------------------------------------
  if      bit > 0 && ((bit+N) > 8)
i = 0
  while i < 8-bit
    bsf    reg,bit+i
i += 1
  endw
i = 0
  while i < N-(8-bit)
    bsf       reg+1+i/8,i-(8*(i/8))
i += 1
    endw
    endif
;------------------------------------
```

The following shows how the related part of the code is assembled for this example. Note that the division $\frac{i}{8}$ is an integer divison. Thus we have following results: $\frac{i}{8}$

$= 0$ [$i = 0, 1, ..., 7$], $\frac{i}{8} = 1$ [$i = 8, 9, ..., 15$], $\frac{i}{8} = 2$ [$i = 16, 17, ..., 23$], ...

```
if      bit > 0 and (bit+N) > 8
    i = 0
        while i < 8 - bit [i = 0, 8 - bit = 8 - 6 = 2]
        bsf     reg,bi t+i = bsf     74h,6+0 = bsf    74h,6
    i = i + 1 = 0 + 1 = 1
        while i < 8 - bit [i = 1, 8 - bit = 8 - 6 = 2]
        bsf     reg,bit+i = bsf    74h,6+1 = bsf    74h,7
        endwhile
    i = 0
        while i < N - (8 - bit) [i = 0, N - (8 - bit) = 6 - (8
        - 6) = 4]
```

$$bsf \quad reg + 1 + \frac{i}{8}, \quad i - \left(8 * \left(\frac{i}{8}\right)\right) = bsf \quad 74h + 1 + \frac{0}{8},$$

$$0 - \left(8 * \left(\frac{0}{8}\right)\right) = bsf \quad 75h,0$$

```
    i = i + 1 = 0 + 1 = 1
        while i < N - (8 - bit) [i = 1, N - (8 - bit) = 6 - (8
        - 6) = 4]
```

$$bsf \quad reg + 1 + \frac{i}{8}, \quad i - \left(8 * \left(\frac{i}{8}\right)\right) = bsf \quad 74h + 1 + \frac{1}{8},$$

$$1 - \left(8 * \left(\frac{1}{8}\right)\right) = bsf \quad 75h,1$$

```
    i = i + 1 = 1 + 1 = 2
        while i < N - (8 - bit) [i = 2, N - (8 - bit) = 6 - (8
        - 6) = 4]
```

$$bsf \quad reg + 1 + \frac{i}{8}, \quad i - \left(8 * \left(\frac{i}{8}\right)\right) = bsf \quad 74h + 1 + \frac{2}{8},$$

$$2 - \left(8 * \left(\frac{2}{8}\right)\right) = bsf \quad 75h,2$$

```
    i = i + 1 = 2 + 1 = 3
        while i < N - (8 - bit) [i = 3, N - (8 - bit) = 6 - (8
        - 6) = 4]
```

$$bsf \quad reg + 1 + \frac{i}{8}, \quad i - \left(8 * \left(\frac{i}{8}\right)\right) = bsf \quad 74h + 1 + \frac{3}{8},$$

$$3 - \left(8 * \left(\frac{3}{8}\right)\right) = bsf \quad 75h,3$$

```
        endwhile
endif
```

The assembled code has 6-bit set f "bsf" assembly instructions for consecutive bits starting from 74h,6 to 75h,3. When I0.0 = 1, the shaded cells in the following table shows the set 6 bits starting from Q0.6 due to the instruction "setBF Q0.6,6":

Address	Name	Bit 7	Bit 6	Bit 5	Bit 4	Bit 3	Bit 2	Bit 1	Bit 0
074h	Q0	Q0.7	Q0.6	Q0.5	Q0.4	Q0.3	Q0.2	Q0.1	Q0.0
075h	Q1	Q1.7	Q1.6	Q1.5	Q1.4	Q1.3	Q1.2	Q1.1	Q1.0

3.30   MACRO "RESETBF" (RESET BIT FIELD)

The symbols of the macro "resetBF" are depicted in Table 3.54. Figures 3.31 and 3.32 show the macro "resetBF" and its flowchart, respectively. This macro has a Boolean enable input variable EN, passed in through "W", the parameter N (the number of consecutive bits to be reset), and a Boolean output variable OUT (reg,bit), showing the first bit to reset. The Boolean enable input variable EN should be loaded into (W,0) before this macro is invoked. When EN = 1, the macro "resetBF" writes a data value of 0 to "N" consecutive bits starting from the address OUT. When EN = 0, the "resetBF" macro is not activated, and the states of "N" consecutive bits starting from address OUT are not changed. This instruction must be located at the end of a PLC rung. Operands for the instruction "resetBF" are shown in Table 3.55. Assumption 1: the operand "reg,bit" can be in any bank. Assumption 2: all consecutive bits are in the same bank.

3.31   EXAMPLES FOR CONTACT AND RELAY-BASED MACROS

Up to now in this chapter, we have seen contact and relay-based macros developed for the PIC16F1847-Based PLC. It is now time to consider eight examples related to these macros. Before you can run the example programs considered here, you are expected to construct your own PIC16F1847-Based PLC hardware by using the necessary PCB files, and by producing your PCBs with their components. For an effective use of examples, all example programs considered in this book (*Hardware and Basic Concepts*) are allocated within the file "PICPLC_PIC16F1847_user_Bsc .inc", which is downloadable from this book's webpage under the downloads section.

TABLE 3.54
The Symbols of the Macro "resetBF"

Ladder Diagram Symbol   Schematic Symbol

inputs & output

EN (Boolean enable input variable, through W) = 0 or 1

N (the number of consecutive bits to be reset) = 1, 2, ..., 256

OUT (the first bit to reset, reg,bit) = 0 or 1

```
resetBF macro  reg,bit, N
         local    i,L0
;-----------------------------------------------------------------------
     if    ((N>0)&&(N<257)) ;if (N>0)&&(N<257) then carry on, else do not compile.
;-----------------------------------------------------------------------
         movwf   Temp_1
         btfss   Temp_1,0
         goto    L0
         banksel reg
;--------------------------------------
     if    bit==0
i=0
   while i < N
         bcf       reg+i/8,i-(8*(i/8))
i += 1
   endw
   endif
;--------------------------------------
     if    bit>0 && ((bit+N)<=8)
i=0
   while i < N
         bcf       reg,bit+i
i += 1
   endw
   endif
;--------------------------------------
     if    bit>0 && ((bit+N)>8)
i=0
   while i < 8–bit
         bcf       reg,bit+i
i += 1
   endw
i=0
   while i < N–(8–bit)
         bcf       reg+1+i/8,i–(8*(i/8))
i += 1
   endw
   endif
L0
;-----------------------------------------------------------------------
   else
   error "N must be a number from 1 to 256"
   endif
;-----------------------------------------------------------------------
         endm
```

FIGURE 3.31   The macro "resetBF".

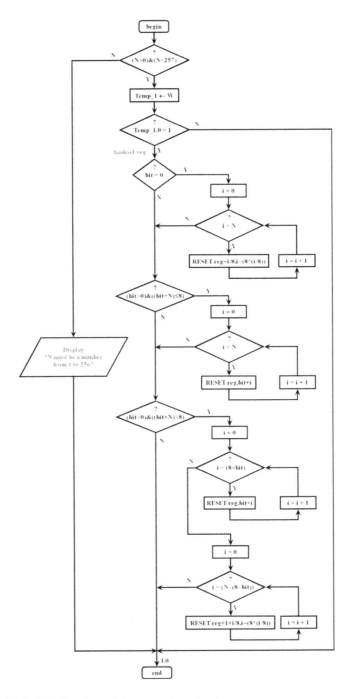

FIGURE 3.32 The flowchart of the macro "resetBF".

TABLE 3.55

Operands for the Instruction "resetBF"

OUT

(reg,bit)	Data type	Operands
Bit	BOOL	Q, M, T_Q, C_Q, C_QD, C_QU, drum_Q, drum_TQ, drum_steps, drum_events, SF, MB, any Boolean variable from SRAM.

```
user_program_1   macro
;
;--- PLC codes to be allocated in the "user_program_1" macro start from here ---
;__Example_Bsc_3.1
          ld        I0.0              ;rung 1          ; (W,0) = I0.0
          out       Q0.0                              ; Q0.0 = (W,0)

          ld_not    I0.1              ;rung 2          ; (W,0) = complement of I0.1
          out       Q0.1                              ; Q0.1 = (W,0)

          ld        I0.2              ;rung 3          ; (W,0) = I0.2
          out       M127.7                            ; M127.7 = (W,0)

          ld        M127.7            ;rung 4          ; (W,0) = M127.7
          out_not   Q0.2                              ; Q0.2 = complement of (W,0)

          ld        I0.3              ;rung 5          ; (W,0) = I0.3
          not                                         ; (W,0) = complement of (W,0)
          out       Q0.3                              ; Q0.3 = (W,0)

          in_out    I0.4,Q0.4         ;rung 6          ; Q0.4 = I0.4

          inv_out   I0.5,Q0.5         ;rung 7          ; Q0.5 = complement of I0.5

          in_out    LOGIC1,Q0.6       ;rung 8          ; Q0.6 = LOGIC1

          in_out    LOGIC0,Q0.7       ;rung 9          ; Q0.7 = LOGIC0

          in_out    T_10ms,Q1.0       ;rung 10         ; Q1.0 = T_10ms

          in_out    re_T10ms,Q1.1     ;rung 11         ; Q1.1 = rising edge of T_10ms

          in_out    T_100ms,Q1.2      ;rung 12         ; Q1.2 = T_100ms

          in_out    re_T100ms,Q1.3    ;rung 13         ; Q1.3 = rising edge of T_100ms

          in_out    T_1s,Q1.4         ;rung 14         ; Q1.4 = T_1s

          in_out    re_T1s,Q1.5       ;rung 15         ; Q1.5 = rising edge of T_1s
;--- PLC codes to be allocated in the "user_program_1" macro end here ----------
          endm
```

FIGURE 3.33   The user program of Example 3.1.

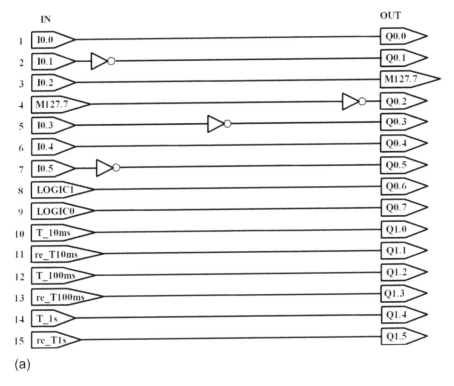

(a)

FIGURE 3.34 The user program of Example 3.1: (a) schematic diagram, (b) ladder diagram.

Initially all example programs are commented out by putting a semicolon ";" in front of each line. When you would like to test one of the example programs, you must uncomment each line of the example program by following the steps shown below:

1. Highlight the block of source lines you want to uncomment by dragging the mouse over these lines with the left mouse button held down. With default coloring in MPLAB X IDE you will now see green characters on a blue background.
2. Release the mouse button.
3. Press Ctrl/Shift/C or press the Alt, S, and M keys in succession, or select "Toggle Comment" from the toolbar Source menu. Now a semicolon will be removed from all selected source lines. With default coloring you will see red characters on a white background.

Then, you can run the project by pressing the symbol ▷ from the toolbar. Next, the MPLAB X IDE will produce the "PICPLC_PIC16F1847.X.production.hex" file for the project. Then the MPLAB X IDE will be connected to the PICkit 3 programmer and finally it will program the PIC16F1847 microcontroller within the CPU board of the PIC16F1847-Based PLC. During these steps, make sure that in the CPU board of the PIC16F1847-Based PLC, the 4PDT switch is in the "PROG" position and the power

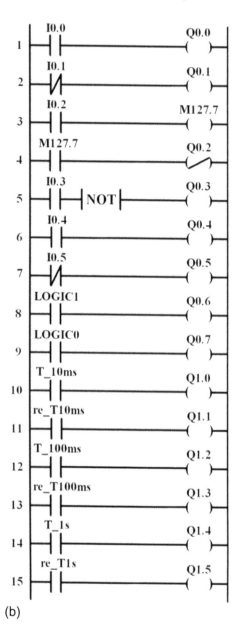

(b)

FIGURE 3.34 Continued

switch is in the "OFF" position. After loading the program file to the PIC16F1847 microcontroller, switch the 4PDT to "RUN" and the power switch to the "ON" position. Finally, you are ready to test the example program. **Warning**: When you finish your study with an example and try to take a look at another example, do not forget to comment the current example program before uncommenting another one. In other words, make sure that only one example program is uncommented and tested at the

```
user_program_1  macro
;--- PLC codes to be allocated in the "user_program_1" macro start from here ---
;__Example_Bsc_3.2

        ld       I0.0    ;rung 1        ; (W,0) = I0.0
        or       I0.1                   ; (W,0)= (W,0) or I0.1
        out      Q0.0                   ; Q0.0 = (W,0)

        ld       I0.0    ;rung 2        ; (W,0) = I0.0
        or       I0.1                   ; (W,0)= (W,0) or I0.1
        or       I0.2                   ; (W,0)= (W,0) or I0.2
        out      Q0.1                   ; Q0.1 = (W,0)

        ld       I0.0    ;rung 3        ; (W,0) = I0.0
        or       I0.1                   ; (W,0)= (W,0) or I0.1
        or       I0.2                   ; (W,0)= (W,0) or I0.2
        or       I0.3                   ; (W,0)= (W,0) or I0.3
        out      Q0.2                   ; Q0.2 = (W,0)

        ld_not   I0.4    ;rung 4        ; (W,0) = complement of I0.4
        or       I0.5                   ; (W,0)= (W,0) or I0.5
        or       I0.6                   ; (W,0)= (W,0) or I0.6
        out      Q0.3                   ; Q0.3 = (W,0)

        ld       I0.4    ;rung 5        ; (W,0) = I0.4
        or       I0.5                   ; (W,0)= (W,0) or I0.5
        or_not   I0.6                   ; (W,0)= (W,0) or complement of I0.6
        out      Q0.4                   ; Q0.4 = (W,0)

        ld       I0.6    ;rung 6        ; (W,0) = I0.6
        or       I0.7                   ; (W,0)= (W,0) or I0.7
        out      Q0.5                   ; Q0.5 = (W,0)

        ld       I0.6    ;rung 7        ; (W,0) = I0.6
        or       I0.7                   ; (W,0)= (W,0) or I0.7
        out_not  Q0.6                   ; Q0.6 = complement of (W,0)

        ld       I0.6    ;rung 8        ; (W,0) = I0.6
        or       I0.7                   ; (W,0)= (W,0) or I0.7
        not                             ; (W,0) = complement of (W,0)
        out      Q0.7                   ; Q0.7 = (W,0)

        ld       I1.0    ;rung 9        ; (W,0) = I1.0
        or       I1.1                   ; (W,0)= (W,0) or I1.1
        or_not   I1.2                   ; (W,0)= (W,0) or complement of I1.2
        or       I1.3                   ; (W,0)= (W,0) or I1.3
        or       I1.4                   ; (W,0)= (W,0) or I1.4
        or       I1.5                   ; (W,0)= (W,0) or I1.5
        or       I1.6                   ; (W,0)= (W,0) or I1.6
        or       I1.7                   ; (W,0)= (W,0) or I1.7
        out      Q1.0                   ; Q1.0 = (W,0)
```

FIGURE 3.35   (*1 of* 2) The user program of Example 3.2.

```
ld        I2.0    ;rung 10      ; (W,0) = I2.0
nor       I2.1                  ; (W,0)= (W,0) nor I2.1
out       Q2.0                  ; Q2.0 = (W,0)

ld        I2.0    ;rung 11      ; (W,0) = I2.0
nor       I2.1                  ; (W,0)= (W,0) nor I2.1
nor       I2.2                  ; (W,0)= (W,0) nor I2.2
out       Q2.1                  ; Q2.1 = (W,0)

ld        I2.0    ;rung 12      ; (W,0) = I2.0
or        I2.1                  ; (W,0)= (W,0) or I2.1
or        I2.2                  ; (W,0)= (W,0) or I2.2
not                            ; (W,0) = complement of (W,0)
out       Q2.2                  ; Q2.2 = (W,0)

ld        I2.0    ;rung 13      ; (W,0) = I2.0
or        I2.1                  ; (W,0)= (W,0) or I2.1
or        I2.2                  ; (W,0)= (W,0) or I2.2
out_not   Q2.3                  ; Q2.3 = complement of (W,0)

ld        I2.0    ;rung 14      ; (W,0) = I2.0
or        I2.1                  ; (W,0)= (W,0) or I2.1
or        I2.2                  ; (W,0)= (W,0) or I2.2
or        I2.3                  ; (W,0)= (W,0) or I2.3
not                            ; (W,0) = complement of (W,0)
out       Q2.4                  ; Q2.4 = (W,0)

ld        I2.0    ;rung 15      ; (W,0) = I2.0
or        I2.1                  ; (W,0)= (W,0) or I2.1
or        I2.2                  ; (W,0)= (W,0) or I2.2
or        I2.3                  ; (W,0)= (W,0) or I2.3
out_not   Q2.5                  ; Q2.5 = complement of (W,0)

;--- PLC codes to be allocated in the "user_program_1" macro end here ----------
endm
```

FIGURE 3.35 Continued

same time. Otherwise, if you somehow leave more than one example uncommented, the example you are trying to test will probably not function as expected since it may try to access the same resources that are being used and changed by other examples.

Please check the accuracy of each program by cross-referencing it with the related macros.

3.31.1 Example 3.1

Example 3.1 shows the usage of the following contact and relay-based macros: "ld", "ld_not", "not", "out", "out_not", "in_out", and "inv_out". The user program of Example 3.1 is shown in Figure 3.33. The schematic diagram and ladder diagram of Example 3.1 are depicted in Figure 3.34(a) and in Figure 3.34(b), respectively. When the project file of the PIC16F1847-Based PLC is open in the MPLAB X IDE, from the file "PICPLC_PIC16F1847_user_Bsc.inc", if you uncomment Example 3.1

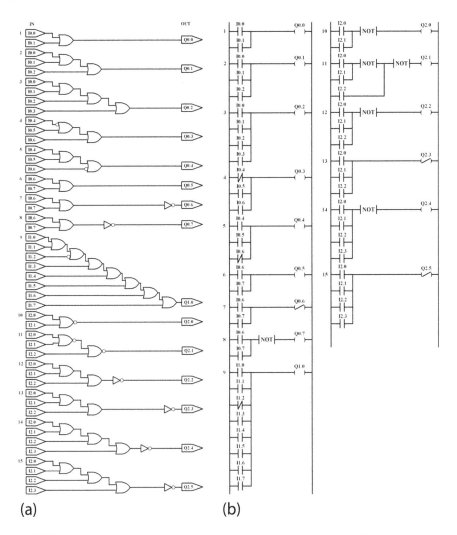

FIGURE 3.36   The user program of Example 3.2: (a) schematic diagram, (b) ladder diagram.

and run the project by pressing the symbol ▷ from the toolbar, then the PIC16F1847 microcontroller within the CPU board of the PIC16F1847-Based PLC will be programmed. After loading the program file to the PIC16F1847 microcontroller, switch the 4PDT to "RUN" and the power switch to the "ON" position. Next you can test the operation of this example.

3.31.2   EXAMPLE 3.2

Example 3.2 shows the usage of the following contact and relay-based macros: "ld", "ld_not", "or", "or_not", "out", and "out_not". The user program of Example 3.2 is shown in Figure 3.35. The schematic diagram and ladder diagram of Example 3.2

```
user_program_1   macro
;--- PLC codes to be allocated in the "user_program_1" macro start from here ---
;__Example_Bsc_3.3
        ld        I0.0     ;rung 1        ; (W,0) = I0.0
        and       I0.1                    ; (W,0)= (W,0) and I0.1
        out       Q0.0                    ; Q0.0 = (W,0)

        ld        I0.0     ;rung 2        ; (W,0) = I0.0
        and       I0.1                    ; (W,0)= (W,0) and I0.1
        and       I0.2                    ; (W,0)= (W,0) and I0.2
        out       Q0.1                    ; Q0.1 = (W,0)

        ld        I0.0     ;rung 3        ; (W,0) = I0.0
        and       I0.1                    ; (W,0)= (W,0) and I0.1
        and       I0.2                    ; (W,0)= (W,0) and I0.2
        and       I0.3                    ; (W,0)= (W,0) and I0.3
        out       Q0.2                    ; Q0.2 = (W,0)

        ld_not    I0.4     ;rung 4        ; (W,0) = complement of I0.4
        and       I0.5                    ; (W,0)= (W,0) and I0.5
        and       I0.6                    ; (W,0)= (W,0) and I0.6
        out       Q0.3                    ; Q0.3 = (W,0)

        ld        I0.4     ;rung 5        ; (W,0) = I0.4
        and       I0.5                    ; (W,0)= (W,0) and I0.5
        and_not I0.6                      ; (W,0)= (W,0) and complement of I0.6
        out       Q0.4                    ; Q0.4 = (W,0)

        ld        I0.6     ;rung 6        ; (W,0) = I0.6
        and       I0.7                    ; (W,0)= (W,0) and I0.7
        out       Q0.5                    ; Q0.5 = (W,0)

        ld        I0.6     ;rung 7        ; (W,0) = I0.6
        and       I0.7                    ; (W,0)= (W,0) and I0.7
        out_not Q0.6                      Q0.6 = complement of (W,0)

        ld        I0.6     ;rung 8        ; (W,0) = I0.6
        and       I0.7                    ; (W,0)= (W,0) and I0.7
        not                               ; (W,0) = complement of (W,0)
        out       Q0.7                    ; Q0.7 = (W,0)

        ld        I1.0     ;rung 9        ; (W,0) = I1.0
        and       I1.1                    ; (W,0)= (W,0) and I1.1
        and_not I1.2                      ; (W,0)= (W,0) and complement of I1.2
        and       I1.3                    ; (W,0)= (W,0) and I1.3
        and       I1.4                    ; (W,0)= (W,0) and I1.4
        and       I1.5                    ; (W,0)= (W,0) and I1.5
        and       I1.6                    ; (W,0)= (W,0) and I1.6
        and       I1.7                    ; (W,0)= (W,0) and I1.7
        out       Q1.0                    ; Q1.0 = (W,0)
```

FIGURE 3.37   (*1 of* 2) The user program of Example 3.3.

```
        ld      I2.0    ;rung 10  ; (W,0) = I2.0
        nand    I2.1              ; (W,0)= (W,0) nand I2.1
        out     Q2.0              ; Q2.0 = (W,0)

        ld      I2.0    ;rung 11  ; (W,0) = I2.0
        nand    I2.1              ; (W,0)= (W,0) nand I2.1
        nand    I2.2              ; (W,0)= (W,0) nand I2.2
        out     Q2.1              ; Q2.1 = (W,0)

        ld      I2.0    ;rung 12  ; (W,0) = I2.0
        and     I2.1              ; (W,0)= (W,0) and I2.1
        and     I2.2              ; (W,0)= (W,0) and I2.2
        not                       ; (W,0) = complement of (W,0)
        out     Q2.2              ; Q2.2 = (W,0)

        ld      I2.0    ;rung 13  ; (W,0) = I2.0
        and     I2.1              ; (W,0)= (W,0) and I2.1
        and     I2.2              ; (W,0)= (W,0) and I2.2
        out_not Q2.3              ; Q2.3 = complement of (W,0)

        ld      I2.0    ;rung 14  ; (W,0) = I2.0
        and     I2.1              ; (W,0)= (W,0) and I2.1
        and     I2.2              ; (W,0)= (W,0) and I2.2
        and     I2.3              ; (W,0)= (W,0) and I2.3
        not                       ; (W,0) = complement of (W,0)
        out     Q2.4              ; Q2.4 = (W,0)

        ld      I2.0    ;rung 15  ; (W,0) = I2.0
        and     I2.1              ; (W,0)= (W,0) and I2.1
        and     I2.2              ; (W,0)= (W,0) and I2.2
        and     I2.3              ; (W,0)= (W,0) and I2.3
        out_not Q2.5              ; Q2.5 = complement of (W,0)

;--- PLC codes to be allocated in the "user_program_1" macro end here ----------
        endm
```

FIGURE 3.37 Continued

are depicted in Figure 3.36(a) and in Figure 3.36(b), respectively. When the project file of the PIC16F1847-Based PLC is open in the MPLAB X IDE, from the file "PICPLC_PIC16F1847_user_Bsc.inc", if you uncomment Example 3.2 and run the project by pressing the symbol ▷ from the toolbar, then the PIC16F1847 microcontroller within the CPU board of the PIC16F1847-Based PLC will be programmed. After loading the program file to the PIC16F1847 microcontroller, switch the 4PDT to "RUN" and the power switch to the "ON" position. Next you can test the operation of this example.

3.31.3 EXAMPLE 3.3

Example 3.3 shows the usage of the following contact and relay-based macros: "ld", "ld_not", "and", "and_not", "nand", "not", "out", and "out_not". The user program of Example 3.3 is shown in Figure 3.37. The schematic diagram and ladder diagram

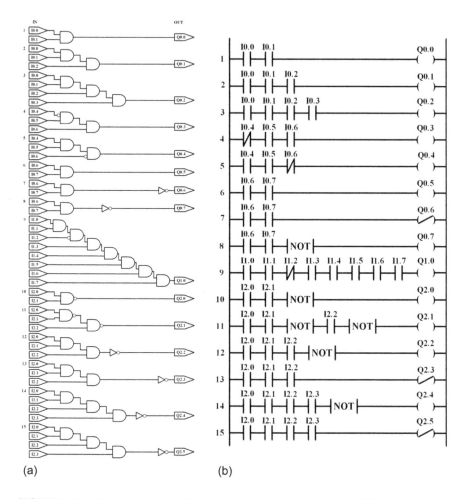

FIGURE 3.38 The user program of Example 3.3: (a) schematic diagram, (b) ladder diagram.

of Example 3.3 are depicted in Figure 3.38(a) and in Figure 3.38(b), respectively. When the project file of the PIC16F1847-Based PLC is open in the MPLAB X IDE, from the file "PICPLC_PIC16F1847_user_Bsc.inc", if you uncomment Example 3.3 and run the project by pressing the symbol ▷ from the toolbar, then the PIC16F1847 microcontroller within the CPU board of the PIC16F1847-Based PLC will be programmed. After loading the program file to the PIC16F1847 microcontroller, switch the 4PDT to "RUN" and the power switch to the "ON" position. Next you can test the operation of this example.

3.31.4 EXAMPLE 3.4

Example 3.4 shows the usage of the following contact and relay-based macros: "ld", "xor", "xor_not", "xnor", "_set", "_reset", "SR", "RS", and "out". The user program of Example 3.4 is shown in Figure 3.39. The schematic diagram and ladder diagram

```
user_program_1   macro
;--- PLC codes to be allocated in the "user_program_1" macro start from here ---
;__Example_Bsc_3.4

        ld      I0.0            ;rung 1          ; (W,0) = I0.0
        xor     I0.1                            ; (W,0)= (W,0) xor I0.1
        out     Q0.0                            ; Q0.0 = (W,0)

        ld      I0.0            ;rung 2          ; (W,0) = I0.0
        xor_not I0.1                            ; (W,0)= (W,0) xor_not I0.1
        out     Q0.1                            ; Q0.1 = (W,0)

        ld      I0.2            ;rung 3          ; (W,0) = I0.2
        xor     I0.3                            ; (W,0)= (W,0) xor I0.3
        out     Q0.2                            ; Q0.2 = (W,0)

        ld      I0.2            ;rung 4          ; (W,0) = I0.2
        xor_not I0.3                            ; (W,0)= (W,0) xor_not I0.3
        out     Q0.3                            ; Q0.3 = (W,0)

        ld      I0.4            ;rung 5          ; (W,0) = I0.4
        xnor    I0.5                            ; (W,0)= (W,0) xnor I0.5
        out     Q0.4                            ; Q0.4 = (W,0)

        ld      I0.6            ;rung 6          ; (W,0) = I0.6
        xnor    I0.7                            ; (W,0)= (W,0) xnor I0.7
        out     Q0.5                            ; Q0.5 = (W,0)

        ld      I1.1            ;rung 7          ; (W,0) = I1.1
        _set    Q1.0                            ; set Q1.0 if (W,0) = 1

        ld      I1.0            ;rung 8          ; (W,0) = I1.0
        _reset  Q1.0                            ; reset Q1.0 if (W,0) = 1

        ld      I1.3            ;rung 9          ; (W,0) = I1.3
        _set    Q1.2                            ; set Q1.2 if (W,0) = 1

        ld      I1.2            ;rung 10         ; (W,0) = I1.2
        _reset  Q1.2                            ; reset Q1.2 if (W,0) = 1

        ld      I1.4            ;rung 11         ; (W,0) = I1.4
        _reset  Q1.4                            ; reset Q1.4 if (W,0) = 1

        ld      I1.5            ;rung 12         ; (W,0) = I1.5
        _set    Q1.4                            ; set Q1.4 if (W,0) = 1

        ld      I1.6            ;rung 13         ; (W,0) = I1.6
        _reset  Q1.6                            ; reset Q1.6 if (W,0) = 1

        ld      I1.7            ;rung 14         ; (W,0) = I1.7
        _set    Q1.6                            ; set Q1.6 if (W,0) = 1
```

FIGURE 3.39   (*1 of* 2) The user program of Example 3.4.

```
    SR      I2.0,I2.1,Q2.0    ;rung 15      ; S=I2.0,R1=I2.1,Q=Q2.0

    SR      I2.2,I2.3,Q2.2    ;rung 16      ; S=I2.2,R1=I2.3,Q=Q2.2

    SR      I2.4,I2.5,Q2.4    ;rung 17      ; S=I2.4,R1=I2.5,Q=Q2.4

    RS      I3.0,I3.1,Q3.0    ;rung 18      ; R=I3.0,S1=I3.1,Q=Q3.0

    RS      I3.2,I3.3,Q3.2    ;rung 19      ; R=I3.2,S1=I3.3,Q=Q3.2

    RS      I3.4,I3.5,Q3.4    ;rung 20      ; R=I3.4,S1=I3.5,Q=Q3.4
  ;
  ;--- PLC codes to be allocated in the "user_program_1" macro end here ---------
      endm
```

FIGURE 3.39 Continued

of Example 3.4 are depicted in Figure 3.40(a) and in Figure 3.40(b), respectively. When the project file of the PIC16F1847-Based PLC is open in the MPLAB X IDE, from the file "PICPLC_PIC16F1847_user_Bsc.inc", if you uncomment Example 3.4 and run the project by pressing the symbol ▷ from the toolbar, then the PIC16F1847 microcontroller within the CPU board of the PIC16F1847-Based PLC will be programmed. After loading the program file to the PIC16F1847 microcontroller, switch the 4PDT to "RUN" and the power switch to the "ON" position. Next you can test the operation of this example.

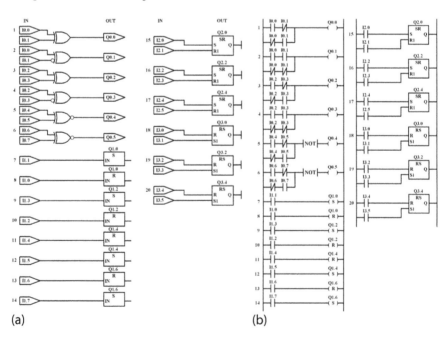

(a)   (b)

FIGURE 3.40 The user program of Example 3.4: (a) schematic diagram, (b) ladder diagram.

```
user_program_1   macro
;--- PLC codes to be allocated in the "user_program_1" macro start from here ---
;__Example_Bsc_3.5

        ld      I0.0            ;rung 1          ; (W,0) = I0.0
        and     I0.1                            ; (W,0) = I0.0 and I0.1
        mid_out M0.0                            ;  M0.0 = I0.0 and I0.1
        and     I0.2                            ; (W,0) = I0.0 and I0.1 and I0.2
        out     Q0.0                            ;  Q0.0 = I0.0 and I0.1 and I0.2

        in_out  M0.0,Q0.2       ;rung 2          ; Q0.2 = M0.0 = I0.0 and I0.1

        ld      I1.0            ;rung 3          ; (W,0) = I1.0
        and     I1.1                            ; (W,0) = I1.0 and I1.1
        and     I1.2                            ; (W,0) = I1.0 and I1.1 and I1.2
        mid_out M33.3                           ;  M33.3 = I1.0 and I1.1 and I1.2
        and     I1.3                            ; (W,0) = I1.0 and I1.1 and I1.2 and I1.3
        out     Q1.0                            ;  Q1.0 = I1.0 and I1.1 and I1.2 and I1.3

        in_out  M33.3,Q1.2      ;rung 4          ; Q1.2 = M33.3 = I1.0 and I1.1 and I1.2

        ld      I2.0            ;rung 5          ; (W,0) = I2.0
        and     I2.1                            ; (W,0) = I2.0 and I2.1
        mid_out_not M2.1                        ;  M2.1 = complement of (I2.0 and I2.1)
        and     I2.2                            ; (W,0) = I2.0 and I2.1 and I2.2
        out     Q2.0                            ;  Q2.0 = I2.0 and I2.1 and I2.2

        in_out  M2.1,Q2.2       ;rung 6          ; Q2.2 = complement of (I2.0 and I2.1)

        ld      I3.0            ;rung 7          ; (W,0) = I3.0
        and     I3.1                            ; (W,0) = I3.0 and I3.1
        and     I3.2                            ; (W,0) = I3.0 and I3.1 and I3.2
        mid_out_not M99.7                       ;  M99.7 = complement of (I3.0 and I3.1 and I3.2)
        and     I3.3                            ; (W,0) = I3.0 and I3.1 and I3.2 and I3.3
        out     Q3.0                            ;  Q3.0 = I3.0 and I3.1 and I3.2 and I3.3

        in_out  M99.7,Q3.2      ;rung 8          ; Q3.2 = complement of (I3.0 and I3.1 and I3.2)
;
;--- PLC codes to be allocated in the "user_program_1" macro end here ----------
        endm
```

FIGURE 3.41   The user program of Example 3.5.

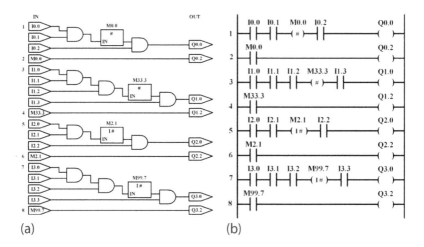

(a)   (b)

FIGURE 3.42   The user program of Example 3.5: (a) schematic diagram, (b) ladder diagram.

```
user_program_1   macro
;--- PLC codes to be allocated in the "user_program_1" macro start from here ---
;__Example_Bsc_3.6
         ld        I0.0            ;rung 1           ; (W,0) = I0.0
         r_edge   M55.3                             ; (W,0) = rising edge of (W,0)
         out       Q0.0                             ; Q0.0 = (W,0)

         ld        I0.3            ;rung 2           ; (W,0) = I0.3
         r_edge   M127.7                            ; (W,0) = rising edge of (W,0)
         out       Q0.3                             ; Q0.3 = (W,0)

         ld_not   I0.6            ;rung 3           ; (W,0) = complement of I0.6
         r_edge   M99.0                             ; (W,0) = rising edge of (W,0)
         out       Q0.6                             ; Q0.6 = (W,0)

         ld        I1.0            ;rung 4           ; (W,0) = I1.0
         f_edge   M0.2                              ; (W,0) = falling edge of (W,0)
         out       Q1.0                             ;  Q1.0 = (W,0)

         ld        I1.3            ;rung 5           ; (W,0) = I1.3
         f_edge   M127.6                            ; (W,0) = falling edge of (W,0)
         out       Q1.3                             ; Q1.3 = (W,0)

         ld_not   I1.6            ;rung 6           ; (W,0) = complement of I1.6
         f_edge   M0.1                              ; (W,0) = falling edge of (W,0)
         out       Q1.6                             ; Q1.6 = (W,0)

         ld        I2.0            ;rung 7           ; (W,0) = I2.0
         r_toggle M2.0,Q2.0                         ; Q2.0 will be toggled everytime
                                                    ; when I2.0 goes from low to high

         ld        I2.3            ;rung 8           ; (W,0) = I2.3
         r_toggle M2.1,Q2.3                         ; Q2.3 will be toggled everytime
                                                    ; when I2.3 goes from low to high

         ld_not   I2.6            ;rung 9           ; (W,0) = complement of I2.6
         r_toggle M2.2,Q2.6                         ; Q2.6 will be toggled everytime
                                                    ; when I2.6 goes from low to high

         ld        I3.0            ;rung 10          ; (W,0) = I3.0
         f_toggle M3.0,Q3.0                         ; Q3.0 will be toggled everytime
                                                    ; when I3.0 goes from high to low

         ld        I3.3            ;rung 11          ; (W,0) = I3.3
         f_toggle M3.1,Q3.3                         ; Q3.3 will be toggled everytime
                                                    ; when I3.3 goes from high to low

         ld_not   I3.6            ;rung 12          ; (W,0) = complement of I3.6
         f_toggle M3.2,Q3.6                         ; Q3.6 will be toggled everytime
                                                    ; when I3.6 goes from high to low
;--- PLC codes to be allocated in the "user_program_1" macro end here ----------
         endm
```

FIGURE 3.43   The user program of Example 3.6.

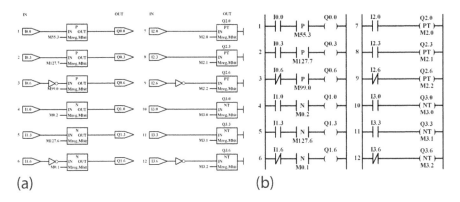

FIGURE 3.44 The user program of Example 3.6: (a) schematic diagram, (b) ladder diagram.

3.31.5 EXAMPLE 3.5

Example 3.5 shows the usage of the following contact and relay-based macros: "ld", "and", "mid_out", "mid_ out_not", "in_out", and "out". The user program of Example 3.5 is shown in Figure 3.41. The schematic diagram and ladder diagram of Example 3.5 are depicted in Figure 3.42(a) and in Figure 3.42(b), respectively. When the project file of the PIC16F1847-Based PLC is open in the MPLAB X IDE, from the file "PICPLC_PIC16F1847_user_Bsc.inc", if you uncomment Example 3.5 and run the project by pressing the symbol ▷ from the toolbar, then the PIC16F1847 microcontroller within the CPU board of the PIC16F1847-Based PLC will be programmed. After loading the program file to the PIC16F1847 microcontroller, switch the 4PDT to "RUN" and the power switch to the "ON" position. Next you can test the operation of this example.

3.31.6 EXAMPLE 3.6

Example 3.6 shows the usage of the following contact and relay-based macros: "ld", "ld_not", "r_edge", "f_edge", "r_toggle", "f_toggle", and "out". The user program of Example 3.6 is shown in Figure 3.43. The schematic diagram and ladder diagram of Example 3.6 are depicted in Figure 3.44(a) and in Figure 3.44(b), respectively. When the project file of the PIC16F1847-Based PLC is open in the MPLAB X IDE, from the file "PICPLC_PIC16F1847_user_Bsc.inc", if you uncomment Example 3.6 and run the project by pressing the symbol ▷ from the toolbar, then the PIC16F1847 microcontroller within the CPU board of the PIC16F1847-Based PLC will be programmed. After loading the program file to the PIC16F1847 microcontroller, switch the 4PDT to "RUN" and the power switch to the "ON" position. Next you can test the operation of this example.

3.31.7 EXAMPLE 3.7

Example 3.7 shows the usage of the following contact and relay-based macros: "ld", "and", "adrs_re", "adrs_fe", "in_out", and "out". The user program of Example 3.7 is

```
user_program_1   macro
;--- PLC codes to be allocated in the "user_program_1" macro start from here ---
;__Example_Bsc_3.7
        ld        I0.0              ;rung 1 ; (W,0) = I0.0
        and       I0.1                      ; (W,0) = (W,0) and I0.1
        and       I0.2                      ; (W,0) = (W,0) and I0.2
        adrs_re I0.3,M0.0                   ; If (I0.0 and I0.1 and I0.2)=1 &
                                            ; the rising edge signal of I0.3 occurs
                                            ; then (W,0)=1, else (W,0)=0.
        out       Q0.0                      ; Q0.0 = (W,0)

        in_out    I1.0,M1.0        ;rung 2 ; M1.0 = I1.0
        in_out    I1.1,M19.1       ;rung 3 ; M19.1 = I1.1
        in_out    I1.2,M49.2       ;rung 4 ; M49.2 = I1.2
        in_out    I1.3,M54.3       ;rung 5 ; M54.3 = I1.3
        in_out    I1.4,M127.6      ;rung 6 ; M127.6 = I1.4

        ld        M1.0             ;rung 7 ; (W,0) = M1.0
        and       M19.1                     ; (W,0) = (W,0) and M19.1
        and       M49.2                     ; (W,0) = (W,0) and M49.2
        adrs_re M54.3,M118.3                ; If (M1.0 and M19.1 and M49.2)=1 &
                                            ; the rising edge signal of M54.3 occurs
                                            ; then (W,0)=1, else (W,0)=1.
        and       M127.6                    ; (W,0) = (W,0) and M127.6
        out       M44.4                     ; M44.4 = (W,0)

        in_out    M44.4,Q1.0       ;rung 8 ; Q1.0 = M44.4

        ld        I2.0             ;rung 9 ; (W,0) = I2.0
        and       I2.1                      ; (W,0) = (W,0) and I2.1
        and       I2.2                      ; (W,0) = (W,0) and I2.2
        adrs_fe I2.3,M2.0                   ; If (I2.0 and I2.1 and I2.2)=1 &
                                            ; the falling edge signal of I2.3 occurs
                                            ; then (W,0)=1, else (W,0)=0.
        out       Q2.0                      ; Q2.0 = (W,0)

        in_out    I3.0,M3.0        ;rung 10; M3.0  = I3.0
        in_out    I3.1,M1.1        ;rung 11; M1.1  = I3.1
        in_out    I3.2,M4.2        ;rung 12; M4.2  = I3.2
        in_out    I3.3,M5.3        ;rung 13; M5.3  = I3.3
        in_out    I3.4,M12.6       ;rung 14; M12.6 = I3.4

        ld        M3.0             ;rung 15; (W,0) = M3.0
        and       M1.1                      ; (W,0) = (W,0) and M1.1
        and       M4.2                      ; (W,0) = (W,0) and M4.2
        adrs_fe M5.3,M11.3                  ; If (M3.0 and M1.1 and M4.2)=1 &
                                            ; the falling edge signal of M5.3 occurs
                                            ; then (W,0)=1, else (W,0)=1.
        and       M12.6                     ; (W,0) = (W,0) and M12.6
        out       M4.4                      ; M4.4 = (W,0)

        in_out    M4.4,Q3.0        ;rung 16; Q3.0 = M4.4
;--- PLC codes to be allocated in the "user_program_1" macro end here ----------
        endm
```

FIGURE 3.45   The user program of Example 3.7.

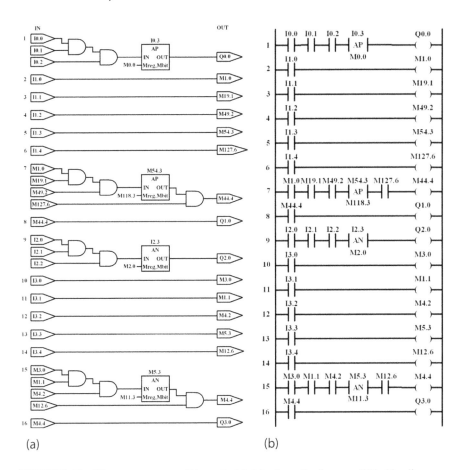

(a)   (b)

FIGURE 3.46 The user program of Example 3.7: (a) schematic diagram, (b) ladder diagram.

shown in Figure 3.45. The schematic diagram and ladder diagram of Example 3.7 are depicted in Figure 3.46(a) and in Figure 3.46(b), respectively. When the project file of the PIC16F1847-Based PLC is open in the MPLAB X IDE, from the file "PICPLC _PIC16F1847_user_Bsc.inc", if you uncomment Example 3.7 and run the project by pressing the symbol ▷ from the toolbar, then the PIC16F1847 microcontroller within the CPU board of the PIC16F1847-Based PLC will be programmed. After loading the program file to the PIC16F1847 microcontroller, switch the 4PDT to "RUN" and the power switch to the "ON" position. Next you can test the operation of this example.

3.31.8 EXAMPLE 3.8

Example 3.8 shows the usage of the following contact and relay-based macros: "ld", "setBF", "resetBF", "in_out", and "move_R" (see Chapter 4 of *Intermediate Concepts*). The user program of Example 3.8 is shown in Figure 3.47. The schematic diagram and ladder diagram of Example 3.8 are depicted in Figure 3.48(a)

```
user_program_1  macro
;--- PLC codes to be allocated in the "user_program_1" macro start from here ---
;__Example_Bsc_3.8

          ld        I0.0         ;rung 1    ; (W,0) = I0.0
          setBF     M32.0,23               ; If (W,0)=1 then set successive
                                           ; 23 bits starting from M32.0.

          ld        I0.1         ;rung 2    ; (W,0) = I0.1
          resetBF   M32.0,17               ; If (W,0)=1 then reset successive
                                           ; 17 bits starting from M32.0.

          in_out    I1.0,M0.0    ;rung 3    ; M0.0  = I1.0

          in_out    I1.7,M127.7  ;rung 4    ; M127.7  = I1.7

          ld        M0.0         ;rung 5    ; (W,0) = M0.0
          setBF     M32.0,256              ; If (W,0)=1 then set successive
                                           ; 256 bits starting from M32.0.

          ld        M127.7       ;rung 6    ; (W,0) = M0.0
          resetBF   M32.0,256              ; If (W,0)=1 then reset successive
                                           ; 256 bits starting from M32.0.

          ld        LOGIC1       ;rung 7    ; (W,0) = LOGIC1
          move_R M32,Q0                    ; Q0 = M32
          move_R M33,Q1                    ; Q1 = M33
          move_R M34,Q2                    ; Q2 = M34
          move_R M63,Q3                    ; Q3 = M63
;
;--- PLC codes to be allocated in the "user_program_1" macro end here ---------
          endm
```

FIGURE 3.47   The user program of Example 3.8.

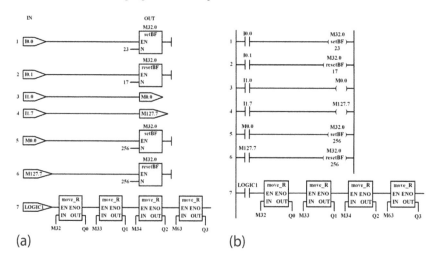

FIGURE 3.48   The user program of Example 3.8: (a) schematic diagram, (b) ladder diagram.

and in Figure 3.48(b), respectively. When the project file of the PIC16F1847-Based PLC is open in the MPLAB X IDE, from the file "PICPLC_PIC16F1847_user_Bsc .inc", if you uncomment Example 3.8 and run the project by pressing the symbol ▷ from the toolbar, then the PIC16F1847 microcontroller within the CPU board of the PIC16F1847-Based PLC will be programmed. After loading the program file to the PIC16F1847 microcontroller, switch the 4PDT to "RUN" and the power switch to the "ON" position. Next you can test the operation of this example.

4 Flip-Flop Macros

INTRODUCTION

In this chapter, the following flip-flop macros are described:

latch1 (D latch with active high enable),

latch0 (D latch with active low enable),

dff_r (rising edge–triggered D flip-flop),

dff_r_SR (rising edge–triggered D flip-flop with active high preset [S] and clear [R] inputs),

dff_f (falling edge–triggered D flip-flop),

dff_f_SR (falling edge–triggered D flip-flop with active high preset [S] and clear [R] inputs),

tff_r (rising edge–triggered T flip-flop),

tff_r_SR (rising edge–triggered T flip-flop with active high preset [S] and clear [R] inputs),

tff_f (falling edge–triggered T flip-flop),

tff_f_SR (falling edge–triggered T flip-flop with active high preset [S] and clear [R] inputs),

jkff_r (rising edge–triggered JK flip-flop),

jkff_r_SR (rising edge–triggered JK flip-flop with active high preset [S] and clear [R] inputs),

jkff_f (falling edge–triggered JK flip-flop),

jkff_f_SR (falling edge–triggered JK flip-flop with active high preset [S] and clear [R] inputs).

The file "PICPLC_PIC16F1847_macros_Bsc.inc", which is downloadable from this book's webpage under the downloads section, contains macros defined for the PIC16F1847-Based PLC explained in this book (*Hardware and Basic Concepts*). Flip-flop macros are defined to operate on Boolean (1-bit) variables. Let us now consider these macros.

4.1 MACRO "LATCH1" (D LATCH WITH ACTIVE HIGH ENABLE)

The symbol and the truth table of the macro "latch1" are depicted in Table 4.1. Figures 4.1 and 4.2 show the macro "latch1" and its flowchart, respectively. This macro has a Boolean enable input EN, passed in through "W", a Boolean input D, passed in through "Dreg,Dbit", and a Boolean output Q, passed out through "Qreg,Qbit". The macro "latch1" defines a D latch function with active high enable. EN (active high enable input) should be loaded into "W" before this macro is run. When EN is OFF (0), no state change is issued for the output Q and it holds its current state. When

---

TABLE 4.1

The Symbol and the Truth Table of the Macro "latch1"

Symbol

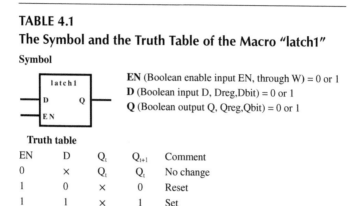

EN (Boolean enable input EN, through W) = 0 or 1

D (Boolean input D, Dreg,Dbit) = 0 or 1

Q (Boolean output Q, Qreg,Qbit) = 0 or 1

Truth table

EN	D	Q_t	Q_{t+1}	Comment
0	×	Q_t	Q_t	No change
1	0	×	0	Reset
1	1	×	1	Set

×: don't care.

```
latch1   macro   Dreg,Dbit, Qreg,Qbit
         local   L1,L2
         andlw   1
         btfsc   STATUS,Z
         goto    L1
         banksel Dreg
         btfss   Dreg,Dbit
         goto    L2
         banksel Qreg
         bsf     Qreg,Qbit
         goto    L1
L2
         banksel Qreg
         bcf     Qreg,Qbit
L1
         endm
```

FIGURE 4.1   The macro "latch1".

EN is ON (1), the output Q is loaded with the state of the input D. Operands for the instruction "latch1" are shown in Table 4.2.

4.2   MACRO "LATCH0" (D LATCH WITH ACTIVE LOW ENABLE)

The symbol and the truth table of the macro "latch0" are depicted in Table 4.3. Figures 4.3 and 4.4 show the macro "latch0" and its flowchart, respectively. This macro has a Boolean enable input EN, passed in through "W", a Boolean input D, passed in through "Dreg,Dbit", and a Boolean output Q, passed out through "Qreg,Qbit". The macro "latch0" defines a D latch function with active low enable. EN (active low enable input) should be loaded into "W" before this macro is run. When EN is ON (1), no state change is issued for the output Q and it holds its current state. When

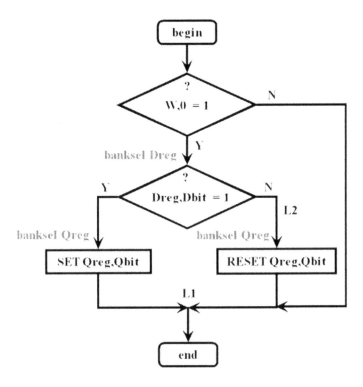

FIGURE 4.2   The flowchart of the macro "latch1".

TABLE 4.2
Operands for the Instruction "latch1"

Input/Output	Data type	Operands
D Dreg,Dbit (Bit)	BOOL	I, Q, M, T_Q, C_Q, C_QD, C_QU, drum_Q, drum_TQ, drum_steps, drum_events, SF, MB, nAI, AI, LOGIC1, LOGIC0, FRSTSCN, SCNOSC, any Boolean variable from SRAM.
Q Qreg,Qbit (Bit)	BOOL	Q, M, T_Q, C_Q, C_QD, C_QU, drum_Q, drum_TQ, drum_steps, drum_events, SF, MB, any Boolean variable from SRAM.

EN is OFF (0), the output Q is loaded with the state of the input D. Operands for the instruction "latch0" are shown in Table 4.4.

4.3   MACRO "DFF_R" (RISING EDGE–TRIGGERED D FLIP-FLOP)

The symbol and the truth table of the macro "dff_r" are depicted in Table 4.5. Figures 4.5 and 4.6 show the macro "dff_r" and its flowchart, respectively. This macro has a Boolean input D, passed in through "Dreg,Dbit", a Boolean clock input C, passed in through "W", a Boolean output Q, passed out through "Qreg,Qbit", and a unique

TABLE 4.3

The Symbol and the Truth Table of the Macro "latch0"

Symbol

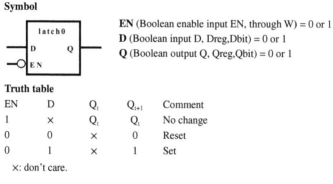

EN (Boolean enable input EN, through W) = 0 or 1
D (Boolean input D, Dreg,Dbit) = 0 or 1
Q (Boolean output Q, Qreg,Qbit) = 0 or 1

Truth table

EN	D	Q_t	Q_{t+1}	Comment
1	×	Q_t	Q_t	No change
0	0	×	0	Reset
0	1	×	1	Set

×: don't care.

```
latch0  macro   Dreg,Dbit, Qreg,Qbit
        local   L1,L2
        andlw   1
        btfss   STATUS,Z
        goto    L1
        banksel Dreg
        btfss   Dreg,Dbit
        goto    L2
        banksel Qreg
        bsf     Qreg,Qbit
        goto    L1
L2
        banksel Qreg
        bcf     Qreg,Qbit
L1
        endm
```

FIGURE 4.3   The macro "latch0".

memory bit "Mreg,Mbit", used to detect the rising edge of the clock input signal C.
The clock input signal C should be loaded into "W" before this macro is run. When
C is ON (1), or OFF (0), or changes its state from ON to OFF (↓), no state change is
issued for the output Q and it holds its current state. When the state of the clock input
signal C is changed from OFF to ON (↑), the output Q is loaded with the state of the
input D. Operands for the instruction "dff_r" are shown in Table 4.6.

4.4   MACRO "DFF_R_SR" (RISING EDGE–TRIGGERED D FLIP-FLOP WITH ACTIVE HIGH PRESET [S] AND CLEAR [R] INPUTS)

The symbol and the truth table of the macro "dff_r_SR" are depicted in Table 4.7.
Figures 4.7 and 4.8 show the macro "dff_r_SR" and its flowchart, respectively. This

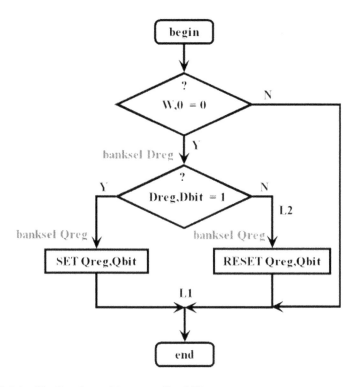

FIGURE 4.4 The flowchart of the macro "latch0".

TABLE 4.4
Operands for the Instruction "latch0"

Input/Output	Data type	Operands
D Dreg,Dbit (Bit)	BOOL	I, Q, M, T_Q, C_Q, C_QD, C_QU, drum_Q, drum_TQ, drum_steps, drum_events, SF, MB, nAI, AI, LOGIC1, LOGIC0, FRSTSCN, SCNOSC, any Boolean variable from SRAM.
Q Qreg,Qbit (Bit)	BOOL	Q, M, T_Q, C_Q, C_QD, C_QU, drum_Q, drum_TQ, drum_steps, drum_events, SF, MB, any Boolean variable from SRAM.

macro has a Boolean input D, passed in through "Dreg,Dbit", a Boolean clock input C, passed in through "W", a Boolean preset input S, passed in through "Sreg,Sbit", a Boolean clear input R, passed in through "Rreg,Rbit", a Boolean output Q, passed out through "Qreg,Qbit", and a unique memory bit "Mreg,Mbit", used to detect the rising edge of the clock input signal C. When R is ON (1), regardless of all other inputs, Q is reset (Q = 0). When R is OFF (0) and S is ON (1), regardless of all other inputs, Q is set (Q = 1). When R and S are both OFF (0), the following operations take place. When C is ON (1), or OFF (0), or changes its state from ON to OFF (↓), no state change is issued for the output Q and it holds its current state. When the state of the clock input signal

TABLE 4.5
The Symbol and the Truth Table of the Macro "dff_r"

Symbol

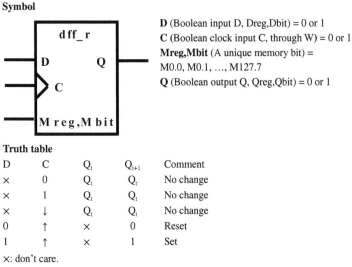

D (Boolean input D, Dreg,Dbit) = 0 or 1
C (Boolean clock input C, through W) = 0 or 1
Mreg,Mbit (A unique memory bit) =
M0.0, M0.1, ..., M127.7
Q (Boolean output Q, Qreg,Qbit) = 0 or 1

Truth table

D	C	Q_t	Q_{t+1}	Comment
×	0	Q_t	Q_t	No change
×	1	Q_t	Q_t	No change
×	↓	Q_t	Q_t	No change
0	↑	×	0	Reset
1	↑	×	1	Set

×: don't care.

```
dff_r    macro    Mreg,Mbit, Dreg,Dbit, Qreg,Qbit
         local    L1,L2
         banksel Mreg
         movwf   Temp_1
         btfss    Temp_1,0
         bsf      Mreg,Mbit      ;Mreg,Mbit = Rising Edge
         btfss    Temp_1,0       ;Detector for rising edge
         goto     L1             ;triggered D flip-flop
         btfss    Mreg,Mbit
         goto     L1
         bcf      Mreg,Mbit
         banksel Dreg
         btfss    Dreg,Dbit
         goto     L2
         banksel Qreg
         bsf      Qreg,Qbit
         goto     L1
L2
         banksel Qreg
         bcf      Qreg,Qbit
L1
         endm
```

FIGURE 4.5   The macro "dff_r".

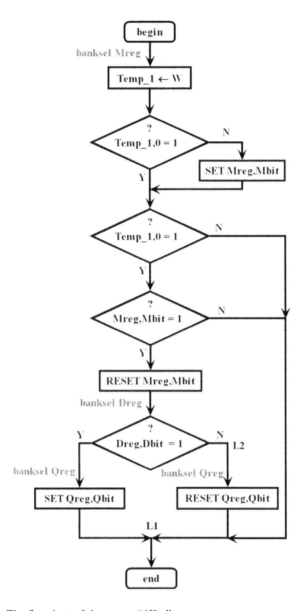

FIGURE 4.6   The flowchart of the macro "dff_r".

C is changed from OFF to ON (↑), the output Q is loaded with the state of the input D. Operands for the instruction "dff_r_SR" are shown in Table 4.8.

4.5   MACRO "DFF_F" (FALLING EDGE–TRIGGERED D FLIP-FLOP)

The symbol and the truth table of the macro "dff_f" are depicted in Table 4.9. Figures 4.9 and 4.10 show the macro "dff_f" and its flowchart, respectively. This macro has

TABLE 4.6
Operands for the Instruction "dff_r"

Input/Output	Data type	Operands
D Dreg,Dbit (Bit)	BOOL	I, Q, M, T_Q, C_Q, C_QD, C_QU, drum_Q, drum_TQ, drum_steps, drum_events, SF, MB, nAI, AI, LOGIC1, LOGIC0, FRSTSCN, SCNOSC, any Boolean variable from SRAM.
Q Qreg,Qbit (Bit)	BOOL	Q, M, T_Q, C_Q, C_QD, C_QU, drum_Q, drum_TQ, drum_steps, drum_events, SF, MB, any Boolean variable from SRAM.

TABLE 4.7
The Symbol and the Truth Table of the Macro "dff_r_SR"

Symbol

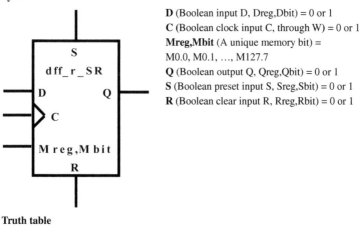

D (Boolean input D, Dreg,Dbit) = 0 or 1
C (Boolean clock input C, through W) = 0 or 1
Mreg,Mbit (A unique memory bit) =
M0.0, M0.1, ..., M127.7
Q (Boolean output Q, Qreg,Qbit) = 0 or 1
S (Boolean preset input S, Sreg,Sbit) = 0 or 1
R (Boolean clear input R, Rreg,Rbit) = 0 or 1

Truth table

S	R	D	C	Q_t	Q_{t+1}	Comment
×	1	×	×	×	0	Reset
1	0	×	×	×	1	Set
0	0	×	0	Q_t	Q_t	No change
0	0	×	1	Q_t	Q_t	No change
0	0	×	↓	Q_t	Q_t	No change
0	0	0	↑	×	0	Reset
0	0	1	↑	×	1	Set

×: don't care.

```
dff_r_SR macro  Mreg,Mbit, Dreg,Dbit, Qreg,Qbit, Sreg,Sbit, Rreg,Rbit
        local     L1,L2,L3
        banksel Rreg
        btfsc     Rreg,Rbit
        goto      L2
        banksel Sreg
        btfsc     Sreg,Sbit
        goto      L3
        ;----------------------;The following codes are the same as "dff_r"
        banksel Mreg      ;plus label L3.
        movwf   Temp_1
        btfss     Temp_1,0
        bsf       Mreg,Mbit    ;Mreg,Mbit = Rising Edge
        btfss     Temp_1,0     ;Detector for rising edge
        goto      L1           ;triggered D flip-flop
        btfss     Mreg,Mbit
        goto      L1
        bcf       Mreg,Mbit
        banksel Dreg
        btfss     Dreg,Dbit
        goto      L2
L3
        banksel Qreg
        bsf       Qreg,Qbit
        goto      L1
L2
        banksel Qreg
        bcf       Qreg,Qbit
L1
        endm
```

FIGURE 4.7 The macro "dff_r_SR".

a Boolean input D, passed in through "Dreg,Dbit", a Boolean clock input C, passed in through "W", a Boolean output Q, passed out through "Qreg,Qbit", and a unique memory bit "Mreg,Mbit", used to detect the falling edge of the clock input signal C. The clock input signal C should be loaded into "W" before this macro is run. When C is ON (1), or OFF (0), or changes its state from OFF to ON (↑), no state change is issued for the output Q and it holds its current state. When the state of the clock input signal C is changed from ON to OFF (↓), the output Q is loaded with the state of the input D. Operands for the instruction "dff_f" are shown in Table 4.10.

4.6 MACRO "DFF_F_SR" (FALLING EDGE–TRIGGERED D FLIP-FLOP WITH ACTIVE HIGH PRESET [S] AND CLEAR [R] INPUTS)

The symbol and the truth table of the macro "dff_f_SR" are depicted in Table 4.11. Figures 4.11 and 4.12 show the macro "dff_f_SR" and its flowchart, respectively. This macro has a Boolean input D, passed in through "Dreg,Dbit", a Boolean clock input

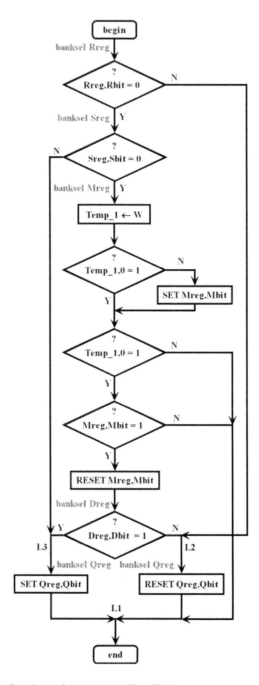

FIGURE 4.8   The flowchart of the macro "dff_r_SR".

TABLE 4.8
Operands for the Instruction "dff_r_SR"

Input/Output	Data type	Operands
D	BOOL	I, Q, M, T_Q, C_Q, C_QD, C_QU, drum_Q,
Dreg,Dbit (Bit)		drum_TQ, drum_steps, drum_events, SF, MB,
S		nAI, AI, LOGIC1, LOGIC0, FRSTSCN,
Sreg,Sbit (Bit)		SCNOSC, any Boolean variable from SRAM.
R		
Rreg,Rbit (Bit)		
Q	BOOL	Q, M, T_Q, C_Q, C_QD, C_QU, drum_Q,
Qreg,Qbit (Bit)		drum_TQ, drum_steps, drum_events, SF, MB,
		any Boolean variable from SRAM.

TABLE 4.9
The Symbol and the Truth Table of the Macro "dff_f"

Symbol

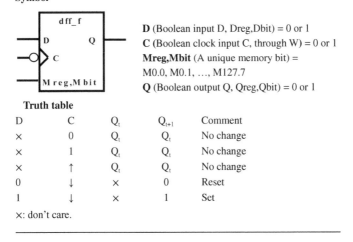

D (Boolean input D, Dreg,Dbit) = 0 or 1
C (Boolean clock input C, through W) = 0 or 1
Mreg,Mbit (A unique memory bit) =
M0.0, M0.1, …, M127.7
Q (Boolean output Q, Qreg,Qbit) = 0 or 1

Truth table

D	C	Q_t	Q_{t+1}	Comment
×	0	Q_t	Q_t	No change
×	1	Q_t	Q_t	No change
×	↑	Q_t	Q_t	No change
0	↓	×	0	Reset
1	↓	×	1	Set

×: don't care.

C, passed in through "W", a Boolean preset input S, passed in through "Sreg,Sbit", a Boolean clear input R, passed in through "Rreg,Rbit", a Boolean output Q, passed out through "Qreg,Qbit", and a unique memory bit "Mreg,Mbit", used to detect the falling edge of the clock input signal C. When R is ON (1), regardless of all other inputs, Q is reset (Q = 0). When R is OFF (0) and S is ON (1), regardless of all other inputs, Q is set (Q = 1). When R and S are both OFF (0), the following operations take place. When C is ON (1), or OFF (0), or changes its state from OFF to ON (↑), no state change is issued for the output Q and it holds its current state. When the state of the clock input signal C is changed from ON to OFF (↓), the output Q is loaded

```
dff_f    macro    Mreg,Mbit, Dreg,Dbit, Qreg,Qbit
         local    L1,L2
         banksel  Mreg
         movwf    Temp_1
         btfsc    Temp_1,0
         bsf      Mreg,Mbit        ;Mreg,Mbit = Falling Edge
         btfsc    Temp_1,0         ;Detector for falling edge
         goto     L1               ;triggered D flip-flop
         btfss    Mreg,Mbit
         goto     L1
         bcf      Mreg,Mbit
         banksel  Dreg
         btfss    Dreg,Dbit
         goto     L2
         banksel  Qreg
         bsf      Qreg,Qbit
         goto     L1
L2
         banksel  Qreg
         bcf      Qreg,Qbit
L1
         endm
```

FIGURE 4.9   The macro "dff_f".

with the state of the input D. Operands for the instruction "dff_f_SR" are shown in Table 4.12.

4.7   MACRO "TFF_R" (RISING EDGE–TRIGGERED T FLIP-FLOP)

The symbol and the truth table of the macro "tff_r" are depicted in Table 4.13. Figures 4.13 and 4.14 show the macro "tff_r" and its flowchart, respectively. This macro has a Boolean input T, passed in through "Treg,Tbit", a Boolean clock input C, passed in through "W", a Boolean output Q, passed out through "Qreg,Qbit", and a unique memory bit "Mreg,Mbit", used to detect the rising edge of the clock input signal C. The clock input signal C should be loaded into "W" before this macro is run. When C is ON (1), or OFF (0), or changes its state from ON to OFF (↓), no state change is issued for the output Q and it holds its current state. When the state of the clock input signal C is changed from OFF to ON (↑), if T = 0, then no state change is issued for the output Q; if T = 1, then the output Q is toggled. Operands for the instruction "tff_r" are shown in Table 4.14.

4.8   MACRO "TFF_R_SR" (RISING EDGE–TRIGGERED T FLIP-FLOP WITH ACTIVE HIGH PRESET [S] AND CLEAR [R] INPUTS)

The symbol and the truth table of the macro "tff_r_SR" are depicted in Table 4.15. Figures 4.15 and 4.16 show the macro "tff_r_SR" and its flowchart, respectively. This macro has a Boolean input T, passed in through "Treg,Tbit", a Boolean clock input

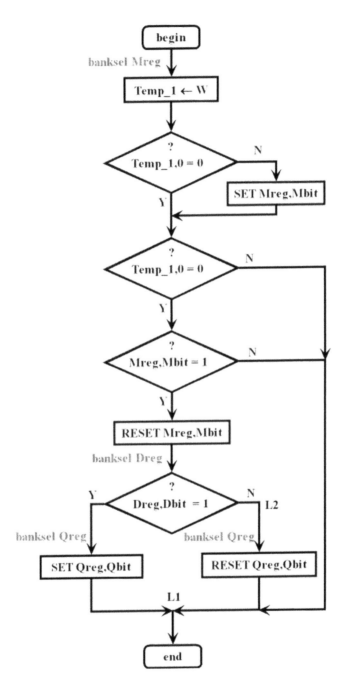

FIGURE 4.10 The flowchart of the macro "dff_f".

TABLE 4.10
Operands for the Instruction "dff_f"

Input/Output	Data type	Operands
D Dreg,Dbit (Bit)	BOOL	I, Q, M, T_Q, C_Q, C_QD, C_QU, drum_Q, drum_TQ, drum_steps, drum_events, SF, MB, nAI, AI, LOGIC1, LOGIC0, FRSTSCN, SCNOSC, any Boolean variable from SRAM.
Q Qreg,Qbit (Bit)	BOOL	Q, M, T_Q, C_Q, C_QD, C_QU, drum_Q, drum_TQ, drum_steps, drum_events, SF, MB, any Boolean variable from SRAM.

TABLE 4.11
The Symbol and the Truth Table of the Macro "dff_f_SR"

Symbol

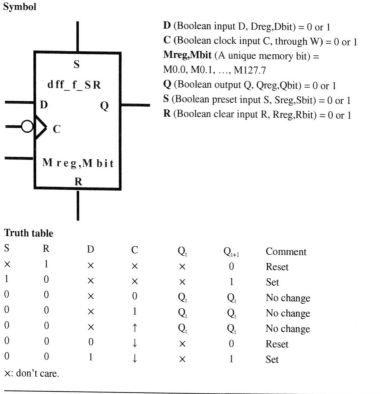

D (Boolean input D, Dreg,Dbit) = 0 or 1
C (Boolean clock input C, through W) = 0 or 1
Mreg,Mbit (A unique memory bit) =
M0.0, M0.1, ..., M127.7
Q (Boolean output Q, Qreg,Qbit) = 0 or 1
S (Boolean preset input S, Sreg,Sbit) = 0 or 1
R (Boolean clear input R, Rreg,Rbit) = 0 or 1

Truth table

S	R	D	C	Q_t	Q_{t+1}	Comment
×	1	×	×	×	0	Reset
1	0	×	×	×	1	Set
0	0	×	0	Q_t	Q_t	No change
0	0	×	1	Q_t	Q_t	No change
0	0	×	↑	Q_t	Q_t	No change
0	0	0	↓	×	0	Reset
0	0	1	↓	×	1	Set

×: don't care.

C, passed in through "W", a Boolean preset input S, passed in through "Sreg,Sbit", a Boolean clear input R, passed in through "Rreg,Rbit", a Boolean output Q, passed out through "Qreg,Qbit", and a unique memory bit "Mreg,Mbit", used to detect the rising edge of the clock input signal C. When R is ON (1), regardless of all other inputs, Q is reset (Q = 0). When R is OFF (0) and S is ON (1), regardless of all other

```
dff_f_SR macro  Mreg,Mbit, Dreg,Dbit, Qreg,Qbit, Sreg,Sbit, Rreg,Rbit
        local    L1,L2,L3
        banksel  Rreg
        btfsc    Rreg,Rbit
        goto     L2
        banksel  Sreg
        btfsc    Sreg,Sbit
        goto     L3
        ;-----------------------;The following codes are the same as "dff_f"
        banksel  Mreg    ;plus label L3.
        movwf    Temp_1
        btfsc    Temp_1,0
        bsf      Mreg,Mbit      ;Mreg,Mbit = Falling Edge
        btfsc    Temp_1,0       ;Detector for falling edge
        goto     L1             ;triggered D flip-flop
        btfss    Mreg,Mbit
        goto     L1
        bcf      Mreg,Mbit
        banksel  Dreg
        btfss    Dreg,Dbit
        goto     L2
L3
        banksel  Qreg
        bsf      Qreg,Qbit
        goto     L1
L2
        banksel  Qreg
        bcf      Qreg,Qbit
L1
        endm
```

FIGURE 4.11   The macro "dff_f_SR".

inputs, Q is set (Q = 1). When R and S are both OFF (0), the following operations take place. When C is ON (1), or OFF (0), or changes its state from ON to OFF (↓), no state change is issued for the output Q and it holds its current state. When the state of the clock input signal C is changed from OFF to ON (↑), if T = 0, then no state change is issued for the output Q; if T = 1, then the output Q is toggled. Operands for the instruction "tff_r_SR" are shown in Table 4.16.

4.9   MACRO "TFF_F" (FALLING EDGE–TRIGGERED T FLIP-FLOP)

The symbol and the truth table of the macro "tff_f" are depicted in Table 4.17. Figures 4.17 and 4.18 show the macro "tff_f" and its flowchart, respectively. This macro has a Boolean input T, passed in through "Treg,Tbit", a Boolean clock input C, passed in through "W", a Boolean output Q, passed out through "Qreg,Qbit", and a unique memory bit "Mreg,Mbit", used to detect the falling edge of the clock input signal C. The clock input signal C should be loaded into "W" before this macro is run. When C is ON (1), or OFF (0), or changes state from OFF to ON (↑), no state change is issued for the output Q and it holds its current state. When the state of clock input signal C is changed from ON to OFF (↓), if T = 0, then no state change is issued for the output Q; if T = 1, then the output Q is toggled. Operands for the instruction "tff_f" are shown in Table 4.18.

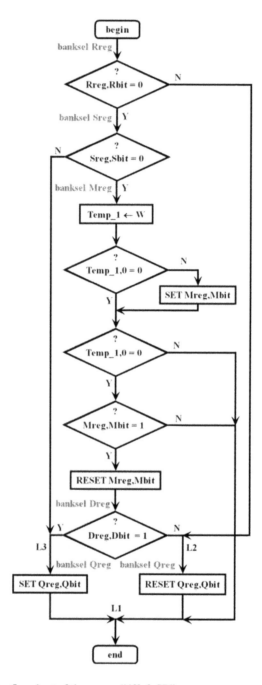

FIGURE 4.12 The flowchart of the macro "dff_f_SR".

TABLE 4.12
Operands for the Instruction "dff_f_SR"

Input/Output	Data type	Operands
D Dreg,Dbit (Bit) S Sreg,Sbit (Bit) R Rreg,Rbit (Bit)	BOOL	I, Q, M, T_Q, C_Q, C_QD, C_QU, drum_Q, drum_TQ, drum_steps, drum_events, SF, MB, nAI, AI, LOGIC1, LOGIC0, FRSTSCN, SCNOSC, any Boolean variable from SRAM.
Q Qreg,Qbit (Bit)	BOOL	Q, M, T_Q, C_Q, C_QD, C_QU, drum_Q, drum_TQ, drum_ steps, drum_events, SF, MB, any Boolean variable from SRAM.

TABLE 4.13
The Symbol and the Truth Table of the Macro "tff_r"

Symbol

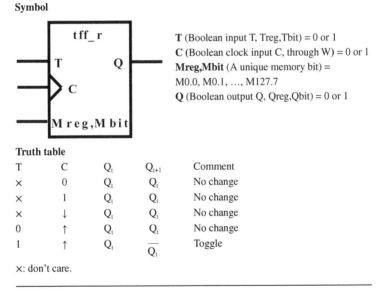

T (Boolean input T, Treg,Tbit) = 0 or 1
C (Boolean clock input C, through W) = 0 or 1
Mreg,Mbit (A unique memory bit) =
M0.0, M0.1, …, M127.7
Q (Boolean output Q, Qreg,Qbit) = 0 or 1

Truth table

T	C	Q_t	Q_{t+1}	Comment
×	0	Q_t	Q_t	No change
×	1	Q_t	Q_t	No change
×	↓	Q_t	Q_t	No change
0	↑	Q_t	Q_t	No change
1	↑	Q_t	$\overline{Q_t}$	Toggle

×: don't care.

4.10   MACRO "TFF_F_SR" (FALLING EDGE–TRIGGERED T FLIP-FLOP WITH ACTIVE HIGH PRESET [S] AND CLEAR [R] INPUTS)

The symbol and the truth table of the macro "tff_f_SR" are depicted in Table 4.19. Figures 4.19 and 4.20 show the macro "tff_f_SR" and its flowchart, respectively. This macro has a Boolean input T, passed in through "Treg,Tbit", a Boolean clock input C, passed in through "W", a Boolean preset input S, passed in through "Sreg,Sbit", a Boolean clear input R, passed in through "Rreg,Rbit", a Boolean output Q, passed

```
tff_r    macro    Mreg,Mbit, Treg,Tbit, Qreg,Qbit
         local    L1,L2
         banksel Mreg
         movwf   Temp_1
         btfss   Temp_1,0
         bsf     Mreg,Mbit        ;Mreg,Mbit = Rising Edge
         btfss   Temp_1,0         ;Detector for rising edge
         goto    L1               ;triggered T flip-flop
         btfss   Mreg,Mbit
         goto    L1
         bcf     Mreg,Mbit
         banksel Treg
         btfss   Treg,Tbit
         goto    L1
         banksel Qreg
         btfsc   Qreg,Qbit
         goto    L2
         bsf     Qreg,Qbit
         goto    L1
L2       bcf     Qreg,Qbit
L1
         endm
```

FIGURE 4.13   The macro "tff_r".

out through "Qreg,Qbit", and a unique memory bit "Mreg,Mbit", used to detect the falling edge of the clock input signal C. When R is ON (1), regardless of all other inputs, Q is reset (Q = 0). When R is OFF (0) and S is ON (1), regardless of all other inputs, Q is set (Q = 1). When R and S are both OFF (0), the following operations take place. When C is ON (1), or OFF (0), or changes state from OFF to ON (↑), no state change is issued for the output Q and it holds its current state. When the state of the clock input signal C is changed from ON to OFF (↓), if T = 0, then no state change is issued for the output Q; if T = 1, then the output Q is toggled. Operands for the instruction "tff_f_SR" are shown in Table 4.20.

4.11   MACRO "JKFF_R" (RISING EDGE–TRIGGERED JK FLIP-FLOP)

The symbol and the truth table of the macro "jkff_r" are depicted in Table 4.21. Figures 4.21 and 4.22 show the macro "jkff_r" and its flowchart, respectively. This macro has a Boolean input J, passed in through "Jreg,Jbit", a Boolean input K, passed in through "Kreg,Kbit", a Boolean clock input C, passed in through "W", a Boolean output Q, passed out through "Qreg,Qbit", and a unique memory bit "Mreg,Mbit", used to detect the rising edge of the clock input signal C.

The clock input signal C should be loaded into "W" before this macro is run. When C is ON (1), or OFF (0), or changes state from ON to OFF (↓), no state change is issued for the output Q and it holds its current state. When the state of clock input signal C is changed from OFF to ON (↑), if JK = 00, then no state change is issued; if JK = 01, then Q is reset; if JK = 10, then Q is set; and finally if JK = 11, then Q is toggled. Operands for the instruction "jkff_r" are shown in Table 4.22.

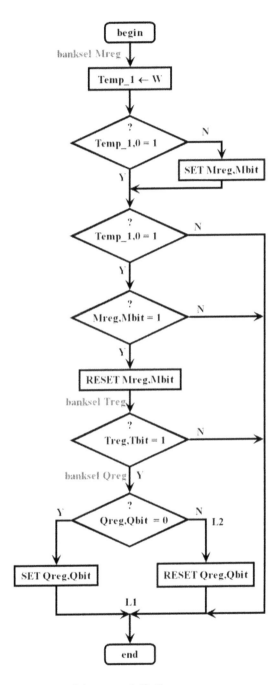

FIGURE 4.14 The flowchart of the macro "tff_r".

TABLE 4.14

Operands for the Instruction "tff_r"

Input/Output	Data type	Operands
T Treg,Tbit (Bit)	BOOL	I, Q, M, T_Q, C_Q, C_QD, C_QU, drum_Q, drum_TQ, drum_steps, drum_events, SF, MB, nAI, AI, LOGIC1, LOGIC0, FRSTSCN, SCNOSC, any Boolean variable from SRAM.
Q Qreg,Qbit (Bit)	BOOL	Q, M, T_Q, C_Q, C_QD, C_QU, drum_Q, drum_TQ, drum_ steps, drum_events, SF, MB, any Boolean variable from SRAM.

TABLE 4.15

The Symbol and the Truth Table of the Macro "tff_r_SR"

Symbol

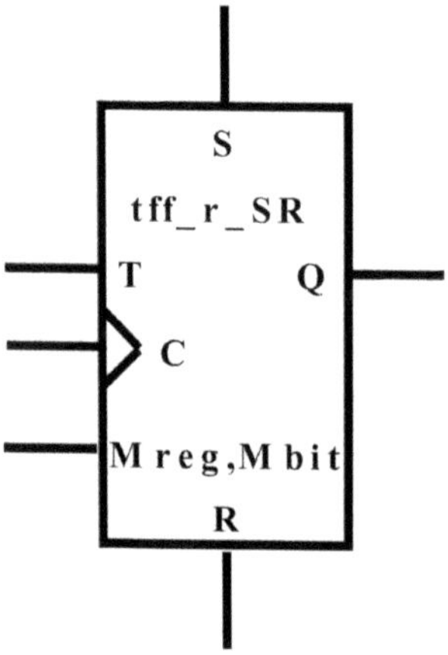

T (Boolean input T, Treg,Tbit) = 0 or 1
C (Boolean clock input C, through W) = 0 or 1
Mreg,Mbit (A unique memory bit) =
M0.0, M0.1, ..., M127.7
Q (Boolean output Q, Qreg,Qbit) = 0 or 1
S (Boolean preset input S, Sreg,Sbit) = 0 or 1
R (Boolean clear input R, Rreg,Rbit) = 0 or 1

Truth table

S	R	T	C	Q_t	Q_{t+1}	Comment
×	1	×	×	×	0	Reset
1	0	×	×	×	1	Set
0	0	×	0	Q_t	Q_t	No change
0	0	×	1	Q_t	Q_t	No change
0	0	×	↓	Q_t	Q_t	No change
0	0	0	↑	Q_t	Q_t	No change
0	0	1	↑	Q_t	$\overline{Q_t}$	Toggle

×: don't care.

```
tff_r_SR macro  Mreg,Mbit, Treg,Tbit, Qreg,Qbit, Sreg,Sbit, Rreg,Rbit
         local    L1,L2,L3
         banksel  Rreg
         btfsc    Rreg,Rbit
         goto     L2
         banksel  Sreg
         btfsc    Sreg,Sbit
         goto     L3
         ;-----------------------;The following codes are the same as "tff_r"
         banksel  Mreg           ;plus label L3.
         movwf    Temp_1
         btfss    Temp_1,0
         bsf      Mreg,Mbit      ;Mreg,Mbit = Rising Edge
         btfss    Temp_1,0       ;Detector for rising edge
         goto     L1             ;triggered T flip-flop
         btfss    Mreg,Mbit
         goto     L1
         bcf      Mreg,Mbit
         banksel  Treg
         btfss    Treg,Tbit
         goto     L1
         banksel  Qreg
         btfsc    Qreg,Qbit
         goto     L2
L3
         banksel  Qreg
         bsf      Qreg,Qbit
         goto     L1
L2
         banksel  Qreg
         bcf      Qreg,Qbit
L1
         endm
```

FIGURE 4.15   The macro "tff_r_SR".

4.12   MACRO "JKFF_R_SR" (RISING EDGE–TRIGGERED JK FLIP-FLOP WITH ACTIVE HIGH PRESET [S] AND CLEAR [R] INPUTS)

The symbol and the truth table of the macro "jkff_r_SR" are depicted in Table 4.23. Figures 4.23 and 4.24 show the macro "jkff_r_SR" and its flowchart, respectively. This macro has a Boolean input J, passed in through "Jreg,Jbit", a Boolean input K, passed in through "Kreg,Kbit", a Boolean clock input C, passed in through "W", a Boolean preset input S, passed in through "Sreg,Sbit", a Boolean clear input R, passed in through "Rreg,Rbit", a Boolean output Q, passed out through "Qreg,Qbit", and a unique memory bit "Mreg,Mbit", used to detect the rising edge of the clock input signal C. When R is ON (1), regardless of all other inputs, Q is reset (Q = 0). When R is OFF (0) and S is ON (1), regardless of all other inputs, Q is set (Q = 1). When R and S are both OFF (0), the following operations take place. When C is ON (1), or OFF (0), or changes state from ON to OFF (↓), no state change is issued for the output Q and it holds its current state. When the state of the clock input signal C is changed from OFF to ON (↑), if JK = 00, then no state change is issued; if JK =

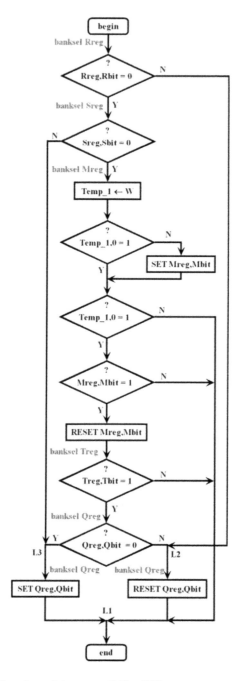

FIGURE 4.16   The flowchart of the macro "tff_r_SR".

TABLE 4.16
Operands for the Instruction "tff_r_SR"

Input/Output	Data type	Operands
T Treg,Tbit (Bit) S Sreg,Sbit (Bit) R Rreg,Rbit (Bit)	BOOL	I, Q, M, T_Q, C_Q, C_QD, C_QU, drum_Q, drum_TQ, drum_steps, drum_events, SF, MB, nAI, AI, LOGIC1, LOGIC0, FRSTSCN, SCNOSC, any Boolean variable from SRAM.
Q Qreg,Qbit (Bit)	BOOL	Q, M, T_Q, C_Q, C_QD, C_QU, drum_Q, drum_TQ, drum_steps, drum_events, SF, MB, any Boolean variable from SRAM.

TABLE 4.17
The Symbol and the Truth Table of the Macro "tff_f"

Symbol

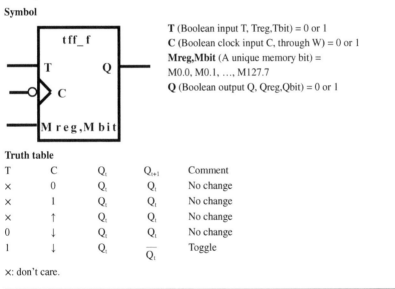

T (Boolean input T, Treg,Tbit) = 0 or 1
C (Boolean clock input C, through W) = 0 or 1
Mreg,Mbit (A unique memory bit) = M0.0, M0.1, …, M127.7
Q (Boolean output Q, Qreg,Qbit) = 0 or 1

Truth table

T	C	Q_t	Q_{t+1}	Comment
×	0	Q_t	Q_t	No change
×	1	Q_t	Q_t	No change
×	↑	Q_t	Q_t	No change
0	↓	Q_t	Q_t	No change
1	↓	Q_t	$\overline{Q_t}$	Toggle

×: don't care.

01, then Q is reset; if JK = 10, then Q is set; and finally if JK = 11, then Q is toggled. Operands for the instruction "jkff_r_SR" are shown in Table 4.24.

4.13   MACRO "JKFF_F" (FALLING EDGE–TRIGGERED JK FLIP-FLOP)

The symbol and the truth table of the macro "jkff_f" are depicted in Table 4.25. Figures 4.25 and 4.26 show the macro "jkff_f" and its flowchart, respectively. This macro has a Boolean input J, passed in through "Jreg,Jbit", a Boolean input K, passed

```
tff_f    macro    Mreg,Mbit, Treg,Tbit, Qreg,Qbit
         local    L1,L2
         banksel  Mreg
         movwf    Temp_1
         btfsc    Temp_1,0
         bsf      Mreg,Mbit      ;Mreg,Mbit = Falling Edge
         btfsc    Temp_1,0       ;Detector for falling edge
         goto     L1             ;triggered T flip-flop
         btfss    Mreg,Mbit
         goto     L1
         bcf      Mreg,Mbit
         banksel  Treg
         btfss    Treg,Tbit
         goto     L1
         banksel  Qreg
         btfsc    Qreg,Qbit
         goto     L2
         bsf      Qreg,Qbit
         goto     L1
L2       bcf      Qreg,Qbit
L1
         endm
```

FIGURE 4.17   The macro "tff_f".

in through "Kreg,Kbit", a Boolean clock input C, passed in through "W", a Boolean output Q, passed out through "Qreg,Qbit", and a unique memory bit "Mreg,Mbit", used to detect the falling edge of the clock input signal C. The clock input signal C should be loaded into "W" before this macro is run. When C is ON (1), or OFF (0), or changes state from OFF to ON (↑), no state change is issued for the output Q and it holds its current state. When the state of clock input signal C is changed from ON to OFF (↓), if JK = 00, then no state change is issued; if JK = 01, then Q is reset; if JK = 10, then Q is set; and finally if JK = 11, then Q is toggled. Operands for the instruction "jkff_f" are shown in Table 4.26.

4.14   MACRO "JKFF_F_SR" (FALLING EDGE–TRIGGERED JK FLIP-FLOP WITH ACTIVE HIGH PRESET [S] AND CLEAR [R] INPUTS)

The symbol and the truth table of the macro "jkff_f_SR" are depicted in Table 4.27. Figures 4.27 and 4.28 show the macro "jkff_f_SR" and its flowchart, respectively. This macro has a Boolean input J, passed in through "Jreg,Jbit", a Boolean input K, passed in through "Kreg,Kbit", a Boolean clock input C, passed in through "W", a Boolean preset input S, passed in through "Sreg,Sbit", a Boolean clear input R, passed in through "Rreg,Rbit", a Boolean output Q, passed out through "Qreg,Qbit", and a unique memory bit "Mreg,Mbit", used to detect the falling edge of the clock input signal C. When R is ON (1), regardless of all other inputs, Q is reset (Q = 0). When R is OFF (0) and S is ON (1), regardless of all other inputs, Q is set (Q = 1).

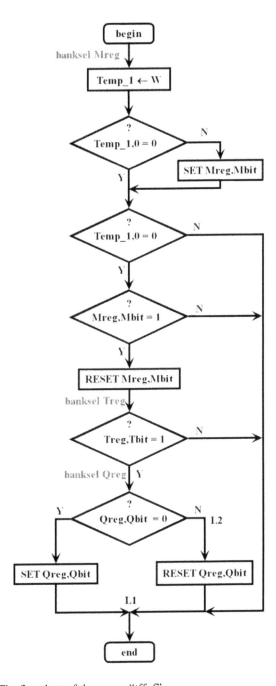

FIGURE 4.18 The flowchart of the macro "tff_f".

TABLE 4.18

Operands for the Instruction "tff_f"

Input/Output	Data type	Operands
T Treg,Tbit (Bit)	BOOL	I, Q, M, T_Q, C_Q, C_QD, C_QU, drum_Q, drum_TQ, drum_steps, drum_events, SF, MB, nAI, AI, LOGIC1, LOGIC0, FRSTSCN, SCNOSC, any Boolean variable from SRAM.
Q Qreg,Qbit (Bit)	BOOL	Q, M, T_Q, C_Q, C_QD, C_QU, drum_Q, drum_TQ, drum_steps, drum_events, SF, MB, any Boolean variable from SRAM.

TABLE 4.19

The Symbol and the Truth Table of the Macro "tff_f_SR"

Symbol

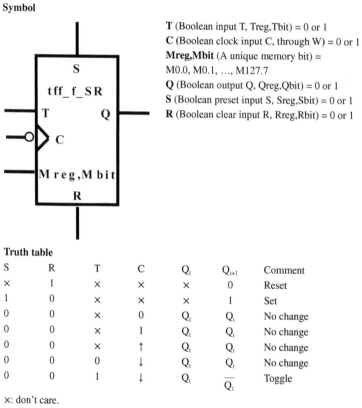

T (Boolean input T, Treg,Tbit) = 0 or 1
C (Boolean clock input C, through W) = 0 or 1
Mreg,Mbit (A unique memory bit) = M0.0, M0.1, ..., M127.7
Q (Boolean output Q, Qreg,Qbit) = 0 or 1
S (Boolean preset input S, Sreg,Sbit) = 0 or 1
R (Boolean clear input R, Rreg,Rbit) = 0 or 1

Truth table

S	R	T	C	Q_t	Q_{t+1}	Comment
×	1	×	×	×	0	Reset
1	0	×	×	×	1	Set
0	0	×	0	Q_t	Q_t	No change
0	0	×	1	Q_t	Q_t	No change
0	0	×	↑	Q_t	Q_t	No change
0	0	0	↓	Q_t	Q_t	No change
0	0	1	↓	Q_t	$\overline{Q_t}$	Toggle

×: don't care.

```
tff_f_SR macro  Mreg,Mbit, Treg,Tbit, Qreg,Qbit, Sreg,Sbit, Rreg,Rbit
        local   L1,L2,L3
        banksel Rreg
        btfsc   Rreg,Rbit
        goto    L2
        banksel Sreg
        btfsc   Sreg,Sbit
        goto    L3
        ;-----------------------;The following codes are the same as "tff_f"
        banksel Mreg     ;plus label L3.
        movwf   Temp_1
        btfsc   Temp_1,0
        bsf     Mreg,Mbit       ;Mreg,Mbit = Falling Edge
        btfsc   Temp_1,0        ;Detector for falling edge
        goto    L1              ;triggered T flip-flop
        btfss   Mreg,Mbit
        goto    L1
        bcf     Mreg,Mbit
        banksel Treg
        btfss   Treg,Tbit
        goto    L1
        banksel Qreg
        btfsc   Qreg,Qbit
        goto    L2
L3
        banksel Qreg
        bsf     Qreg,Qbit
        goto    L1
L2
        banksel Qreg
        bcf     Qreg,Qbit
L1
        endm
```

FIGURE 4.19   The macro "tff_f_SR".

When R and S are both OFF (0), the following operations take place. When C is ON (1), or OFF (0), or changes state from OFF to ON (↑), no state change is issued for the output Q and it holds its current state. When the state of the clock input signal C is changed from ON to OFF (↓), if JK = 00, then no state change is issued; if JK = 01, then Q is reset; if JK = 10, then Q is set; and finally if JK = 11, then Q is toggled. Operands for the instruction "jkff_f_SR" are shown in Table 4.28.

4.15   EXAMPLES FOR FLIP-FLOP MACROS

Up to now in this chapter, we have seen flip-flop macros developed for the PIC16F1847-Based PLC. It is now time to consider some examples related to these macros. Before you can run the example programs considered here, you are expected to construct your own PIC16F1847-Based PLC hardware by using the necessary PCB files and by producing your PCBs with their components. For an effective use of examples, all example programs considered in this book are allocated within the file "PICPLC_PIC16F1847_user_Bsc.inc", which is downloadable from this book's webpage under the downloads section. Initially all example programs are commented out by putting

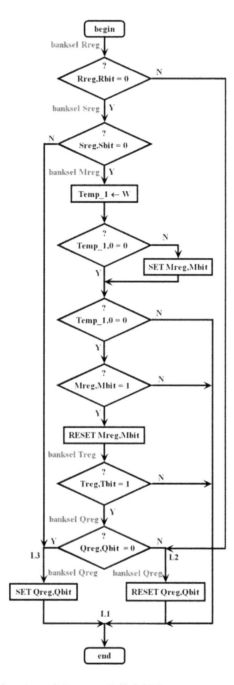

FIGURE 4.20   The flowchart of the macro "tff_f_SR".

TABLE 4.20
Operands for the Instruction "tff_f_SR"

Input/Output	Data type	Operands
T	BOOL	I, Q, M, T_Q, C_Q, C_QD, C_QU, drum_Q, drum_TQ,
Treg,Tbit (Bit)		drum_steps, drum_events, SF, MB, nAI, AI, LOGIC1,
S		LOGIC0, FRSTSCN, SCNOSC, any Boolean variable from
Sreg,Sbit (Bit)		SRAM.
R		
Rreg,Rbit (Bit)		
Q	BOOL	Q, M, T_Q, C_Q, C_QD, C_QU, drum_Q, drum_TQ, drum_
Qreg,Qbit (Bit)		steps, drum_events, SF, MB, any Boolean variable from SRAM.

TABLE 4.21
The Symbol and the Truth Table of the Macro "jkff_r"

Symbol

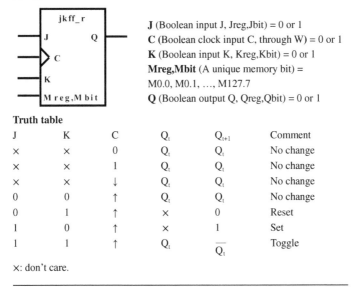

J (Boolean input J, Jreg,Jbit) = 0 or 1
C (Boolean clock input C, through W) = 0 or 1
K (Boolean input K, Kreg,Kbit) = 0 or 1
Mreg,Mbit (A unique memory bit) =
M0.0, M0.1, ..., M127.7
Q (Boolean output Q, Qreg,Qbit) = 0 or 1

Truth table

J	K	C	Q_t	Q_{t+1}	Comment
×	×	0	Q_t	Q_t	No change
×	×	1	Q_t	Q_t	No change
×	×	↓	Q_t	Q_t	No change
0	0	↑	Q_t	Q_t	No change
0	1	↑	×	0	Reset
1	0	↑	×	1	Set
1	1	↑	Q_t	$\overline{Q_t}$	Toggle

×: don't care.

a semicolon ";" in front of each line. When you would like to test one of the example programs, you must uncomment each line of the example program by following the steps shown below:

1. Highlight the block of source lines you want to uncomment by dragging the mouse over these lines with the left mouse button held down. With default coloring in MPLAB X IDE you will now see green characters on a blue background.

```
jkff_r    macro    Mreg,Mbit, Jreg,Jbit, Kreg,Kbit, Qreg,Qbit
          local    L1,L2,L3,L4
          banksel  Mreg
          movwf    Temp_1
          btfss    Temp_1,0
          bsf      Mreg,Mbit        ;Mreg,Mbit = Rising Edge
          btfss    Temp_1,0         ;Detector for rising edge
          goto     L1               ;triggered JK flip-flop
          btfss    Mreg,Mbit
          goto     L1
          bcf      Mreg,Mbit
          banksel  Jreg
          btfss    Jreg,Jbit
          goto     L4               ;if j=0 then goto L4
          banksel  Kreg
          btfss    Kreg,Kbit
          goto     L3               ;if j=1&k=0 then SET Qreg,Qbit (goto L3)
          banksel  Qreg             ;
          btfsc    Qreg,Qbit        ;if j=1&k=1
          goto     L2               ;then TOGGLE
          goto     L3               ;Qreg,Qbit
L4
          banksel  Kreg
          btfss    Kreg,Kbit
          goto     L1               ;if j=0&k=0 then NO CHANGE (goto L1)
          goto     L2               ;if j=0&k=1 then RESET Qreg,Qbit
L3
          banksel  Qreg
          bsf      Qreg,Qbit
          goto     L1
L2
          banksel  Qreg
          bcf      Qreg,Qbit
L1
          endm
```

FIGURE 4.21   The macro "jkff_r".

2. Release the mouse button.
3. Press Ctrl/Shift/C or press the Alt, S, and M keys in succession, or select "Toggle Comment" from the toolbar Source menu. Now a semicolon will be removed from all selected source lines. With default coloring you will see red characters on a white background.

Then, you can run the project by pressing the symbol ▷ from the toolbar. Next, the MPLAB X IDE will produce the "PICPLC_PIC16F1847.X.production.hex" file for the project. Then the MPLAB X IDE will be connected to the PICkit 3 programmer and finally it will program the PIC16F1847 microcontroller within the CPU board of the PIC16F1847-Based PLC. During these steps, make sure that in the CPU board of the PIC16F1847-Based PLC, the 4PDT switch is in the "PROG" position and the power switch is in the "OFF" position. After loading the program file to the PIC16F1847 microcontroller, switch the 4PDT to "RUN" and the power switch to the "ON" position. Finally, you are ready to test the example program. **Warning**: when you finish your study with an example and try to take a look at another example, do not forget to comment the current example program before uncommenting

```
jkff_r_SR macro Mreg,Mbit, Jreg,Jbit, Kreg,Kbit, Qreg,Qbit, Sreg,Sbit, Rreg,Rbit
            local    L1,L2,L3,L4
            banksel Rreg
            btfsc    Rreg,Rbit
            goto     L2
            banksel Sreg
            btfsc    Sreg,Sbit
            goto     L3
            ;----------------------;The following codes are the same as "jkff_r"
            banksel Mreg
            movwf   Temp_1
            btfss    Temp_1,0
            bsf      Mreg,Mbit       ;Mreg,Mbit = Rising Edge
            btfss    Temp_1,0        ;Detector for rising edge
            goto     L1              ;triggered JK flip-flop
            btfss    Mreg,Mbit
            goto     L1
            bcf      Mreg,Mbit
            banksel Jreg
            btfss    Jreg,Jbit
            goto     L4              ;if j=0 then goto L4
            banksel Kreg
            btfss    Kreg,Kbit
            goto     L3              ;if j=1&k=0 then SET Qreg,Qbit (goto L3)
            banksel Qreg             ;
            btfsc    Qreg,Qbit       ;if j=1&k=1
            goto     L2              ;then TOGGLE
            goto     L3              ;Qreg,Qbit
L4
            banksel Kreg
            btfss    Kreg,Kbit
            goto     L1              ;if j=0&k=0 then NO CHANGE (goto L1)
            goto     L2              ;if j=0&k=1 then RESET Qreg,Qbit
L3
            banksel Qreg
            bsf      Qreg,Qbit
            goto     L1
L2
            banksel Qreg
            bcf      Qreg,Qbit
L1
            endm
```

FIGURE 4.23   The macro "jkff_r_SR".

another one. In other words, make sure that only one example program is uncommented and tested at the same time. Otherwise, if you somehow leave more than one example uncommented, the example you are trying to test will probably not function as expected since it may try to access the same resources that are being used and changed by other examples.

Please check the accuracy of each program by cross-referencing it with the related macros.

4.15.1   EXAMPLE 4.1

Example 4.1 shows the usage of the following flip-flop macros: "latch1", "latch0", "r_edge", and "f_edge". The user program of Example 4.1 is shown in Figure 4.29. The schematic diagram and ladder diagram of Example 4.1 are depicted in Figure 4.30(a) and in Figure 4.30(b), respectively. When the project file of the PIC16F1847-Based

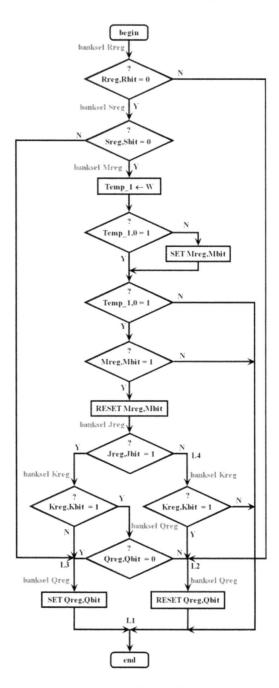

FIGURE 4.24 The flowchart of the macro "jkff_r_SR".

TABLE 4.24
Operands for the Instruction "jkff_r_SR"

Input/Output	Data type	Operands
J Jreg,Jbit (Bit) K Kreg,Kbit (Bit) S Sreg,Sbit (Bit) R Rreg,Rbit (Bit)	BOOL	I, Q, M, T_Q, C_Q, C_QD, C_QU, drum_Q, drum_TQ, drum_steps, drum_events, SF, MB, nAI, AI, LOGIC1, LOGIC0, FRSTSCN, SCNOSC, any Boolean variable from SRAM.
Q Qreg,Qbit (Bit)	BOOL	Q, M, T_Q, C_Q, C_QD, C_QU, drum_Q, drum_TQ, drum_steps, drum_events, SF, MB, any Boolean variable from SRAM.

TABLE 4.25
The Symbol and the Truth Table of the Macro "jkff_f"

Symbol

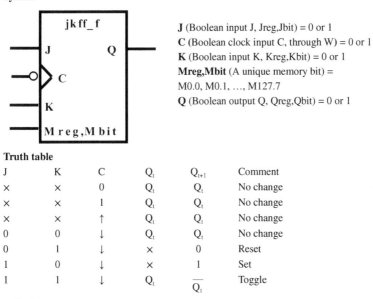

J (Boolean input J, Jreg,Jbit) = 0 or 1
C (Boolean clock input C, through W) = 0 or 1
K (Boolean input K, Kreg,Kbit) = 0 or 1
Mreg,Mbit (A unique memory bit) = M0.0, M0.1, …, M127.7
Q (Boolean output Q, Qreg,Qbit) = 0 or 1

Truth table

J	K	C	Q_t	Q_{t+1}	Comment
×	×	0	Q_t	Q_t	No change
×	×	1	Q_t	Q_t	No change
×	×	↑	Q_t	Q_t	No change
0	0	↓	Q_t	Q_t	No change
0	1	↓	×	0	Reset
1	0	↓	×	1	Set
1	1	↓	Q_t	$\overline{Q_t}$	Toggle

×: don't care.

```
jkff_f    macro   Mreg,Mbit ,Jreg,Jbit, Kreg,Kbit, Qreg,Qbit
          local   L1,L2,L3,L4
          banksel Mreg
          movwf   Temp_1
          btfsc   Temp_1,0
          bsf     Mreg,Mbit       ;Mreg,Mbit = Falling Edge
          btfsc   Temp_1,0        ;Detector for falling edge
          goto    L1              ;triggered JK flip-flop
          btfss   Mreg,Mbit
          goto    L1
          bcf     Mreg,Mbit
          banksel Jreg
          btfss   Jreg,Jbit
          goto    L4              ;if j=0 then goto L4
          banksel Kreg
          btfss   Kreg,Kbit
          goto    L3              ;if j=1&k=0 then SET Qreg,Qbit (goto L3)
          banksel Qreg            ;
          btfsc   Qreg,Qbit       ;if j=1&k=1
          goto    L2              ;then TOGGLE
          goto    L3              ;Qreg,Qbit
L4
          banksel Kreg
          btfss   Kreg,Kbit
          goto    L1              ;if j=0&k=0 then NO CHANGE (goto L1)
          goto    L2              ;if j=0&k=1 then RESET Qreg,Qbit
L3
          banksel Qreg
          bsf     Qreg,Qbit
          goto    L1
L2
          banksel Qreg
          bcf     Qreg,Qbit
L1
          endm
```

FIGURE 4.25 The macro "jkff_f".

PLC is open in the MPLAB X IDE, from the file "PICPLC_PIC16F1847_user_Bsc .inc", if you uncomment Example 4.1 and run the project by pressing the symbol ▷ from the toolbar, then the PIC16F1847 microcontroller within the CPU board of the PIC16F1847-Based PLC will be programmed. After loading the program file to the PIC16F1847 microcontroller, switch the 4PDT to "RUN" and the power switch to the "ON" position. Next you can test the operation of this example.

Basically, this example is dedicated to macros "latch1" and "latch0". In addition, observe that in rung 15 we obtain a rising edge–triggered D flip-flop by using an "r_edge" and a "latch1" macro. Similarly, in rung 16 we obtain a falling edge–triggered D flip-flop by using an "f_edge" and a "latch1" macro.

4.15.2   EXAMPLE 4.2

Example 4.2 shows the usage of the following flip-flop macros: "dff_r", "dff_r_SR", "dff_f", and "dff_f_SR". The user program of Example 4.2 is shown in Figure 4.31. The schematic diagram and ladder diagram of Example 4.2 are depicted in Figure 4.32(a) and in Figure 4.32(b), respectively. When the project file of the

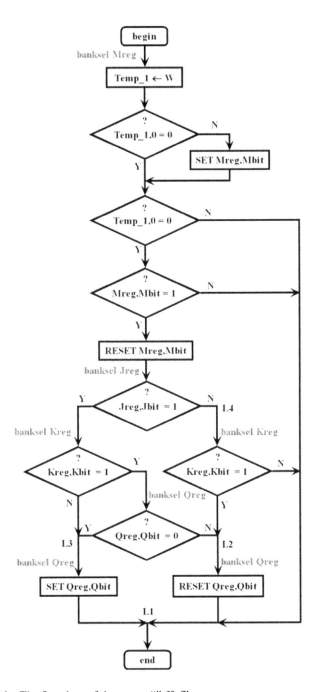

FIGURE 4.26 The flowchart of the macro "jkff_f".

TABLE 4.26
Operands for the Instruction "jkff_f"

Input/Output	Data type	Operands
J Jreg,Jbit (Bit) K Kreg,Kbit (Bit)	BOOL	I, Q, M, T_Q, C_Q, C_QD, C_QU, drum_Q, drum_TQ, drum_steps, drum_events, SF, MB, nAI, AI, LOGIC1, LOGIC0, FRSTSCN, SCNOSC, any Boolean variable from SRAM.
Q Qreg,Qbit (Bit)	BOOL	Q, M, T_Q, C_Q, C_QD, C_QU, drum_Q, drum_TQ, drum_ steps, drum_events, SF, MB, any Boolean variable from SRAM.

TABLE 4.27
The Symbol and the Truth Table of the Macro "jkff_f_SR"

Symbol

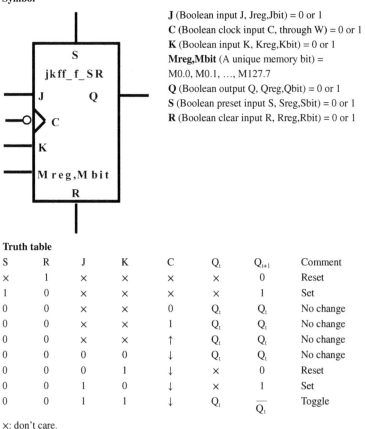

J (Boolean input J, Jreg,Jbit) = 0 or 1
C (Boolean clock input C, through W) = 0 or 1
K (Boolean input K, Kreg,Kbit) = 0 or 1
Mreg,Mbit (A unique memory bit) =
M0.0, M0.1, ..., M127.7
Q (Boolean output Q, Qreg,Qbit) = 0 or 1
S (Boolean preset input S, Sreg,Sbit) = 0 or 1
R (Boolean clear input R, Rreg,Rbit) = 0 or 1

Truth table

S	R	J	K	C	Q_t	Q_{t+1}	Comment
×	1	×	×	×	×	0	Reset
1	0	×	×	×	×	1	Set
0	0	×	×	0	Q_t	Q_t	No change
0	0	×	×	1	Q_t	Q_t	No change
0	0	×	×	↑	Q_t	Q_t	No change
0	0	0	0	↓	Q_t	Q_t	No change
0	0	0	1	↓	×	0	Reset
0	0	1	0	↓	×	1	Set
0	0	1	1	↓	Q_t	$\overline{Q_t}$	Toggle

×: don't care.

```
jkff_f_SR macro Mreg,Mbit, Jreg,Jbit, Kreg,Kbit, Qreg,Qbit, Sreg,Sbit, Rreg,Rbit
      local   L1,L2,L3,L4
      banksel Rreg
      btfsc   Rreg,Rbit
      goto    L2
      banksel Sreg
      btfsc   Sreg,Sbit
      goto    L3
      ;----------------------;The following codes are the same as "jkff_f"
      banksel Mreg
      movwf   Temp_1
      btfsc   Temp_1,0
      bsf     Mreg,Mbit      ;Mreg,Mbit = Falling Edge
      btfsc   Temp_1,0       ;Detector for falling edge
      goto    L1             ;triggered JK flip-flop
      btfss   Mreg,Mbit
      goto    L1
      bcf     Mreg,Mbit
      banksel Jreg
      btfss   Jreg,Jbit
      goto    L4             ;if j=0 then goto L4
      banksel Kreg
      btfss   Kreg,Kbit
      goto    L3             ;if j=1&k=0 then SET Qreg,Qbit (goto L3)
      banksel Qreg           ;
      btfsc   Qreg,Qbit      ;if j=1&k=1
      goto    L2             ;then TOGGLE
      goto    L3             ;Qreg,Qbit
L4
      banksel Kreg
      btfss   Kreg,Kbit
      goto    L1             ;if j=0&k=0 then NO CHANGE (goto L1)
      goto    L2             ;if j=0&k=1 then RESET Qreg,Qbit
L3
      banksel Qreg
      bsf     Qreg,Qbit
      goto    L1
L2
      banksel Qreg
      bcf     Qreg,Qbit
L1
      endm
```

FIGURE 4.27   The macro "jkff_f_SR".

PIC16F1847-Based PLC is open in the MPLAB X IDE, from the file "PICPLC_PIC16F1847_user_Bsc.inc", if you uncomment Example 4.2 and run the project by pressing the symbol ▷ from the toolbar, then the PIC16F1847 microcontroller within the CPU board of the PIC16F1847-Based PLC will be programmed. After loading the program file to the PIC16F1847 microcontroller, switch the 4PDT to "RUN" and the power switch to the "ON" position. Next you can test the operation of this example. Rungs 4, 5, and 6 show how to obtain a rising edge–triggered D flip-flop with **active low** preset (S) and clear (R) inputs. The same technique can be used to obtain a falling edge–triggered D flip-flop with **active low** preset (S) and clear (R) inputs.

4.15.3   Example 4.3

Example 4.3 shows the usage of the following flip-flop macros: "tff_r", "tff_r_SR", "tff_f", and "tff_f_SR". The user program of Example 4.3 is shown in Figure

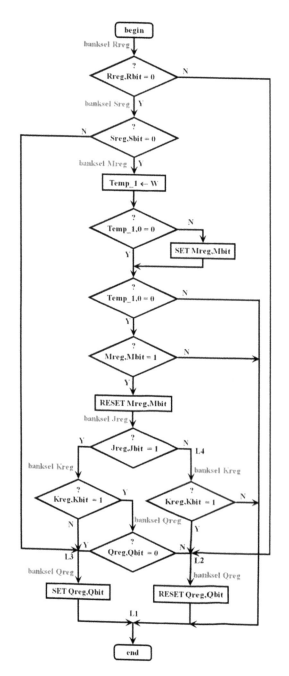

FIGURE 4.28 The flowchart of the macro "jkff_f_SR".

TABLE 4.28

Operands for the Instruction "jkff_f_SR"

Input/Output	Data type	Operands
J	BOOL	I, Q, M, T_Q, C_Q, C_QD, C_QU, drum_Q, drum_TQ, drum_steps,
Jreg,Jbit (Bit)		drum_events, SF, MB, nAI, AI, LOGIC1, LOGIC0, FRSTSCN,
K		SCNOSC, any Boolean variable from SRAM.
Kreg,Kbit (Bit)		
S		
Sreg,Sbit (Bit)		
R		
Rreg,Rbit (Bit)		
Q	BOOL	Q, M, T_Q, C_Q, C_QD, C_QU, drum_Q, drum_TQ, drum_steps,
Qreg,Qbit (Bit)		drum_events, SF, MB, any Boolean variable from SRAM.

4.33. The schematic diagram and ladder diagram of Example 4.3 are depicted in Figure 4.34(a) and in Figure 4.34(b), respectively. When the project file of the PIC16F1847-Based PLC is open in the MPLAB X IDE, from the file "PICPLC _PIC16F1847_user_Bsc.inc", if you uncomment Example 4.3 and run the project by pressing the symbol ▷ from the toolbar, then the PIC16F1847 microcontroller within the CPU board of the PIC16F1847-Based PLC will be programmed. After loading the program file to the PIC16F1847 microcontroller, switch the 4PDT to "RUN" and the power switch to the "ON" position. Next you can test the operation of this example. Rungs 4, 5, and 6 show how to obtain a rising edge–triggered T flip-flop with **active low** preset (S) and clear (R) inputs. The same technique can be used to obtain a falling edge–triggered T flip-flop with **active low** preset (S) and clear (R) inputs.

4.15.4 Example 4.4

Example 4.4 shows the usage of the following flip-flop macros: "jkff_r", "jkff_r_SR", "jkff_f", and "jkff_f_SR". The user program of Example 4.4 is shown in Figure 4.35. The schematic diagram and ladder diagram of Example 4.4 are depicted in Figure 4.36(a) and in Figure 4.36(b), respectively. When the project file of the PIC16F1847-Based PLC is open in the MPLAB X IDE, from the file "PICPLC_PIC16F1847_user_Bsc.inc", if you uncomment Example 4.4 and run the project by pressing the symbol ▷ from the toolbar, then the PIC16F1847 microcontroller within the CPU board of the PIC16F1847-Based PLC will be programmed. After loading the program file to the PIC16F1847 microcontroller, switch the 4PDT to "RUN" and the power switch to the "ON" position. Next you can test the operation of this example. Rungs 4, 5, and 6 show how to obtain a rising edge–triggered JK flip-flop with **active low** preset (S) and clear (R) inputs. The same technique can be used to obtain a falling edge–triggered JK flip-flop with **active low** preset (S) and clear (R) inputs.

```
user_program_1   macro
;--- PLC codes to be allocated in the "user_program_1" macro start from here ---
;__Example_Bsc_4.1
        ld        I0.0              ;rung 1   ; EN = I0.0
                  ;D ,Q                       ; D = I0.1
        latch1    I0.1,Q0.0                   ; Q = Q0.0

        ld        I0.2              ;rung 2   ; EN = I0.2
                  ;D ,Q                       ; D = I0.3
        latch1    I0.3,Q0.2                   ; Q = Q0.2

        ld_not    I0.4              ;rung 3   ; EN = complement of I0.4
                  ;D ,Q                       ; D = I0.5
        latch1    I0.5,Q0.4                   ; Q = Q0.4

        in_out    I0.6,M0.0         ;rung 4   ; M0.0 = I0.6
                                              ;
        in_out    I0.7,M127.7       ;rung 5   ; M127.7 = I0.7

        ld        M0.0              ;rung 6   ; EN = M0.0 = I0.6
                  ;D ,Q                       ; D = M127.7 = I0.7
        latch1    M127.7,M44.4                ; Q = M44.4 = Q0.6

        in_out    M44.4,Q0.6        ;rung 7   ; Q0.6 = M44.4

        ld        I1.0              ;rung 8   ; EN = I1.0
                  ;D ,Q                       ; D = I1.1
        latch0    I1.1,Q1.0                   ; Q = Q1.0

        ld        I1.2              ;rung 9   ; EN = I1.2
                  ;D ,Q                       ; D = I1.3
        latch0    I1.3,Q1.2                   ; Q = Q1.2

        ld_not    I1.4              ;rung 10  ; EN = complement of I1.4
                  ;D ,Q                       ; D = I1.5
        latch0    I1.5,Q1.4                   ; Q = Q1.4
        in_out    I1.6,M1.0         ;rung 11  ; M1.0 = I1.6

        in_out    I1.7,M12.7        ;rung 12  ; M12.7 = I1.7

        ld        M1.0              ;rung 13  ; EN = M1.0 = I1.6
                  ;D ,Q                       ; D = M12.7 = I1.7
        latch0    M12.7,M4.4                  ; Q = M4.4 = Q1.6

        in_out    M4.4,Q1.6         ;rung 14  ; Q1.6 = M4.4

        ld        I2.0              ;rung 15  ; (W,0) = I2.0
        r_edge    M99.0                       ; EN = (W,0) = rising edge of (W,0)
                  ;D ,Q                       ; D = I2.1
        latch1    I2.1,Q2.0                   ; Q = Q2.0

        ld        I2.2              ;rung 16  ; (W,0) = I2.2
        f_edge    M99.1                       ; EN = (W,0) = falling edge of (W,0)
                  ;D ,Q                       ; D = I2.3
        latch1    I2.3,Q2.2                   ; Q = Q2.2
;--- PLC codes to be allocated in the "user_program_1" macro end here ----------
        endm
```

FIGURE 4.29 The user program of Example 4.1.

4.15.5 EXAMPLE 4.5: 4-BIT ASYNCHRONOUS UP COUNTER

Example 4.5 explains the implementation of a 4-bit asynchronous up counter by using the PIC16F1847-Based PLC. A 4-bit asynchronous up counter designed by using falling edge–triggered T flip-flops is shown in Figure 4.37. A 4-bit binary up counter counts sequence from 0000 to 1111 (from 0 to 15 in decimal). Clock inputs of all flip-flops are cascaded and the T input of each flip-flop is connected to HIGH. This means that in the 4-bit asynchronous up counter, the flip-flops are connected in toggle mode and they will toggle at each negative-going edge (falling

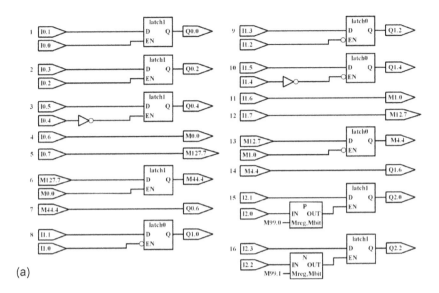

FIGURE 4.30 (a) The user program of Example 4.1: schematic diagram.

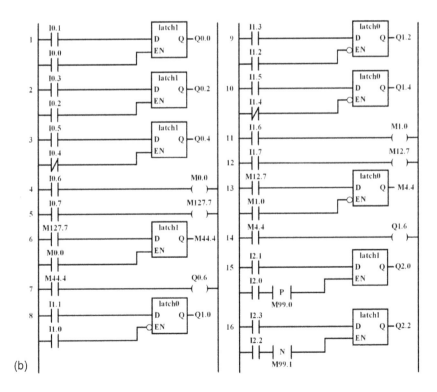

FIGURE 4.30 (b) The user program of Example 4.1: ladder diagram.

```
user_program_1  macro
;--- PLC codes to be allocated in the "user_program_1" macro start from here ---
;__Example_Bsc_4.2

        ld        I0.0                      ;rung 1 ;clock signal C = I0.0
        dff_r     M0.0,I0.1,Q0.0                    ;D=I0.1, Q=O0.0

        ld        I0.2                      ;rung 2 ;clock signal C = I0.2
        dff_r     M0.1,I0.3,Q0.2                    ;D=I0.3, Q=O0.2

        ld        I1.0                      ;rung 3 ;clock signal C = I1.0
        dff_r_SR  M0.2,I1.1,Q1.0,I1.2,I1.3          ;D=I1.1, Q=Q1.0, S=I1.2, R=I1.3

        inv_out   I1.6,M1.6                 ;rung 4 ; M1.6 = complement of I1.6

        inv_out   I1.7,M1.7                 ;rung 5 ; M1.7 = complement of I1.7

        ld        I1.4                      ;rung 6 ;clock signal C = I1.4
        dff_r_SR  M0.3,I1.5,Q1.2,M1.6,M1.7          ;D=I1.5, Q=Q1.2, S=M1.6, R=M1.7

        ld        I2.0                      ;rung 7 ;clock signal C = I2.0
        dff_f     M0.4,I2.1,Q2.0                    ;D=I2.1, Q=O2.0
        ld        I2.2                      ;rung 8 ;clock signal C = I2.2
        dff_f     M0.5,I2.3,Q2.2                    ;D=I2.3, Q=O2.2
        ld        I3.0                      ;rung 9 ;clock signal C = I3.0
        dff_f_SR  M0.6,I3.1,Q3.0,I3.2,I3.3          ;D=I3.1, Q=Q3.0, S=I3.2, R=I3.3

        ld        I3.4                      ;rung 10;clock signal C = I3.4
        dff_f_SR  M0.7,I3.5,Q3.2,I3.6,I3.7          ;D=I3.5, Q=Q3.2, S=I3.6, R=I3.7
;
;--- PLC codes to be allocated in the "user_program_1" macro end here ----------
        endm
```

FIGURE 4.31   The user program of Example 4.2.

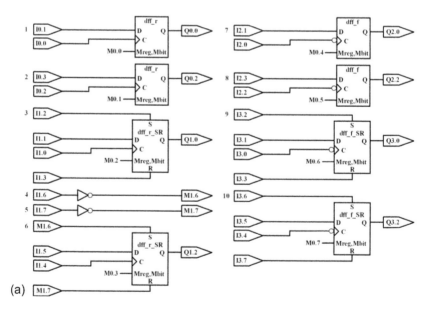

(a)

FIGURE 4.32   (a) The user program of Example 4.2: schematic diagram.

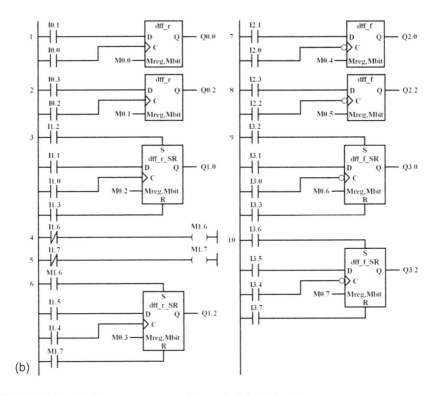

FIGURE 4.32   (b) The user program of Example 4.2: ladder diagram.

edge) of their respective clock signal. The clock input is connected to the first flip-flop. The other flip-flops receive the clock signal input from the Q output of the previous flip-flop.

Figure 4.38 shows the logic diagram of the 4-bit asynchronous up counter implemented by using the PIC16F1847-Based PLC. T flip-flops are implemented by using "tff_f" macros. The CP clock signal can be applied through I0.1 or it can be fixed by using the T_1s reference timing signal. The choice between these two clock sources is made by using a 2-to-1 multiplexer "Mux 2x1". When the select input I0.0 of "Mux 2x1" is 0 (or 1, respectively), the CP clock input is I0.1 (or T_1s, respectively).

The user program of Example 4.5 (4-bit asynchronous up counter) is shown in Figure 4.39. The schematic diagram of the user program of Example 4.5 is depicted in Figure 4.40. When the project file of the PIC16F1847-Based PLC is open in the MPLAB X IDE, from the file "PICPLC_PIC16F1847_user_Bsc.inc", if you uncomment Example 4.5 and run the project by pressing the symbol ▷ from the toolbar, then the PIC16F1847 microcontroller within the CPU board of the PIC16F1847-Based PLC will be programmed. After loading the program file to the PIC16F1847 microcontroller, switch the 4PDT to "RUN" and the power switch to the "ON" position. Next you can test the operation of this example. Table 4.29 shows the 4-bit binary count up sequence.

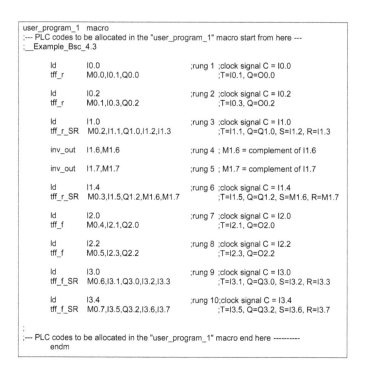

```
user_program_1   macro
;--- PLC codes to be allocated in the "user_program_1" macro start from here ---
;__Example_Bsc_4.3

        ld        I0.0                    ;rung 1 ;clock signal C = I0.0
        tff_r     M0.0,I0.1,Q0.0                  ;T=I0.1, Q=O0.0

        ld        I0.2                    ;rung 2 ;clock signal C = I0.2
        tff_r     M0.1,I0.3,Q0.2                  ;T=I0.3, Q=O0.2

        ld        I1.0                    ;rung 3 ;clock signal C = I1.0
        tff_r_SR  M0.2,I1.1,Q1.0,I1.2,I1.3        ;T=I1.1, Q=Q1.0, S=I1.2, R=I1.3

        inv_out   I1.6,M1.6               ;rung 4 ; M1.6 = complement of I1.6

        inv_out   I1.7,M1.7               ;rung 5 ; M1.7 = complement of I1.7

        ld        I1.4                    ;rung 6 ;clock signal C = I1.4
        tff_r_SR  M0.3,I1.5,Q1.2,M1.6,M1.7        ;T=I1.5, Q=Q1.2, S=M1.6, R=M1.7

        ld        I2.0                    ;rung 7 ;clock signal C = I2.0
        tff_f     M0.4,I2.1,Q2.0                  ;T=I2.1, Q=O2.0

        ld        I2.2                    ;rung 8 ;clock signal C = I2.2
        tff_f     M0.5,I2.3,Q2.2                  ;T=I2.3, Q=O2.2

        ld        I3.0                    ;rung 9 ;clock signal C = I3.0
        tff_f_SR  M0.6,I3.1,Q3.0,I3.2,I3.3        ;T=I3.1, Q=Q3.0, S=I3.2, R=I3.3

        ld        I3.4                    ;rung 10;clock signal C = I3.4
        tff_f_SR  M0.7,I3.5,Q3.2,I3.6,I3.7        ;T=I3.5, Q=Q3.2, S=I3.6, R=I3.7

;
;--- PLC codes to be allocated in the "user_program_1" macro end here ----------
        endm
```

FIGURE 4.33   The user program of Example 4.3.

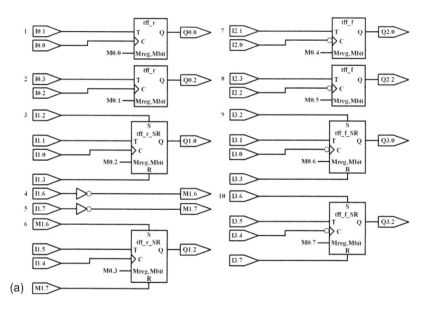

(a)

FIGURE 4.34   (a) The user program of Example 4.3: schematic diagram.

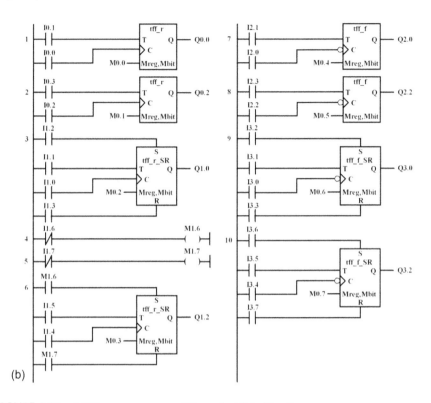

FIGURE 4.34   (b) The user program of Example 4.3: ladder diagram.

4.15.6 EXAMPLE 4.6: 4-BIT ASYNCHRONOUS DOWN COUNTER

Example 4.6 explains the implementation of a 4-bit asynchronous down counter by using the PIC16F1847-Based PLC. A 4-bit asynchronous down counter designed by using falling edge–triggered T flip-flops is shown in Figure 4.41. A 4-bit binary down counter counts sequence from 1111 to 0000 (from 15 to 0 in decimal). Clock inputs of all flip-flops are cascaded and the T input of each flip-flop is connected to HIGH. This means that in the 4-bit asynchronous down counter, the flip-flops are connected in toggle mode and they will toggle at each negative-going edge (falling edge) of their respective clock signal. The clock input is connected to the first flip-flop. The other flip-flops receive the clock signal input from the $\overline{Q}$ output of the previous flip-flop.

Figure 4.42 shows the logic diagram of the 4-bit asynchronous down counter implemented by using the PIC16F1847-Based PLC. T flip-flops are implemented by using "tff_f" macros. The CP clock signal can be applied through I0.1 or it can be fixed by using a T_1s reference timing signal. The choice between these two clock sources is made by using a 2-to-1 multiplexer "Mux 2x1". When the select input I0.0 of "Mux 2x1" is 0 (or 1, respectively), the CP clock input is I0.1 (or T_1s, respectively).

```
user_program_1  macro
;--- PLC codes to be allocated in the "user_program_1" macro start from here ---
;__Example_Bsc_4.4

        ld      I0.0                        ;rung 1 ;clock signal C = I0.0
        jkff_r  M0.0,I0.1,I0.2,Q0.0                ;J=I0.1, K=I0.2, Q=Q0.0

        ld      I0.3                        ;rung 2 ;clock signal C = I0.3
        jkff_r  M0.1,I0.4,I0.5,Q0.2                ;J=I0.4, K=I0.5, Q=Q0.2

        ld      I0.6                        ;rung 3 ;clock signal C = I0.6
        jkff_r_SR M0.2,I1.0,I1.1,Q1.0,I1.2,I1.3
                                            ;J=I1.0, K=I1.1, Q=Q1.0, S=I1.2, R=I1.3

        inv_out  I1.6,M1.6                  ;rung 4 ; M1.6 = complement of I1.6

        inv_out  I1.7,M1.7                  ;rung 5 ; M1.7 = complement of I1.7

        ld      I0.7                        ;rung 6 ;clock signal C = I0.7
        jkff_r_SR M0.3,I1.4,I1.5,Q1.2,M1.6,M1.7
                                            ;J=I1.4, K=I1.5, Q=Q1.2, S=M1.6, R=M1.7

        ld      I2.0                        ;rung 7 ;clock signal C = I2.0
        jkff_f  M0.4,I2.1,I2.2,Q2.0                ;J=I2.1, K=I2.2, Q=Q2.0

        ld      I2.3                        ;rung 8 ;clock signal C = I2.3
        jkff_f  M0.5,I2.4,I2.5,Q2.2                ;J=I2.4, K=I2.5, Q=Q2.2

        ld      I2.6                        ;rung 9 ;clock signal C = I2.6
        jkff_f_SR M0.6,I3.0,I3.1,Q3.0,I3.2,I3.3
                                            ;J=I3.0, K=I3.1, Q=Q3.0, S=I3.2, R=I3.3

        ld      I2.7                        ;rung 10;clock signal C = I2.7
        jkff_f_SR M0.7,I3.4,I3.5,Q3.2,I3.6,I3.7
                                            ;J=I3.4, K=I3.5, Q=Q3.2, S=I3.6, R=I3.7
;--- PLC codes to be allocated in the "user_program_1" macro end here ----------
        endm
```

FIGURE 4.35   The user program of Example 4.4.

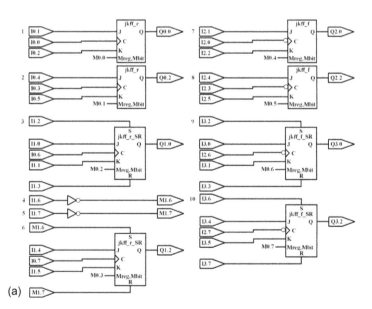

(a)

FIGURE 4.36   (a) The user program of Example 4.4: schematic diagram.

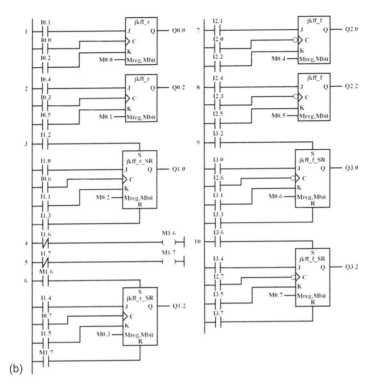

(b)

FIGURE 4.36   (b) The user program of Example 4.4: ladder diagram.

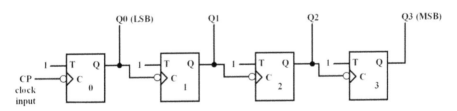

FIGURE 4.37   The logic diagram of the 4-bit asynchronous up counter designed by using falling edge–triggered T flip-flops.

The user program of Example 4.6 (4-bit asynchronous down counter) is shown in Figure 4.43. The schematic diagram of the user program of Example 4.6 is depicted in Figure 4.44. When the project file of the PIC16F1847-Based PLC is open in the MPLAB X IDE, from the file "PICPLC_PIC16F1847_user_Bsc.inc", if you uncomment Example 4.6 and run the project by pressing the symbol ▷ from the toolbar, then the PIC16F1847 microcontroller within the CPU board of the PIC16F1847-Based PLC will be programmed. After loading the program file to the PIC16F1847 microcontroller, switch the 4PDT to "RUN" and the power switch to the "ON" position. Next you can test the operation of this example. Table 4.30 shows the 4-bit binary count down sequence.

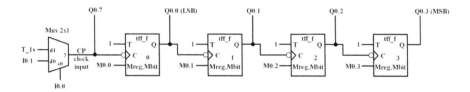

FIGURE 4.38 The logic diagram of the 4-bit asynchronous up counter implemented by using the PIC16F1847-Based PLC.

```
user_program_1   macro
;
;--- PLC codes to be allocated in the "user_program_1" macro start from here ---
;__Example_Bsc_4.5
;
;_____ 4-bit asynchronous up counter _____
;
                        ;s0,i1,i0,rego,bito          ;rung 1
            mux_2_1  I0.0,T_1s,I0.1,Q0.7

            ld        Q0.7                            ;rung 2
                     ;Mreg,Mbit,Treg,Tbit,Qreg,Qbit
            tff_f     M0.0,LOGIC1,Q0.0

            ld        Q0.0                            ;rung 3
                     ;Mreg,Mbit,Treg,Tbit,Qreg,Qbit
            tff_f     M0.1,LOGIC1,Q0.1

            ld        Q0.1                            ;rung 4
                     ;Mreg,Mbit,Treg,Tbit,Qreg,Qbit
            tff_f     M0.2,LOGIC1,Q0.2

            ld        Q0.2                            ;rung 5
                     ;Mreg,Mbit,Treg,Tbit,Qreg,Qbit
            tff_f     M0.3,LOGIC1,Q0.3

;--- PLC codes to be allocated in the "user_program_1" macro end here ----------
            endm
```

FIGURE 4.39 The user program of Example 4.5 (4-bit asynchronous up counter).

4.15.7 Example 4.7: Asynchronous Decade Counter

Example 4.7 explains the implementation of an asynchronous decade counter by using the PIC16F1847-Based PLC. The binary counters have 2^n states, where n is the number of flip-flops. Counters with fewer than 2^n states are designed with truncated sequences. These sequences are achieved by forcing the counter to recycle before going through all of its normal states. A common modulus for counters with truncated sequences is 10. A counter with 10 states in its sequence is called a decade counter. An asynchronous decade counter designed by using falling edge–triggered T flip-flops is shown in Figure 4.45. An asynchronous decade counter counts sequence from 0000 to 1001 (from 0 to 9 in decimal). Clock inputs of all flip-flops are cascaded and the T input of each flip-flop is connected to HIGH. This means that in the asynchronous decade counter, the flip-flops are connected in toggle mode and they will toggle at each negative-going edge (falling edge) of their respective clock signal. The clock input is connected to the first flip-flop. The other flip-flops receive

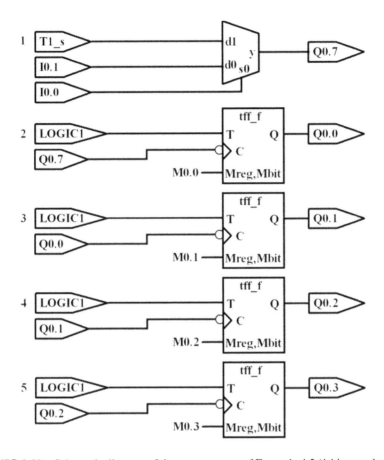

FIGURE 4.40 Schematic diagram of the user program of Example 4.5 (4-bit asynchronous up counter).

the clock signal input from the Q output of the previous flip-flop. Once the counter counts to 10 (1010), all the flip-flops are cleared by means of the AND gate. Notice that only Q1 and Q3 are used to decode the count of 10. This is called partial decoding, as none of the other states (0 to 9) have both Q1 and Q3 HIGH at the same time.

Figure 4.46 shows the logic diagram of the asynchronous decade counter implemented by using the PIC16F1847-Based PLC. JK flip-flops are implemented by using "jkff_f_SR" macros. The 2-input AND gate is implemented by using an "and" macro. The CP clock signal can be applied through I0.1 or it can be fixed by using a T_1s reference timing signal. The choice between these two clock sources is made by using a 2-to-1 multiplexer "Mux 2x1". When the select input I0.0 of "Mux 2x1" is 0 (or 1, respectively), the CP clock input is I0.1 (or T_1s, respectively).

The user program of Example 4.7 (asynchronous decade counter) is shown in Figure 4.47. The schematic diagram of the user program of Example 4.7 is depicted in Figure 4.48. When the project file of the PIC16F1847-Based PLC is open in the MPLAB X IDE, from the file "PICPLC_PIC16F1847_user_Bsc.inc", if you

TABLE 4.29

4-Bit Binary Count Up Sequence

CP Clock input	Q3 (MSB) Q0.3	Q2 Q0.2	Q1 Q0.1	Q0 (LSB) Q0.0	Decimal Count Value
	0	0	0	0	0
1	0	0	0	1	1
2	0	0	1	0	2
3	0	0	1	1	3
4	0	1	0	0	4
5	0	1	0	1	5
6	0	1	1	0	6
7	0	1	1	1	7
8	1	0	0	0	8
9	1	0	0	1	9
10	1	0	1	0	10
11	1	0	1	1	11
12	1	1	0	0	12
13	1	1	0	1	13
14	1	1	1	0	14
15	1	1	1	1	15

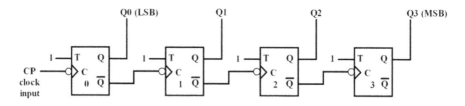

FIGURE 4.41 The logic diagram of the 4-bit asynchronous down counter designed by using falling edge–triggered T flip-flops.

uncomment Example 4.7 and run the project by pressing the symbol ▷ from the toolbar, then the PIC16F1847 microcontroller within the CPU board of the PIC16F1847-Based PLC will be programmed. After loading the program file to the PIC16F1847 microcontroller, switch the 4PDT to "RUN" and the power switch to the "ON" position. Next you can test the operation of this example. Table 4.31 shows the count up sequence for the decade counter.

4.15.8 EXAMPLE 4.8: 4-BIT ASYNCHRONOUS UP/DOWN COUNTER

Example 4.8 explains the implementation of a 4-bit asynchronous up/down counter by using the PIC16F1847-Based PLC. Up/down counters, also known as bidirectional counters, are capable of counting in either the up direction or the down

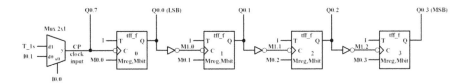

FIGURE 4.42 The logic diagram of the 4-bit asynchronous down counter implemented by using the PIC16F1847-Based PLC.

```
user_program_1  macro
;
;--- PLC codes to be allocated in the "user_program_1" macro start from here ---
;__Example_Bsc_4.6
;
;_____ 4-bit asynchronous down counter _____
;
                    ;s0,i1,i0,rego,bito          ;rung 1
            mux_2_1  I0.0,T_1s,I0.1,Q0.7

            ld       Q0.7                         ;rung 2
                     ;Mreg,Mbit,Treg,Tbit,Qreg,Qbit
            tff_f    M0.0,LOGIC1,Q0.0

            inv_out Q0.0,M1.0                     ;rung 3

            ld       M1.0                         ;rung 4
                     ;Mreg,Mbit,Treg,Tbit,Qreg,Qbit
            tff_f    M0.1,LOGIC1,Q0.1

            inv_out Q0.1,M1.1                     ;rung 5

            ld       M1.1                         ;rung 6
                     ;Mreg,Mbit,Treg,Tbit,Qreg,Qbit
            tff_f    M0.2,LOGIC1,Q0.2

            inv_out Q0.2,M1.2                     ;rung 7

            ld       M1.2                         ;rung 8
                     ;Mreg,Mbit,Treg,Tbit,Qreg,Qbit
            tff_f    M0.3,LOGIC1,Q0.3

;--- PLC codes to be allocated in the "user_program_1" macro end here ----------
            endm
```

FIGURE 4.43 The user program of Example 4.6 (4-bit asynchronous down counter).

direction through any given count sequence. A 4-bit asynchronous up/down counter designed by using falling edge–triggered T flip-flops is shown in Figure 4.49. In this circuit, the input K is used to control the count direction. When K = 0, the counter counts up with each falling edge of the CP clock input signal. In this case, the EXOR gate between flip-flops FF0 and FF1 will connect the non-inverted output (Q0) of FF0 into the clock input of FF1. Q1 of FF1 will be connected through the second EXOR gate into the clock input of FF2. Similarly, Q2 of FF2 will be connected through the third EXOR gate into the clock input of FF3. Thus the counter will count up. When K = 1, the counter counts down with each falling edge of the CP clock input signal. In this case, the EXOR gate between flip-flops FF0 and FF1 will

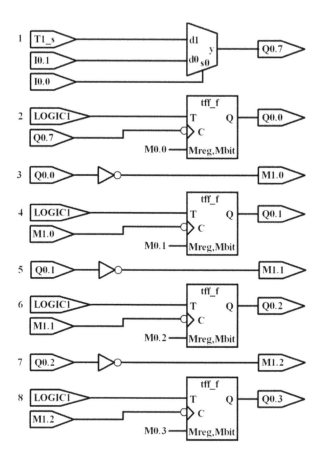

FIGURE 4.44   Schematic diagram of the user program of Example 4.6 (4-bit asynchronous down counter).

connect the inverted output $\left(\overline{Q0}\right)$ of FF0 into the clock input of FF1. $\left(\overline{Q1}\right)$ of FF1 will be connected through the second EXOR gate into the clock input of FF2. Similarly, $\left(\overline{Q2}\right)$ of FF2 will be connected through the third EXOR gate into the clock input of FF3. Thus the counter will count down. When K = 0, the counter counts up sequence from 0000 to 1111 (from 0 to 15 in decimal). When K = 1, the counter counts down sequence from 1111 to 0000 (from 15 to 0 in decimal). The T input of each flip-flop is connected to HIGH. This means that flip-flops are connected in toggle mode and they will toggle at each negative-going edge (falling edge) of their respective clock signal. The main drawback of these types of asynchronous up/down counters is that during normal operation, assuming that CP is 0 or 1, when the state of K is changed from LOW to HIGH or vice versa, this also acts as a clock input signal, and outputs of the flip-flops are changed accordingly, thus destroying the current count value at that instant.

Figure 4.50 shows the logic diagram of the 4-bit asynchronous up/down counter implemented by using the PIC16F1847-Based PLC. T flip-flops are implemented by

TABLE 4.30

4-Bit Binary Count Down Sequence

CP Clock input	Q3 (MSB) Q0.3	Q2 Q0.2	Q1 Q0.1	Q0 (LSB) Q0.0	Decimal Count Value
	0	0	0	0	0
1	1	1	1	1	15
2	1	1	1	0	14
3	1	1	0	1	13
4	1	1	0	0	12
5	1	0	1	1	11
6	1	0	1	0	10
7	1	0	0	1	9
8	1	0	0	0	8
9	0	1	1	1	7
10	0	1	1	0	6
11	0	1	0	1	5
12	0	0	0	0	4
13	0	0	1	1	3
14	0	0	1	0	2
15	0	0	0	1	1

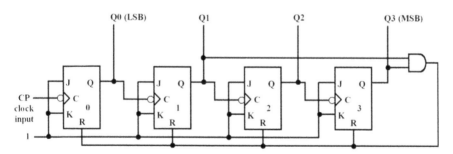

FIGURE 4.45 The logic diagram of the asynchronous decade counter designed by using falling edge–triggered JK flip-flops.

using "tff_f" macros. 2-input EXOR gates are implemented by using "xor" macros. The CP clock signal can be applied through I0.1 or it can be fixed by using a T_1s reference timing signal. The choice between these two clock sources is made by using a 2-to-1 multiplexer "Mux 2x1". When the select input I0.0 of "Mux 2x1" is 0 (or 1, respectively), the CP clock input is I0.1 (or T_1s, respectively).

The user program of Example 4.8 (4-bit asynchronous up/down counter) is shown in Figure 4.51. The schematic diagram of the user program of Example 4.8 is depicted in Figure 4.52. When the project file of the PIC16F1847-Based PLC is open in the MPLAB X IDE, from the file "PICPLC_PIC16F1847_user_Bsc.inc", if you uncomment Example 4.8 and run the project by pressing the symbol ▷ from the toolbar,

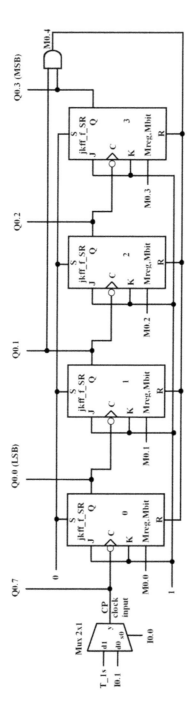

FIGURE 4.46   The logic diagram of the asynchronous decade counter implemented by using the PIC16F1847-Based PLC.

```
user_program_1   macro
;
;--- PLC codes to be allocated in the "user_program_1" macro start from here ---
;__Example_Bsc_4.7
;
;_____ Asynchronous decade counter _____
;
                 ;s0,i1,i0,rego,bito              ;rung 1
          mux_2_1 I0.0,T_1s,I0.1,Q0.7

          ld        Q0.1                          ;rung 2
          and       Q0.3
          out       M0.4

          ld        Q0.2                          ;rung 3
                    ;Mreg,Mbit,J,K,Q,S,R
          jkff_f_SR M0.3,LOGIC1,LOGIC1,Q0.3,LOGIC0,M0.4

          ld        Q0.1                          ;rung 4
                    ;Mreg,Mbit,J,K,Q,S,R
          jkff_f_SR M0.2,LOGIC1,LOGIC1,Q0.2,LOGIC0,M0.4

          ld        Q0.0                          ;rung 5
                    ;Mreg,Mbit,J,K,Q,S,R
          jkff_f_SR M0.1,LOGIC1,LOGIC1,Q0.1,LOGIC0,M0.4

          ld        Q0.7                          ;rung 6
                    ;Mreg,Mbit,J,K,Q,S,R
          jkff_f_SR M0.0,LOGIC1,LOGIC1,Q0.0,LOGIC0,M0.4

;--- PLC codes to be allocated in the "user_program_1" macro end here ----------
          endm
```

FIGURE 4.47   The user program of Example 4.7 (asynchronous decade counter).

then the PIC16F1847 microcontroller within the CPU board of the PIC16F1847-Based PLC will be programmed. After loading the program file to the PIC16F1847 microcontroller, switch the 4PDT to "RUN" and the power switch to the "ON" position. Next you can test the operation of this example.

4.15.9   EXAMPLE 4.9: SYNCHRONOUS DECADE COUNTER

Example 4.9 explains the implementation of a synchronous decade counter by using the PIC16F1847-Based PLC. In synchronous counters, the clock input of all individual flip-flops within the counter are all clocked together at the same time by the same clock signal. From the previous asynchronous counter examples, it can be seen that the Q or $\overline{Q}$ output of one counter stage is connected directly to the clock input of the next counter stage and so on along the chain. The result of this is that the asynchronous counter suffers from what is known as "propagation delay", in which the timing signal is delayed a fraction through each flip-flop. However, with the synchronous counter, the external clock signal is connected to the clock input of every individual flip-flop within the counter so that all flip-flops are clocked together simultaneously (in parallel) at the same time, giving a fixed time relationship. In other words, changes in the output occur in "synchronization" with the clock signal. The result of this synchronization is that all the individual output bits change state exactly at the same time in response to the common clock signal with no ripple effect and therefore, no propagation delay. The design of a synchronous counter using flip-flops requires a number of steps to follow. However, in this book we are

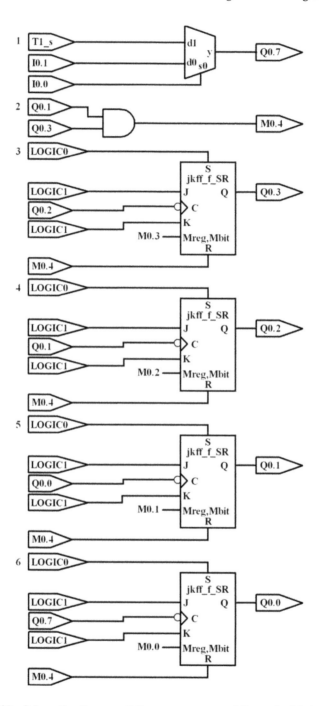

FIGURE 4.48 Schematic diagram of the user program of Example 4.7 (asynchronous decade counter).

TABLE 4.31

Count Up Sequence for the Decade Counter

CP Clock input	Q3 (MSB) Q0.3	Q2 Q0.2	Q1 Q0.1	Q0 (LSB) Q0.0	Decimal Count Value
	0	0	0	0	0
1	0	0	0	1	1
2	0	0	1	0	2
3	0	0	1	1	3
4	0	1	0	0	4
5	0	1	0	1	5
6	0	1	1	0	6
7	0	1	1	1	7
8	1	0	0	0	8
9	1	0	0	1	9

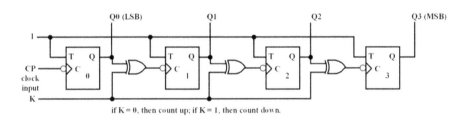

if K = 0, then count up; if K = 1, then count down.

FIGURE 4.49 The logic diagram of the 4-bit asynchronous up/down counter designed by using falling edge–triggered T flip-flops.

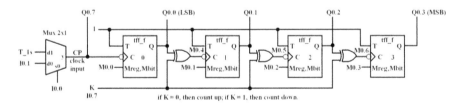

if K = 0, then count up; if K = 1, then count down.

FIGURE 4.50 The logic diagram of the 4-bit asynchronous up/down counter implemented by using the PIC16F1847-Based PLC.

not interested in the design steps of synchronous counters. Therefore, we just use the resulting flip-flop input functions of a particular counter designed previously, and then we implement the counter by using the PIC16F1847-Based PLC. The design of a synchronous decade counter with four JK flip-flops, namely FF0, FF1, FF2, and FF3, results in four sets of input functions as shown below:

$$J0 = K0 = 1, J1 = \overline{Q3}.Q0, K1 = Q0, J2 = K2 = Q1.Q0, J3 = Q2.Q1.Q0, K3 = Q0.$$

```
user_program_1  macro
;
;--- PLC codes to be allocated in the "user_program_1" macro start from here ---
;__Example_Bsc_4.8
;
;_____ 4-bit asynchronous up/down counter _____
;
                ;s0,i1,i0,rego,bito          ;rung 1
        mux_2_1 I0.0,T_1s,I0.1,Q0.7

        ld      Q0.0                        ;rung 2
        xor     I0.7
        out     M0.4

        ld      Q0.1                        ;rung 3
        xor     I0.7
        out     M0.5

        ld      Q0.2                        ;rung 4
        xor     I0.7
        out     M0.6

        ld      Q0.7                        ;rung 5
                ;Mreg,Mbit,Treg,Tbit,Qreg,Qbit
        tff_f   M0.0,LOGIC1,Q0.0

        ld      M0.4                        ;rung 6
                ;Mreg,Mbit,Treg,Tbit,Qreg,Qbit
        tff_f   M0.1,LOGIC1,Q0.1

        ld      M0.5                        ;rung 7
                ;Mreg,Mbit,Treg,Tbit,Qreg,Qbit
        tff_f   M0.2,LOGIC1,Q0.2

        ld      M0.6                        ;rung 8
                ;Mreg,Mbit,Treg,Tbit,Qreg,Qbit
        tff_f   M0.3,LOGIC1,Q0.3
;
;--- PLC codes to be allocated in the "user_program_1" macro end here ----------
        endm
```

FIGURE 4.51   The user program of Example 4.8 (4-bit asynchronous up/down counter).

A synchronous decade counter designed by using falling edge–triggered JK flip-flops, based on the above flip-flop input functions, is shown in Figure 4.53. A synchronous decade counter counts sequence from 0000 to 1001 (from 0 to 9 in decimal). Figure 4.54 shows the logic diagram of the synchronous decade counter implemented by using the PIC16F1847-Based PLC. JK flip-flops are implemented by using "jkff_f" macros. Two 2-input AND gates and the 3-input AND gate are all implemented by using "and" macros. The CP clock signal can be applied through I0.1 or it can be fixed by using a T_1s reference timing signal. The choice between these two clock sources is made by using a 2-to-1 multiplexer "Mux 2x1". When the select input I0.0 of "Mux 2x1" is 0 (or 1, respectively), the CP clock input is I0.1 (or T_1s, respectively).

The user program of Example 4.9 (synchronous decade counter) is shown in Figure 4.55. The schematic diagram of the user program of Example 4.9 is depicted in Figure 4.56. When the project file of the PIC16F1847-Based PLC is open in the MPLAB X IDE, from the file "PICPLC_PIC16F1847_user_Bsc.inc", if you uncomment Example 4.9 and run the project by pressing the symbol ▷ from the toolbar, then the PIC16F1847 microcontroller within the CPU board of the PIC16F1847-Based PLC will be programmed. After loading the program file to the PIC16F1847

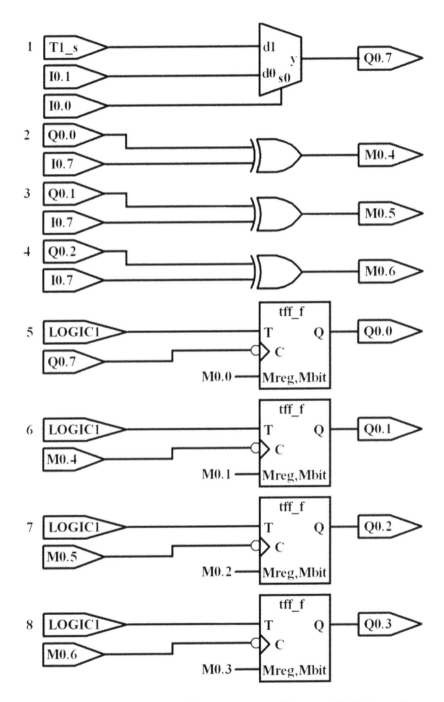

FIGURE 4.52 Schematic diagram of the user program of Example 4.8 (4-bit asynchronous up/down counter).

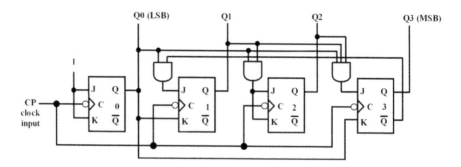

FIGURE 4.53 The logic diagram of the synchronous decade counter designed by using falling edge–triggered JK flip-flops.

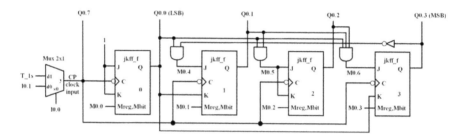

FIGURE 4.54 The logic diagram of the synchronous decade counter implemented by using the PIC16F1847-Based PLC.

microcontroller, switch the 4PDT to "RUN" and the power switch to the "ON" position. Next you can test the operation of this example. Table 4.31 shows the count up sequence of the decade counter.

4.15.10 Example 4.10: 4-Bit Synchronous Up/Down Counter

Example 4.10 explains the implementation of a 4-bit synchronous up/down counter by using the PIC16F1847-Based PLC. The design of a 4-bit synchronous **up/down** counter with four JK flip-flops, namely FF0, FF1, FF2, and FF3, is accomplished in three steps. In the first step, the design of a 4-bit synchronous **up** counter with four JK flip-flops results in four sets of input functions as shown below:

For a 4-bit synchronous **up** counter:

$$J0 = K0 = 1, J1 = K1 = Q0, J2 = K2 = Q1.Q0, J3 = K3 = Q2.Q1.Q0$$

In the second step, the design of a 4-bit synchronous **down** counter with four JK flip-flops results in four sets of input functions as shown below:

For a 4-bit synchronous **down** counter:

$$J0 = K0 = 1, J1 = K1 = \overline{Q0}, J2 = K2 = \overline{Q1}.\overline{Q0}, J3 = K3 = \overline{Q2}.\overline{Q1}.\overline{Q0}$$

```
user_program_1  macro
;
;--- PLC codes to be allocated in the "user_program_1" macro start from here ---
;__Example_Bsc_4.9
;
;_____ Synchronous decade counter _____
;
                        ;s0,i1,i0,rego,bito              ;rung 1
        mux_2_1         I0.0,T_1s,I0.1,Q0.7

        ld              Q0.0                            ;rung 2
        and_not         Q0.3
        out             M0.4

        ld              Q0.0                            ;rung 3
        and             Q0.1
        out             M0.5

        ld              Q0.0                            ;rung 4
        and             Q0.1
        and             Q0.2
        out             M0.6

        ld              Q0.7                            ;rung 5
                        ;Mreg,Mbit,J,K,Q
        jkff_f          M0.3,M0.6,Q0.0,Q0.3

        ld              Q0.7                            ;rung 6
                        ;Mreg,Mbit,J,K,Q
        jkff_f          M0.2,M0.5,M0.5,Q0.2

        ld              Q0.7                            ;rung 7
                        ;Mreg,Mbit,J,K,Q
        jkff_f          M0.1,M0.4,Q0.0,Q0.1

        ld              Q0.7                            ;rung 8
                        ;Mreg,Mbit,J,K,Q
        jkff_f          M0.0,LOGIC1,LOGIC1,Q0.0
;
;--- PLC codes to be allocated in the "user_program_1" macro end here ----------
        endm
```

FIGURE 4.55   The user program of Example 4.9 (synchronous decade counter).

In the third and final step, we come up with one function for each J and K flip-flop input of four flip-flops by using the above functions together with the mode control input M, in such a way that when M = 1, the 4-bit synchronous up/down counter will count up and when M = 0, the 4-bit synchronous up/down counter will count down. As a result, we obtain the following four sets of flip-flop input functions.

For a 4-bit synchronous **up/down** counter:

$$J0 = K0 = 1, J1 = K1 = M.Q0 + \overline{M}.\overline{Q0}, J2 = K2 = M.Q1.Q0 + \overline{M}.\overline{Q1}.\overline{Q0},$$

$$J3 = K3 = M.Q2.Q1.Q0 + \overline{M}.\overline{Q2}.\overline{Q1}.\overline{Q0}.$$

A 4-bit synchronous up/down counter designed by using falling edge–triggered JK flip-flops, based on the above flip-flop input functions, is shown in Figure 4.57. When M = 1, the 4-bit synchronous up/down counter counts sequence from 0000 to 1111 (from 0 to 15 in decimal). When M = 0, the 4-bit synchronous up/down counter counts sequence from 1111 to 0000 (from 15 to 0 in decimal). When the count direction is changed by altering the state of M from LOW to HIGH or vice versa, the count operation will continue in the other direction without destroying the current

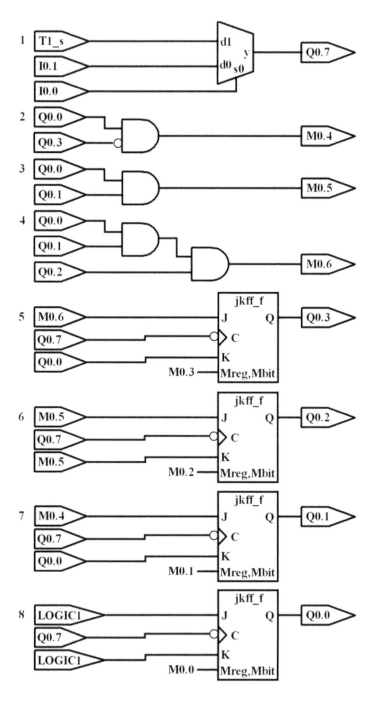

FIGURE 4.56 Schematic diagram of the user program of Example 4.9 (synchronous decade counter).

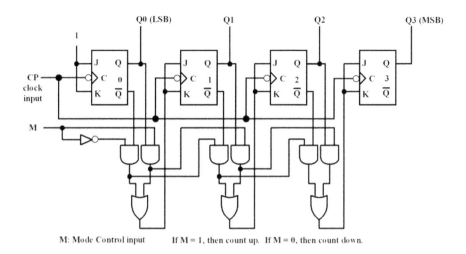

FIGURE 4.57 The logic diagram of the 4-bit synchronous up/down counter designed by using falling edge–triggered JK flip-flops.

count value at that instant. This is the main drawback of asynchronous up/down counters, as explained in Example 4.8.

Figure 4.58 shows the logic diagram of the 4-bit synchronous up/down counter implemented by using the PIC16F1847-Based PLC. JK flip-flops are implemented by using "jkff_f" macros. 2-input AND gates are implemented by using "and" macros and 2-input OR gates are implemented by using "or" macros. The CP clock signal can be applied through I0.1 or it can be fixed by using a T_1s reference timing signal. The choice between these two clock sources is made by using a 2-to-1 multiplexer "Mux 2x1". When the select input I0.0 of "Mux 2x1" is 0 (or 1, respectively), the CP clock input is I0.1 (or T_1s, respectively).

The user program of Example 4.10 (synchronous decade counter) is shown in Figure 4.59. The schematic diagram of the user program of Example 4.10 is depicted in Figure 4.60. When the project file of the PIC16F1847-Based PLC is open in the

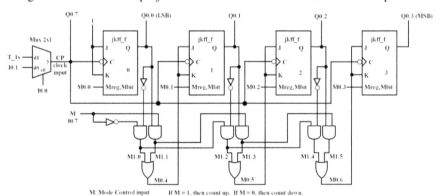

FIGURE 4.58 The logic diagram of the 4-bit synchronous up/down counter implemented by using the PIC16F1847-Based PLC.

```
user_program_1  macro
;--- PLC codes to be allocated in the "user_program_1" macro start from here ---
;__Example_Bsc_4.10
;_____ 4-bit synchronous up/down counter _____
                        ;s0,i1,i0,rego,bito              ;rung 1
        mux_2_1         I0.0,T_1s,I0.1,Q0.7

        ld_not          I0.7                             ;rung 2
        and_not         Q0.0
        out             M1.0

        ld              I0.7                             ;rung 3
        and             Q0.0
        out             M1.1

        ld              M1.0                             ;rung 4
        or              M1.1
        out             M0.4

        ld              M1.0                             ;rung 5
        and_not         Q0.1
        out             M1.2

        ld              M1.1                             ;rung 6
        and             Q0.1
        out             M1.3

        ld              M1.2                             ;rung 7
        or              M1.3
        out             M0.5

        ld              M1.2                             ;rung 8
        and_not         Q0.2
        out             M1.4

        ld              M1.3                             ;rung 9
        and             Q0.2
        out             M1.5

        ld              M1.4                             ;rung 10
        or              M1.5
        out             M0.6

        ld              Q0.7                             ;rung 11
                        ;Mreg,Mbit,J,K,Q
        jkff_f          M0.0,LOGIC1,LOGIC1,Q0.0

        ld              Q0.7                             ;rung 12
        jkff_f          M0.1,M0.4,M0.4,Q0.1

        ld              Q0.7                             ;rung 13
        jkff_f          M0.2,M0.5,M0.5,Q0.2

        ld              Q0.7                             ;rung 14
        jkff_f          M0.3,M0.6,M0.6,Q0.3
;--- PLC codes to be allocated in the "user_program_1" macro end here ----------
        endm
```

FIGURE 4.59 The user program of Example 4.10 (4-bit synchronous up/down counter).

MPLAB X IDE, from the file "PICPLC_PIC16F1847_user_Bsc.inc", if you uncomment Example 4.10 and run the project by pressing the symbol ▷ from the toolbar, then the PIC16F1847 microcontroller within the CPU board of the PIC16F1847-Based PLC will be programmed. After loading the program file to the PIC16F1847 microcontroller, switch the 4PDT to "RUN" and the power switch to the "ON" position. Next you can test the operation of this example.

4.15.11 EXAMPLE 4.11: 4-BIT SERIAL-IN, PARALLEL-OUT SHIFT RIGHT REGISTER

Example 4.11 explains the implementation of a 4-bit serial-in, parallel-out (SIPO) shift right register by using the PIC16F1847-Based PLC. Shift registers consist of

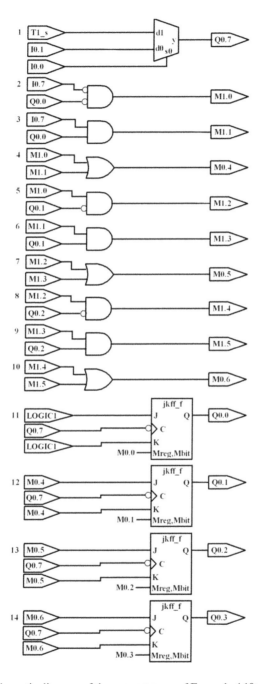

FIGURE 4.60 Schematic diagram of the user program of Example 4.10 (4-bit synchronous up/down counter).

flip-flops and are important in applications involving the storage of data in a digital system. A register, unlike a counter, has no specified sequence of states, except in certain specialised applications. A register, in general, is used solely for storing and shifting data (1s and 0s) entered into it from an external source.

The logic diagram of the 4-bit SIPO shift right register designed by using rising edge–triggered D flip-flops is shown in Figure 4.61. Table 4.32 depicts the function table of the 4-bit SIPO shift right register shown in Figure 4.61. A sample timing diagram and a sample output table for the 4-bit SIPO shift right register are also provided in Figure 4.62 and Table 4.33, respectively.

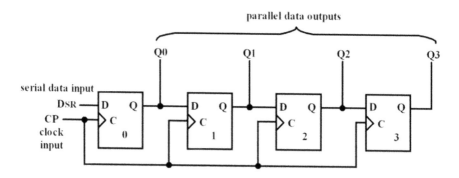

FIGURE 4.61   The logic diagram of the 4-bit SIPO shift right register designed by using rising edge–triggered D flip-flops.

TABLE 4.32
The Function Table of the 4-Bit SIPO Shift Right Register Shown in Figure 4.61

	Inputs		Parallel Data Outputs			
Operating Mode	DSR	CP	Q0	Q1	Q2	Q3
Serial Shift Right	l	↑	L	q_0	q_1	q_2
	h	↑	H	q_0	q_1	q_2

DSR: Serial Data (Shift Right) Input
CP: Clock Input (Active HIGH Going Edge)
Q0, Q1, Q2, Q3: Parallel Data Outputs

L = LOW Voltage Level, H = HIGH Voltage Level, × = Don't Care
l = LOW voltage level one set-up time prior to the LOW to HIGH clock transition
h = HIGH voltage level one set-up time prior to the LOW to HIGH clock transition
q_n = Lower case letters indicate the state of the referenced output one set-up time prior to the LOW to HIGH clock transition
↑ = LOW to HIGH clock transition

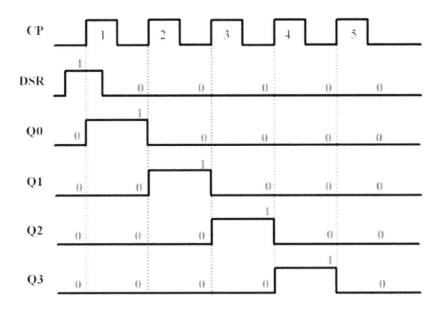

FIGURE 4.62   Sample timing diagram for the 4-bit SIPO shift right register shown in Figure 4.61.

TABLE 4.33

Sample Output Table for the 4-Bit SIPO Shift Right Register Shown in Figure 4.61

Inputs		Parallel Data Outputs			
DSR	CP	Q0 (Q0.0)	Q1 (Q0.1)	Q2 (Q0.2)	Q3 (Q0.3)
1	1st ↑	1	0	0	0
0	2nd ↑	0	1	0	0
0	3rd ↑	0	0	1	0
0	4th ↑	0	0	0	1
0	5th ↑	0	0	0	0

Figure 4.63 shows the logic diagram of the 4-bit SIPO shift right register imple-mented by using the PIC16F1847-Based PLC. D flip-flops are implemented by using "dff_r" macros. The CP clock signal can be applied through I0.1 or it can be fixed by using a T_1s reference timing signal. The choice between these two clock sources is made by using a 2-to-1 multiplexer "Mux 2x1". When the select input I0.0 of "Mux 2x1" is 0 (or 1, respectively), the CP clock input is I0.1 (or T_1s, respectively).

The user program of Example 4.11.1 (4-bit SIPO shift right register) is shown in Figure 4.64. The schematic diagram of the user program of Example 4.11.1 is depicted in Figure 4.65. When the project file of the PIC16F1847-Based PLC is

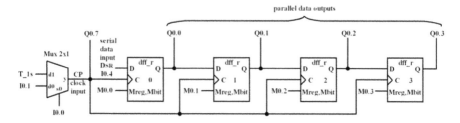

FIGURE 4.63   The logic diagram of the 4-bit SIPO shift right register implemented by using the PIC16F1847-Based PLC.

```
user_program_1  macro
;--- PLC codes to be allocated in the "user_program_1" macro start from here ---
;__Example_Bsc_4.11.1
;
;_____ 4-bit serial-input-parallel-output (SIPO) shift right register _____
;_____ This implementation suffers from the avalanche effect problem. _____
;
                    ;s0,i1,i0,rego,bito          ;rung 1
        mux_2_1     I0.0,T_1s,I0.1,Q0.7

        ld          Q0.7                         ;rung 2
                    ;Mreg,Mbit,D,Q
        dff_r       M0.0,I0.4,Q0.0

        ld          Q0.7                         ;rung 3
                    ;Mreg,Mbit,D,Q
        dff_r       M0.1,Q0.0,Q0.1

        ld          Q0.7                         ;rung 4
                    ;Mreg,Mbit,D,Q
        dff_r       M0.2,Q0.1,Q0.2

        ld          Q0.7                         ;rung 5
                    ;Mreg,Mbit,D,Q
        dff_r       M0.3,Q0.2,Q0.3
;
;--- PLC codes to be allocated in the "user_program_1" macro end here ----------
        endm
```

FIGURE 4.64   The user program of Example 4.11.1 (4-bit SIPO shift right register).

open in the MPLAB X IDE, from the file "PICPLC_PIC16F1847_user_Bsc.inc", if you uncomment Example 4.11.1 and run the project by pressing the symbol ▷ from the toolbar, then the PIC16F1847 microcontroller within the CPU board of the PIC16F1847-Based PLC will be programmed. After loading the program file to the PIC16F1847 microcontroller, switch the 4PDT to "RUN" and the power switch to the "ON" position. Next you can test the operation of this example. When we test Example 4.11.1, while DSR = 1 and CP = ↑, we face with the fact that all outputs are as shown below:

Inputs		Parallel Data Outputs			
DSR	CP	Q0	Q1	Q2	Q3
1	1st ↑	1	1	1	1

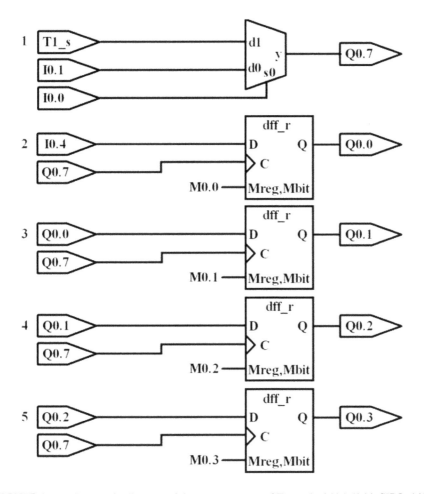

FIGURE 4.65 Schematic diagram of the user program of Example 4.11.1 (4-bit SIPO shift right register).

This is not the result we would like to obtain from the 4-bit SIPO shift right register when we compare it with the results provided in Table 4.33. This result is due to a problem called *avalanche effect*. Now, firstly, let us see how we ended up with such an outcome. Secondly, we consider a solution to this problem in this example.

In the user program of Example 4.11.1, shown in Figure 4.64, rungs from 2 to 5 implement flip-flops from 0 to 3, respectively. Therefore, flip-flop outputs Q0.0, Q0.1, Q0.2, and Q0.3 are evaluated sequentially from rungs 2 to 5, respectively, one by one. The evaluation results of their states either as 0 or as 1 are stored in SRAM immediately upon being evaluated. In the beginning, if we have DSR = 1 and CP = ↑, then the following operations take place in one PLC scan time: in rung 2, since D = DSR (I0.4) = 1 and CP = ↑, Q0.0 = 1. Next, rung 3 is handled by the PLC: since D = Q0.0 = 1 and CP = ↑, Q0.1 = 1. Next, rung 4 is handled by the PLC: since D = Q0.1 = 1 and CP = ↑, Q0.2 = 1. Finally, rung 5 is handled by the PLC: since D

```
user_program_1  macro
;--- PLC codes to be allocated in the "user_program_1" macro start from here ---
;__Example_Bsc_4.11.2
;
;_____ 4-bit serial-input-parallel-output (SIPO) shift right register _____
;
                        ;s0,i1,i0,rego,bito            ;rung 1
          mux_2_1       I0.0,T_1s,I0.1,Q0.7

          ld            Q0.7                           ;rung 2
                        ;Mreg,Mbit,D,Q
          dff_r         M0.3,Q0.2,Q0.3

          ld            Q0.7                           ;rung 3
                        ;Mreg,Mbit,D,Q
          dff_r         M0.2,Q0.1,Q0.2

          ld            Q0.7                           ;rung 4
                        ;Mreg,Mbit,D,Q
          dff_r         M0.1,Q0.0,Q0.1

          ld            Q0.7                           ;rung 5
                        ;Mreg,Mbit,D,Q
          dff_r         M0.0,I0.4,Q0.0
;
;--- PLC codes to be allocated in the "user_program_1" macro end here ----------
          endm
```

FIGURE 4.66   The user program of Example 4.11.2 (4-bit SIPO shift right register).

= $Q0.2 = 1$ and CP = ↑, $Q0.3 = 1$. It can easily be seen that this result is due to the sequential scanning of the user program rungs from top to bottom. To avoid this problem in this case, we can reorder the rungs and solve the problem, as can be seen from the user program of Example 4.11.2 (4-bit SIPO shift right register) shown in Figure 4.66. The schematic diagram of the user program of Example 4.11.2 is depicted in Figure 4.67. In this case, rungs from 2 to 5 implement flip-flops from 3 to 0, respectively. Therefore, flip-flop outputs Q0.3, Q0.2, Q0.1, and Q0.0 are evaluated sequentially from rungs 2 to 5, respectively, one by one. This new order of rungs solves the avalanche effect problem in this example. However, this is not the unique solution for all avalanche effect problems we may face in a PLC program. In some examples, as seen in Example 4.21, reordering the PLC rungs is not enough to tackle this problem.

When the project file of the PIC16F1847-Based PLC is open in the MPLAB X IDE, from the file "PICPLC_PIC16F1847_user_Bsc.inc", if you uncomment Example 4.11.2 and run the project by pressing the symbol ▷ from the toolbar, then the PIC16F1847 microcontroller within the CPU board of the PIC16F1847-Based PLC will be programmed. After loading the program file to the PIC16F1847 microcontroller, switch the 4PDT to "RUN" and the power switch to the "ON" position. Next you can test the operation of this example. It can be seen that this time, the 4-bit SIPO shift right register will function as expected.

4.15.12   Example 4.12: 4-Bit Serial-in, Serial-out Shift Right Register

Example 4.12 explains the implementation of a 4-bit serial-in, serial-out (SISO) shift right register by using the PIC16F1847-Based PLC. The logic diagram of the 4-bit SISO shift right register designed by using rising edge–triggered D flip-flops

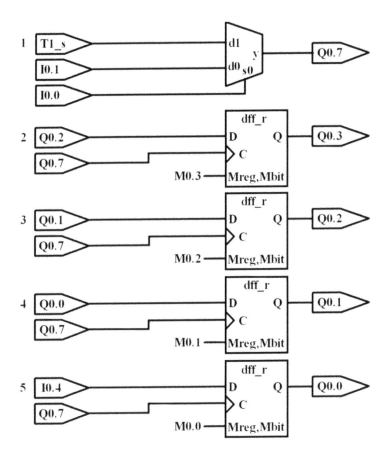

FIGURE 4.67   Schematic diagram of the user program of Example 4.11.2.

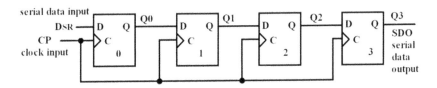

FIGURE 4.68   The logic diagram of the 4-bit SISO shift right register designed by using rising edge–triggered D flip-flops.

is shown in Figure 4.68. Table 4.34 depicts the function table of the 4-bit SISO shift right register shown in Figure 4.68. The structure of this register is identical to that of the 4-bit SIPO shift right register considered in the previous section. The only difference here is that Qn outputs, i.e., outputs Q0, Q1, and Q2, are internal to the register and cannot be observed from outside of the register directly. In this register, only serial data output (SDO) Q3 can be observed from outside of the register.

TABLE 4.34
The Function Table of the 4-Bit SISO Shift Right Register Shown in Figure 4.68

Operating Mode	Inputs		Qn Outputs			SDO
	DSR	CP	Q0	Q1	Q2	Q3
Serial Shift Right	l	↑	L	q_0	q_1	q_2
	h	↑	H	q_0	q_1	q_2

DSR: Serial Data (Shift Right) Input
CP: Clock Input (Active HIGH Going Edge)
Q0, Q1, Q2: Qn Outputs
Q3: Serial Data Output

L = LOW Voltage Level, H = HIGH Voltage Level, × = Don't Care
l = LOW voltage level one set-up time prior to the LOW to HIGH clock transition
h = HIGH voltage level one set-up time prior to the LOW to HIGH clock transition
q_n = Lower case letters indicate the state of the referenced output one set-up time prior to the LOW to HIGH clock transition
↑ = LOW to HIGH clock transition

Figure 4.69 shows the logic diagram of the 4-bit SISO shift right register implemented by using the PIC16F1847-Based PLC. D flip-flops are implemented by using "dff_r" macros. The CP clock signal can be applied through I0.1 or it can be fixed by using a T_1s reference timing signal. The choice between these two clock sources is made by using a 2-to-1 multiplexer "Mux 2x1". When the select input I0.0 of "Mux 2x1" is 0 (or 1, respectively), the CP clock input is I0.1 (or T_1s, respectively).

The user program of Example 4.12 (4-bit SISO shift right register) is shown in Figure 4.70. The schematic diagram of the user program of Example 4.12 is depicted in Figure 4.71. When the project file of the PIC16F1847-Based PLC is open in the MPLAB X IDE, from the file "PICPLC_PIC16F1847_user_Bsc.inc", if you uncomment Example 4.12 and run the project by pressing the symbol ▷ from the toolbar, then the PIC16F1847 microcontroller within the CPU board of the PIC16F1847-Based PLC will be programmed. After loading the program file to the PIC16F1847

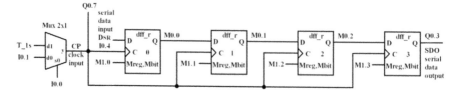

FIGURE 4.69   The logic diagram of the 4-bit SISO shift right register implemented by using the PIC16F1847-Based PLC.

```
user_program_1  macro
;--- PLC codes to be allocated in the "user_program_1" macro start from here ---
;__Example_Bsc_4.12

;_____ 4-bit serial-input- serial-output (SISO) shift right register _____
;
                            ;s0,i1,i0,rego,bito           ;rung 1
        mux_2_1             I0.0,T_1s,I0.1,Q0.7

        ld                  Q0.7                          ;rung 2
                            ;Mreg,Mbit,D,Q
        dff_r               M1.3,M0.2,Q0.3

        ld                  Q0.7                          ;rung 3
                            ;Mreg,Mbit,D,Q
        dff_r               M1.2,M0.1,M0.2

        ld                  Q0.7                          ;rung 4
                            ;Mreg,Mbit,D,Q
        dff_r               M1.1,M0.0,M0.1

        ld                  Q0.7                          ;rung 5
                            ;Mreg,Mbit,D,Q
        dff_r               M1.0,I0.4,M0.0
;
;--- PLC codes to be allocated in the "user_program_1" macro end here ----------
        endm
```

FIGURE 4.70   The user program of Example 4.12 (4-bit SISO shift right register).

microcontroller, switch the 4PDT to "RUN" and the power switch to the "ON" position. Next you can test the operation of this example.

4.15.13   EXAMPLE 4.13: 4-BIT SERIAL-IN, PARALLEL-OUT
SHIFT RIGHT OR SHIFT LEFT REGISTER

Example 4.13 explains the implementation of a 4-bit SIPO shift right or shift left
register by using the PIC16F1847-Based PLC. The logic diagram of the 4-bit SIPO
shift right or shift left register designed by using rising edge–triggered D flip-flops
is shown in Figure 4.72. Table 4.35 depicts the function table of the 4-bit SIPO shift
right or shift left register shown in Figure 4.72. The 4-bit SIPO shift right register
considered in Example 4.11 is capable of shifting the serial input data in one direction only. In this example, the shift operation is possible in both directions from left
to right or from right to left. In each case, the serial data is taken from a different
serial input data source. When the register shifts left (S = 0), the serial input data is
taken from the input DSL. Likewise, when the register shifts right (S = 1), the serial
input data is taken from the input DSR.

Figure 4.73 shows the logic diagram of the 4-bit SIPO shift right or shift left register implemented by using the PIC16F1847-Based PLC. D flip-flops are implemented
by using "dff_r" macros. 2-to-1 multiplexers are implemented by using "mux_2_1"
macros (see Chapter 4 of *Intermediate Concepts*). The CP clock signal can be applied
through I0.1 or it can be fixed by using a T_1s reference timing signal. The choice
between these two clock sources is made by using a 2-to-1 multiplexer "Mux 2x1".
When the select input I0.0 of "Mux 2x1" is 0 (or 1, respectively), the CP clock input
is I0.1 (or T_1s, respectively).

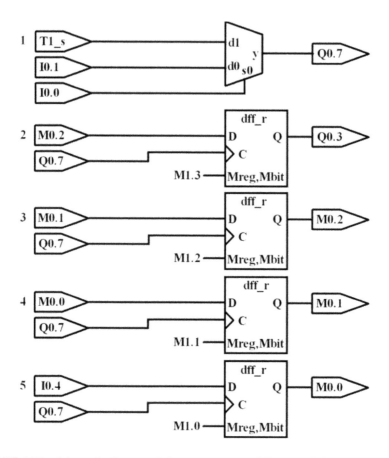

FIGURE 4.71 Schematic diagram of the user program of Example 4.12 (4-bit SISO shift right register).

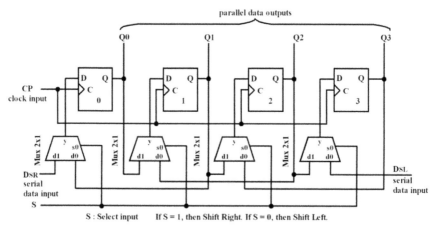

FIGURE 4.72 The logic diagram of the 4-bit SIPO shift right or shift left register designed by using rising edge–triggered D flip-flops.

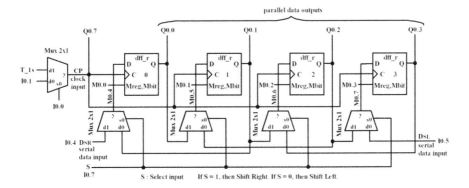

FIGURE 4.73 The logic diagram of the 4-bit SIPO shift right or shift left register implemented by using the PIC16F1847-Based PLC.

The user program of Example 4.13 (4-bit SIPO shift right or shift left register) is shown in Figure 4.74. The schematic diagram of the user program of Example 4.13 is depicted in Figure 4.75. When the project file of the PIC16F1847-Based PLC is open in the MPLAB X IDE, from the file "PICPLC_PIC16F1847_user _Bsc.inc", if you uncomment Example 4.13 and run the project by pressing the

```
user_program_1   macro
;--- PLC codes to be allocated in the "user_program_1" macro start from here ---
;__Example_Bsc_4.13
;
;___ 4-bit serial-input-parallel-output (SIPO) shift right or shift left register ___
;
                    ;s0,i1,i0,rego,bito          ;rung 1
        mux_2_1     I0.0,T_1s,I0.1,Q0.7

                    ;s0,i1,i0,rego,bito          ;rung 2
        mux_2_1     I0.7,I0.4,Q0.1,M0.4

                    ;s0,i1,i0,rego,bito          ;rung 3
        mux_2_1     I0.7,Q0.0,Q0.2,M0.5

                    ;s0,i1,i0,rego,bito          ;rung 4
        mux_2_1     I0.7,Q0.1,Q0.3,M0.6

                    ;s0,i1,i0,rego,bito          ;rung 5
        mux_2_1     I0.7,Q0.2,I0.5,M0.7

        ld          Q0.7                         ;rung 6
                    ;Mreg,Mbit,D,Q
        dff_r       M0.3,M0.7,Q0.3

        ld          Q0.7                         ;rung 7
                    ;Mreg,Mbit,D,Q
        dff_r       M0.2,M0.6,Q0.2

        ld          Q0.7                         ;rung 8
                    ;Mreg,Mbit,D,Q
        dff_r       M0.1,M0.5,Q0.1

        ld          Q0.7                         ;rung 9
                    ;Mreg,Mbit,D,Q
        dff_r       M0.0,M0.4,Q0.0
;
;--- PLC codes to be allocated in the "user_program_1" macro end here ----------
        endm
```

FIGURE 4.74 The user program of Example 4.13 (4-bit SIPO shift right or shift left register).

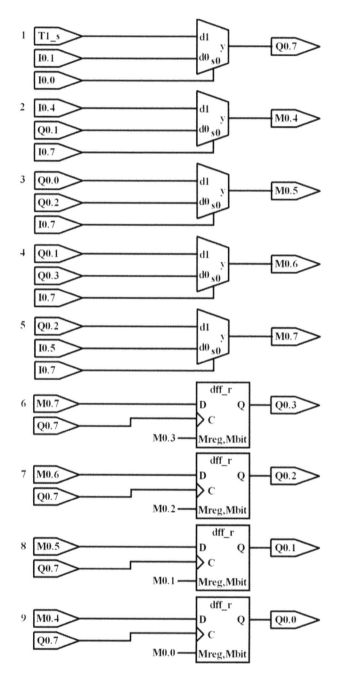

FIGURE 4.75 Schematic diagram of the user program of Example 4.13 (4-bit SIPO shift right or shift left register).

TABLE 4.35

The Function Table of the 4-Bit SIPO Shift Right or Shift Left Register Shown in Figure 4.72

		Inputs			Parallel Data Outputs			
Operating Mode	S	CP	DSR	DSL	Q0	Q1	Q2	Q3
Serial Shift Left	L	↑	×	l	q_1	q_2	q_3	L
	L	↑	×	h	q_1	q_2	q_3	H
Serial Shift Right	H	↑	l	×	L	q_0	q_1	q_2
	H	↑	h	×	H	q_0	q_1	q_2

S: Select Input

CP: Clock Input (Active HIGH Going Edge)

DSR: Serial Data (Shift Right) Input

DSL: Serial Data (Shift Left) Input

Q0, Q1, Q2, Q3: Parallel Data Outputs

L = LOW Voltage Level, H = HIGH Voltage Level, × = Don't Care

l = LOW voltage level one set-up time prior to the LOW to HIGH clock transition

h = HIGH voltage level one set-up time prior to the LOW to HIGH clock transition

q_n = Lower case letters indicate the state of the referenced output one set-up time prior to the LOW to HIGH clock transition

↑ = LOW to HIGH clock transition

symbol ▷ from the toolbar, then the PIC16F1847 microcontroller within the CPU board of the PIC16F1847-Based PLC will be programmed. After loading the program file to the PIC16F1847 microcontroller, switch the 4PDT to "RUN" and the power switch to the "ON" position. Next you can test the operation of this example (Table 4.35).

4.15.14   EXAMPLE 4.14: 4-BIT PARALLEL-IN, SERIAL-OUT SHIFT RIGHT REGISTER

Example 4.14 explains the implementation of a 4-bit parallel-in, serial-out (PISO) shift right register by using the PIC16F1847-Based PLC. The logic diagram of the 4-bit PISO shift right register designed by using rising edge–triggered D flip-flops is shown in Figure 4.76. Table 4.36 depicts the function table of the 4-bit PISO shift right register shown in Figure 4.76. When the Shift /Load select input S/$\overline{L}$ = 0, the data in the parallel inputs P0, P1, P2, and P3 is loaded in the register (i.e., in the flip-flops Q0, Q1, Q2, and Q3, respectively) with the rising edge of the clock signal (parallel load). When the Shift /Load select input S/$\overline{L}$ = 1, the data stored in the register is shifted right with the rising edge of the clock signal (shift right). In this case the serial input is taken from the DSR input.

Figure 4.77 shows the logic diagram of the 4-bit PISO shift right register implemented by using the PIC16F1847-Based PLC. D flip-flops are implemented by using

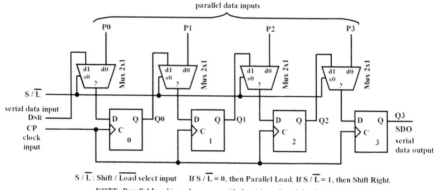

FIGURE 4.76   The logic diagram of the 4-bit PISO shift right register designed by using rising edge–triggered D flip-flops.

TABLE 4.36
The Function Table of the 4-Bit PISO Shift Right Register Shown in Figure 4.76

Operating Mode	Inputs							Qn Outputs			SDO
	$S/\bar{L}$	CP	DSR	P0	P1	P2	P3	Q0	Q1	Q2	Q3
Parallel	L	↑	×	1	1	1	1	L	L	L	L
Load	L	↑	×	1	1	1	h	L	L	L	H
	L	↑	×	.	.	.	.	.	.	.	.
	L	↑	×	h	h	h	h	H	H	H	H
Serial Shift	H	↑	1	×	×	×	×	L	q_0	q_1	q_2
	H	↑	h	×	×	×	×	H	q_0	q_1	q_2

$S/\bar{L}$: Shift / Load Select Input
CP: Clock Input (Active HIGH Going Edge)
DSR: Serial Data (Shift Right) Input
P0, P1, P2, P3: Parallel Data Inputs
Q0, Q1, Q2: Qn Outputs
Q3 (SDO): Serial Data Output from the Last Stage
L = LOW Voltage Level, H = HIGH Voltage Level, × = Don't Care
l = LOW voltage level one set-up time prior to the LOW to HIGH clock transition
h = HIGH voltage level one set-up time prior to the LOW to HIGH clock transition
q_n = Lower case letters indicate the state of the referenced output one set-up time prior to the LOW to HIGH clock transition
↑ = LOW to HIGH clock transition

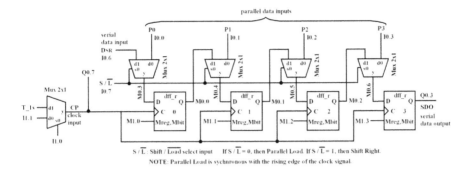

FIGURE 4.77   The logic diagram of the 4-bit PISO shift right register implemented by using the PIC16F1847-Based PLC.

"dff_r" macros. 2-to-1 multiplexers are implemented by using "mux_2_1" macros (see Chapter 4 of *Intermediate Concepts*). The CP clock signal can be applied through I0.1 or it can be fixed by using a T_1s reference timing signal. The choice between these two clock sources is made by using a 2-to-1 multiplexer "Mux 2x1". When the select input I0.0 of "Mux 2x1" is 0 (or 1, respectively), the CP clock input is I0.1 (or T_1s, respectively).

The user program of Example 4.14 (4-bit PISO shift right register) is shown in Figure 4.78. The schematic diagram of the user program of Example 4.14 is depicted in Figure 4.79. When the project file of the PIC16F1847-Based PLC is open in the MPLAB X IDE, from the file "PICPLC_PIC16F1847_user_Bsc.inc", if you uncomment Example 4.14 and run the project by pressing the symbol ▷ from the toolbar, then the PIC16F1847 microcontroller within the CPU board of the PIC16F1847-Based PLC will be programmed. After loading the program file to the PIC16F1847 microcontroller, switch the 4PDT to "RUN" and the power switch to the "ON" position. Next you can test the operation of this example.

4.15.15   Example 4.15: 4-Bit Parallel-in, Parallel-out Register

Example 4.15 explains the implementation of a 4-bit parallel-in, parallel-out (PIPO) register by using the PIC16F1847-Based PLC. The logic diagram of the 4-bit PIPO register designed by using rising edge–triggered D flip-flops is shown in Figure 4.80. Table 4.37 depicts the function table of the 4-bit PIPO register shown in Figure 4.80. The data in the parallel inputs P0, P1, P2, and P3 are loaded in the register (i.e., in the flip-flops Q0, Q1, Q2, and Q3, respectively) with the rising edge of the clock signal (parallel load).

Figure 4.81 shows the logic diagram of the 4-bit PIPO register implemented by using the PIC16F1847-Based PLC. D flip-flops are implemented by using "dff_r" macros. The CP clock signal can be applied through I0.1 or it can be fixed by using a T_1s reference timing signal. The choice between these two clock sources is made by using a 2-to-1 multiplexer "Mux 2x1". When the select input I0.0 of "Mux 2x1" is 0 (or 1, respectively), the CP clock input is I0.1 (or T_1s, respectively).

The user program of Example 4.15 (4-bit PIPO register) is shown in Figure 4.82. The schematic diagram of the user program of Example 4.15 is depicted in Figure

```
user_program_1  macro
;--- PLC codes to be allocated in the "user_program_1" macro start from here ---
;__Example_Bsc_4.14
;
;___ 4-bit parallel-in-serial-out (PISO) shift right register _____

                    ;s0,i1,i0,rego,bito              ;rung 1
      mux_2_1       I1.0,T_1s,I1.1,Q0.7

                    ;s0,i1,i0,rego,bito              ;rung 2
      mux_2_1       I0.7,I0.6,I0.0,M0.3

                    ;s0,i1,i0,rego,bito              ;rung 3
      mux_2_1       I0.7,M0.0,I0.1,M0.4

                    ;s0,i1,i0,rego,bito              ;rung 4
      mux_2_1       I0.7,M0.1,I0.2,M0.5

                    ;s0,i1,i0,rego,bito              ;rung 5
      mux_2_1       I0.7,M0.2,I0.3,M0.6

      ld            Q0.7                             ;rung 6
                    ;Mreg,Mbit,D,Q
      dff_r         M1.3,M0.6,Q0.3

      ld            Q0.7                             ;rung 7
                    ;Mreg,Mbit,D,Q
      dff_r         M1.2,M0.5,M0.2

      ld            Q0.7                             ;rung 8
                    ;Mreg,Mbit,D,Q
      dff_r         M1.1,M0.4,M0.1

      ld            Q0.7                             ;rung 9
                    ;Mreg,Mbit,D,Q
      dff_r         M1.0,M0.3,M0.0
;
;--- PLC codes to be allocated in the "user_program_1" macro end here ----------
      endm
```

FIGURE 4.78   The user program of Example 4.14 (4-bit PISO shift right register).

4.83. When the project file of the PIC16F1847-Based PLC is open in the MPLAB X IDE, from the file "PICPLC_PIC16F1847_user_Bsc.inc", if you uncomment Example 4.15 and run the project by pressing the symbol ▷ from the toolbar, then the PIC16F1847 microcontroller within the CPU board of the PIC16F1847-Based PLC will be programmed. After loading the program file to the PIC16F1847 microcontroller, switch the 4PDT to "RUN" and the power switch to the "ON" position. Next you can test the operation of this example.

4.15.16   EXAMPLE 4.16: 74164 8-BIT SERIAL-IN, PARALLEL-OUT SHIFT REGISTER

Example 4.16 explains the implementation of the 74164 8-bit serial-in, parallel-out (SIPO) register by using the PIC16F1847-Based PLC. The logic diagram of the 74164 8-bit SIPO register designed by using rising edge–triggered D flip-flops is shown in Figure 4.84. Table 4.38 depicts the function table of the 74164 8-bit SIPO register shown in Figure 4.84. The 74164 is an edge-triggered 8-bit shift register with serial data entry and an output from each of the eight stages. Data is entered serially through one of two inputs (A or B); either of these inputs can be used as an active HIGH enable for data entry through the other input. An unused input must be tied HIGH, or both inputs connected together. Each LOW–HIGH transition on the clock

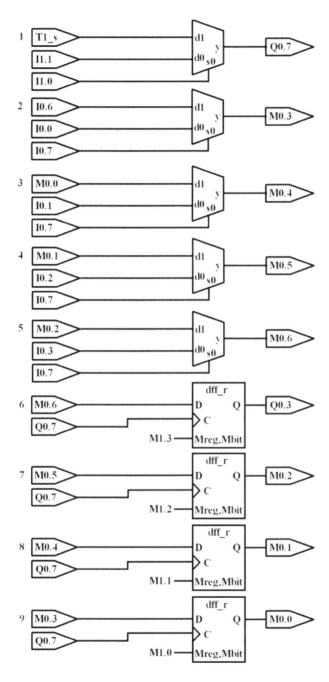

FIGURE 4.79 Schematic diagram of the user program of Example 4.14 (4-bit PISO shift right register).

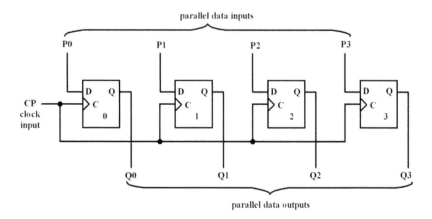

parallel data inputs

FIGURE 4.80 The logic diagram of the 4-bit PIPO register designed by using rising edge–triggered D flip-flops.

TABLE 4.37

The Function Table of the 4-Bit PIPO Register Shown in Figure 4.80

Operating Mode	Inputs					Parallel Data Outputs			
	CP	P0	P1	P2	P3	Q0	Q1	Q2	Q3
Parallel Load	↑	l	l	l	l	L	L	L	L
	↑	l	l	l	h	L	L	L	H
	↑	l	l	h	l	L	L	H	L
	↑	l	l	h	h	L	L	H	H
	↑	l	h	l	l	L	H	L	L
	↑	l	h	l	h	L	H	L	H
	↑	l	h	h	l	L	H	H	L
	↑	l	h	h	h	L	H	H	H
	↑	h	l	l	l	H	L	L	L
	↑	h	l	l	h	H	L	L	H
	↑	h	l	h	l	H	L	H	L
	↑	h	l	h	h	H	L	H	H
	↑	h	h	l	l	H	H	L	L
	↑	h	h	l	h	H	H	L	H
	↑	h	h	h	l	H	H	H	L
	↑	h	h	h	h	H	H	H	H

CP: Clock Input (Active HIGH Going Edge)

P0, P1, P2, P3: Parallel Data Inputs

Q0, Q1, Q2, Q3: Parallel Data Outputs

L = LOW Voltage Level

H = HIGH Voltage Level

l = LOW voltage level one set-up time prior to the LOW to HIGH clock transition

h = HIGH voltage level one set-up time prior to the LOW to HIGH clock transition

↑ = LOW to HIGH clock transition

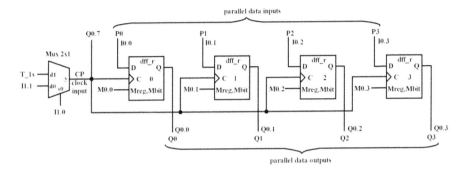

FIGURE 4.81 The logic diagram of the 4-bit PIPO register implemented by using the PIC16F1847-Based PLC.

```
user_program_1  macro
;--- PLC codes to be allocated in the "user_program_1" macro start from here ---
;__Example_Bsc_4.15
;
;____ 4-bit parallel-input-parallel-output (PIPO) register _____

                       ;s0,i1,i0,rego,bito       ;rung 1
        mux_2_1        I1.0,T_1s,I1.1,Q0.7

        ld             Q0.7                      ;rung 2
                       ;Mreg,Mbit,D,Q
        dff_r          M0.3,I0.3,Q0.3

        ld             Q0.7                      ;rung 3
                       ;Mreg,Mbit,D,Q
        dff_r          M0.2,I0.2,Q0.2

        ld             Q0.7                      ;rung 4
                       ;Mreg,Mbit,D,Q
        dff_r          M0.1,I0.1,Q0.1

        ld             Q0.7                      ;rung 5
                       ;Mreg,Mbit,D,Q
        dff_r          M0.0,I0.0,Q0.0
;
;--- PLC codes to be allocated in the "user_program_1" macro end here ----------
        endm
```

FIGURE 4.82 The user program of Example 4.15 (4-bit PIPO register).

(CP) input shifts data one place to the right and enters into Q0 the logical AND of the two data inputs (A•B) that existed before the rising clock edge. A LOW level on the Master Reset $(\overline{MR})$ input overrides all other inputs and clears the register asynchronously, forcing all Q outputs LOW.

Figure 4.85 shows the logic diagram of the 74164 8-bit SIPO register implemented by using the PIC16F1847-Based PLC. D flip-flops are implemented by using "dff_r_ SR" macros. The 2-input AND gate is implemented by using the "and" macro. The CP clock signal can be applied through I0.1 or it can be fixed by using a T_1s reference timing signal. The choice between these two clock sources is made by using a 2-to-1 multiplexer "Mux 2x1". When the select input I0.0 of "Mux 2x1" is 0 (or 1, respectively), the CP clock input is I0.1 (or T_1s, respectively).

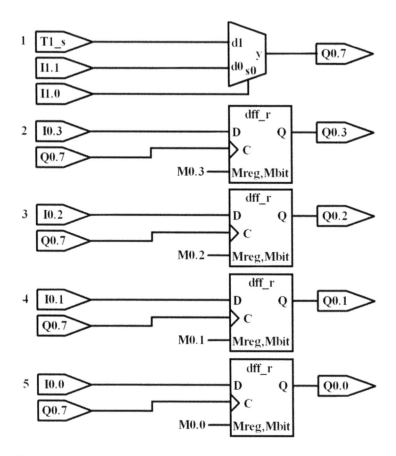

FIGURE 4.83 Schematic diagram of the user program of Example 4.15 (4-bit PIPO register).

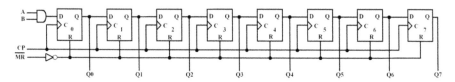

FIGURE 4.84 The logic diagram of the 74164 8-bit SIPO shift register designed by using rising edge–triggered D flip-flops.

The user program of Example 4.16 (74164 8-bit SIPO register) is shown in Figure 4.86. The schematic diagram of the user program of Example 4.16 is depicted in Figure 4.87. When the project file of the PIC16F1847-Based PLC is open in the MPLAB X IDE, from the file "PICPLC_PIC16F1847_user_Bsc.inc", if you uncomment Example 4.16 and run the project by pressing the symbol ▷ from the toolbar, then the PIC16F1847 microcontroller within the CPU board of the PIC16F1847-Based PLC will be programmed. After loading the program file to the PIC16F1847 microcontroller, switch the 4PDT to "RUN" and the power switch to the "ON" position. Next you can test the operation of this example.

TABLE 4.38

The Function Table of the 74164 8-Bit SIPO Shift Register Shown in Figure 4.84

	Inputs				Parallel Data Outputs	
Operating Mode	$\overline{MR}$	CP	A	B	Q0	Q1, Q2, ..., Q7
Reset (Clear)	L	×	×	×	L	L, L, ..., L
Shift	H	↑	l	l	L	$q_0, q_1, ..., q_6$
	H	↑	l	h	L	$q_0, q_1, ..., q_6$
	H	↑	h	l	L	$q_0, q_1, ..., q_6$
	H	↑	h	h	H	$q_0, q_1, ..., q_6$

$\overline{MR}$: Asynchronous Master Reset Input (Active LOW)

CP: Clock Input (Active HIGH Going Edge)

A, B: Data Inputs

Q0, Q1, Q2, Q3, Q4, Q5, Q6, Q7: Parallel Data Outputs

L = LOW Voltage Level

H = HIGH Voltage Level

× = Don't Care

l = LOW voltage level one set-up time prior to the LOW to HIGH clock transition

h = HIGH voltage level one set-up time prior to the LOW to HIGH clock transition

q_n = Lower case letters indicate the state of the referenced output one set-up time prior to the LOW to HIGH clock transition

↑ = LOW to HIGH clock transition

4.15.17   EXAMPLE 4.17: 74165 8-BIT PARALLEL-IN, SERIAL-OUT SHIFT REGISTER

Example 4.17 explains the implementation of the 74165 8-bit parallel-in, serial-out (PISO) register by using the PIC16F1847-Based PLC. The logic diagram of the 74165 8-bit PISO register designed by using rising edge–triggered D flip-flops is shown in Figure 4.88. Table 4.39 depicts the function table of the 74165 8-bit PISO register shown in Figure 4.88. The 74165 is an 8-bit parallel-load or serial-in shift register with complementary serial outputs (Q7 and $\overline{Q7}$) available from the last stage. When parallel load $(\overline{PL})$ input is LOW, parallel data from the P0 to P7 inputs are loaded into the register **asynchronously**. When $(\overline{PL})$ is HIGH, data enters the register serially at the DS input and shifts one place to the right (Q0 $\longrightarrow$ Q1 $\longrightarrow$ Q2, etc.) with each positive-going clock transition. This feature allows parallel-to-serial converter expansion by tying the Q7 output to the DS input of the succeeding stage. The clock input is a gated OR structure which allows one input to be used as an active LOW clock enable $(\overline{CE})$ input. The LOW–HIGH transition of input $\overline{CE}$ should only take place while CP is HIGH for predictable operation. Either the CP or the $\overline{CE}$ should be HIGH before the LOW–HIGH transition of $\overline{PL}$ to prevent shifting the data when $\overline{PL}$ is activated.

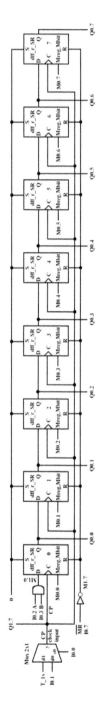

FIGURE 4.85   The logic diagram of the 74164 8-bit SIPO shift register implemented by using the PIC16F1847-Based PLC.

```
user_program_1   macro
;--- PLC codes to be allocated in the "user_program_1" macro start from here ---
;__Example_Bsc_4.16
;
;____ 74164 8-bit serial-input-parallel-output (PIPO) shift register _____

                        ;s0,i1,i0,rego,bito              ;rung 1
          mux_2_1       I0.0,T_1s,I0.1,Q1.7

          ld            I0.2                             ;rung 2
          and           I0.3
          out           M1.0

          inv_out       I0.7,M1.7                        ;rung 3

          ld            Q1.7                             ;rung 4
                        ;Mreg,Mbit,D,Q,S,R
          dff_r_SR      M0.7,Q0.6,Q0.7,LOGIC0,M1.7

          ld            Q1.7                             ;rung 5
                        ;Mreg,Mbit,D,Q,S,R
          dff_r_SR      M0.6,Q0.5,Q0.6,LOGIC0,M1.7

          ld            Q1.7                             ;rung 6
                        ;Mreg,Mbit,D,Q,S,R
          dff_r_SR      M0.5,Q0.4,Q0.5,LOGIC0,M1.7

          ld            Q1.7                             ;rung 7
                        ;Mreg,Mbit,D,Q,S,R
          dff_r_SR      M0.4,Q0.3,Q0.4,LOGIC0,M1.7

          ld            Q1.7                             ;rung 8
                        ;Mreg,Mbit,D,Q,S,R
          dff_r_SR      M0.3,Q0.2,Q0.3,LOGIC0,M1.7

          ld            Q1.7                             ;rung 9
                        ;Mreg,Mbit,D,Q,S,R
          dff_r_SR      M0.2,Q0.1,Q0.2,LOGIC0,M1.7

          ld            Q1.7                             ;rung 10
                        ;Mreg,Mbit,D,Q,S,R
          dff_r_SR      M0.1,Q0.0,Q0.1,LOGIC0,M1.7

          ld            Q1.7                             ;rung 11
                        ;Mreg,Mbit,D,Q,S,R
          dff_r_SR      M0.0,M1.0,Q0.0,LOGIC0,M1.7
;
;--- PLC codes to be allocated in the "user_program_1" macro end here ----------
          endm
```

FIGURE 4.86   The user program of Example 4.16 (74164 8-bit SIPO shift register).

Figure 4.89 shows the logic diagram of the 74165 8-bit PISO register implemented by using the PIC16F1847-Based PLC. D flip-flops are implemented by using "dff_r_SR" macros. 2-input AND gates are implemented by using either the "and" macro or the "and_not" macro. The 3-input NAND gate is implemented by using the following macros: "ld_not", "and_not", "and", and "out_not". The CP clock signal can be applied through I1.1 or it can be fixed by using a T_1s reference timing signal. The choice between these two clock sources is made by using a 2-to-1 multiplexer "Mux 2x1". When the select input I1.0 of "Mux 2x1" is 0 (or 1, respectively), the CP clock input is I1.1 (or T_1s, respectively).

The user program of Example 4.17 (74165 8-bit PISO register) is shown in Figure 4.90. The schematic diagram of the user program of Example 4.17 is depicted in Figure 4.91. When the project file of the PIC16F1847-Based PLC is open in the MPLAB X IDE, from the file "PICPLC_PIC16F1847_user_Bsc.inc", if you uncomment

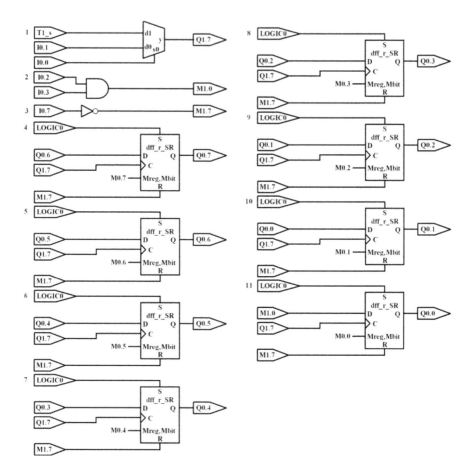

FIGURE 4.87   Schematic diagram of the user program of Example 4.16 (74164 8-bit SIPO shift register).

Example 4.17 and run the project by pressing the symbol ▷ from the toolbar, then the PIC16F1847 microcontroller within the CPU board of the PIC16F1847-Based PLC will be programmed. After loading the program file to the PIC16F1847 microcontroller, switch the 4PDT to "RUN" and the power switch to the "ON" position. Next you can test the operation of this example.

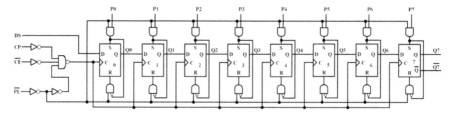

FIGURE 4.88   The logic diagram of the 74165 8-bit PISO shift register designed by using rising edge–triggered D flip-flops.

TABLE 4.39
The Function Table of the 74165 8-Bit PISO Shift Register Shown in Figure 4.88

Operating Mode	Inputs					Qn Outputs		Outputs	
	$\overline{PL}$	$\overline{CE}$	CP	DS	P0, P1, ..., P7	Q0	Q1, ..., Q6	Q7	$\overline{Q7}$
Parallel Load	L	×	×	×	l, l, l, l, l, l, l, l	L	L, ..., L	L	H
	L	×	×	×	l, l, l, l, l, l, l, h	L	L, ..., L	H	L
	L	×	×	×	., ., ., ., ., ., ., .	.	., ..., .	.	.
	L	×	×	×	h, h, h, h, h, h, h, h	H	H, ..., H	H	L
Serial Shift	H	L	↑	l	×	L	$q_0, ..., q_5$	q_6	$\overline{q_6}$
	H	L	↑	h	×	H	$q_0, ..., q_5$	q_6	$\overline{q_6}$
Hold	H	H	×	×	×	q_0	$q_1, ..., q_6$	q_7	$\overline{q_7}$

$\overline{PL}$: Asynchronous Parallel Load Input (Active LOW)

$\overline{CE}$: Clock Enable Input (Active LOW)

CP: Clock Input (Active HIGH Going Edge)

DS: Serial Data Input

P0, P1, P2, P3, P4, P5, P6, P7: Parallel Data Inputs

Q0, Q1, Q2, Q3, Q4, Q5, Q6: Qn Outputs

Q7: Serial Data Output from the Last Stage

$\overline{Q7}$: Complementary Output from the Last Stage

L = LOW Voltage Level, H = HIGH Voltage Level, × = Don't Care

l = LOW voltage level one set-up time prior to the LOW to HIGH clock transition

h = HIGH voltage level one set-up time prior to the LOW to HIGH clock transition

q_n = Lower case letters indicate the state of the referenced output one set-up time prior to the LOW to HIGH clock transition

↑ = LOW to HIGH clock transition

4.15.18   EXAMPLE 4.18: 74194 4-BIT BIDIRECTIONAL UNIVERSAL SHIFT REGISTER

Example 4.18 explains the implementation of the 74194 4-bit bidirectional universal shift register by using the PIC16F1847-Based PLC. The logic diagram of the 74194 4-bit bidirectional universal shift register designed by using rising edge–triggered D flip-flops is shown in Figure 4.92. Table 4.40 depicts the function table of the 74194 4-bit bidirectional universal shift register shown in Figure 4.92. When $\overline{MR} = 0$, regardless of all other inputs the register is cleared and therefore all outputs are forced to 0. When the master reset input is not active, i.e., when $\overline{MR} = 1$, S1 and S0

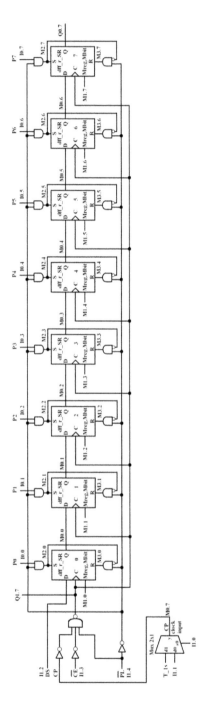

FIGURE 4.89   The logic diagram of the 74165 8-bit PISO shift register implemented by using the PIC16F1847-Based PLC.

```
user_program_1   macro
;--- PLC codes to be allocated in the "user_program_1" macro start from here ---
;__Example_Bsc_4.17
;___  74165 8-bit parallel-input-serial-output shift register _____
                       ;s0,i1,i0,rego,bito                 ;rung 1
      mux_2_1          I1.0,T_1s,I1.1,M0.7

      ld_not           M0.7                                ;rung 2
      and_not          I1.3
      and              I1.4
      out_not          Q1.7

      ld_not           I1.4                                ;rung 3
      and              I0.0
      out              M2.0

      ld_not           I1.4                                ;rung 4
      and              I0.1
      out              M2.1

      ld_not           I1.4                                ;rung 5
      and              I0.2
      out              M2.2

      ld_not           I1.4                                ;rung 6
      and              I0.3
      out              M2.3

      ld_not           I1.4                                ;rung 7
      and              I0.4
      out              M2.4

      ld_not           I1.4                                ;rung 8
      and              I0.5
      out              M2.5

      ld_not           I1.4                                ;rung 9
      and              I0.6
      out              M2.6

      ld_not           I1.4                                ;rung 10
      and              I0.7
      out              M2.7

      ld_not           I1.4                                ;rung 11
      and_not          M2.0
      out              M3.0

      ld_not           I1.4                                ;rung 12
      and_not          M2.1
      out              M3.1
```

FIGURE 4.90 (*1 of* 2) The user program of Example 4.17 (74165 8-bit PISO shift register).

inputs are used to select the operation mode. As can be seen from Table 4.40, there are four types of modes: hold, shift right, shift left, and parallel load. When inputs S1 and S0 are 00, the outputs (Q0, Q1, Q2, and Q3) hold their previous values (hold). When inputs S1 and S0 are 01, the data stored in the register is shifted right with the rising edge of the clock signal (shift right). In this case the serial input is taken from the DSR input. When inputs S1 and S0 are 10, the data stored in the register is shifted

```
    ld_not        I1.4                          ;rung 13
    and_not       M2.2
    out           M3.2

    ld_not        I1.4                          ;rung 14
    and_not       M2.3
    out           M3.3

    ld_not        I1.4                          ;rung 15
    and_not       M2.4
    out           M3.4

    ld_not        I1.4                          ;rung 16
    and_not       M2.5
    out           M3.5

    ld_not        I1.4                          ;rung 17
    and_not       M2.6
    out           M3.6

    ld_not        I1.4                          ;rung 18
    and_not       M2.7
    out           M3.7

    ld            Q1.7                          ;rung 19
                  ;Mreg,Mbit,D,Q,S,R
    dff_r_SR      M1.7,M0.6,Q0.7,M2.7,M3.7

    ld            Q1.7                          ;rung 20
                  ;Mreg,Mbit,D,Q,S,R
    dff_r_SR      M1.6,M0.5,M0.6,M2.6,M3.6

    ld            Q1.7                          ;rung 21
    dff_r_SR      M1.5,M0.4,M0.5,M2.5,M3.5

    ld            Q1.7                          ;rung 22
    dff_r_SR      M1.4,M0.3,M0.4,M2.4,M3.4

    ld            Q1.7                          ;rung 23
    dff_r_SR      M1.3,M0.2,M0.3,M2.3,M3.3

    ld            Q1.7                          ;rung 24
    dff_r_SR      M1.2,M0.1,M0.2,M2.2,M3.2

    ld            Q1.7                          ;rung 25
    dff_r_SR      M1.1,M0.0,M0.1,M2.1,M3.1

    ld            Q1.7                          ;rung 26
    dff_r_SR      M1.0,I1.2,M0.0,M2.0,M3.0
  ;
  ;--- PLC codes to be allocated in the "user_program_1" macro end here ----------
    endm
```

FIGURE 4.90 Continued

left with the rising edge of the clock signal (shift left). In this case the serial input is taken from the DSL input. When inputs S1 and S0 are 11, the data in the parallel inputs P0, P1, P2, and P3 are loaded in the register (i.e., in the flip-flops Q0, Q1, Q2, and Q3, respectively) with the rising edge of the clock signal (parallel load). In the shift right mode (and shift left mode, respectively), Q3 (and Q0, respectively) can be used as the serial data output.

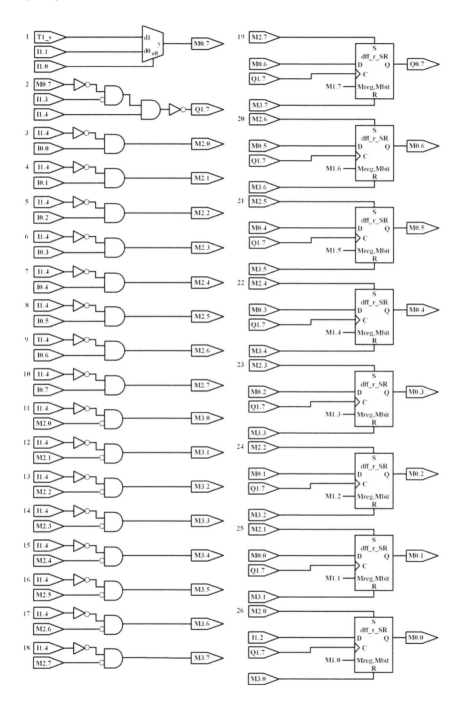

FIGURE 4.91 Schematic diagram of the user program of Example 4.17 (74165 8-bit PISO shift register).

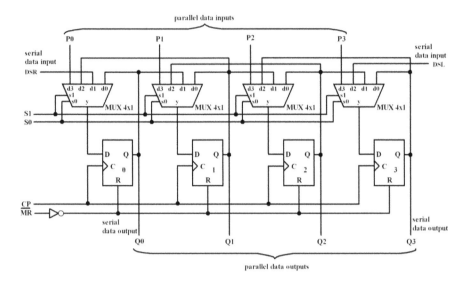

FIGURE 4.92 The logic diagram of the 74194 4-bit bidirectional universal shift register designed by using rising edge–triggered D flip-flops.

Figure 4.93 shows the logic diagram of the 74194 4-bit bidirectional universal shift register implemented by using the PIC16F1847-Based PLC. D flip-flops are implemented by using "dff_r_SR" macros. 4-to-1 multiplexers are implemented by using "mux_4_1" macros (see Chapter 4 of *Intermediate Concepts*). The CP clock signal can be applied through I1.1 or it can be fixed by using a T_1s reference timing signal. The choice between these two clock sources is made by using a 2-to-1 multi-plexer "Mux 2x1". When the select input I1.0 of "Mux 2x1" is 0 (or 1, respectively), the CP clock input is I1.1 (or T_1s, respectively).

The user program of Example 4.18 (74194 4-bit bidirectional universal register) is shown in Figure 4.94. The schematic diagram of the user program of Example 4.18 is depicted in Figure 4.95. When the project file of the PIC16F1847-Based PLC is open in the MPLAB X IDE, from the file "PICPLC_PIC16F1847_user_Bsc .inc", if you uncomment Example 4.18 and run the project by pressing the symbol ▷ from the toolbar, then the PIC16F1847 microcontroller within the CPU board of the PIC16F1847-Based PLC will be programmed. After loading the program file to the PIC16F1847 microcontroller, switch the 4PDT to "RUN" and the power switch to the "ON" position. Next you can test the operation of this example.

4.15.19   EXAMPLE 4.19: 74595 8-BIT SERIAL-IN, SERIAL- OR PARALLEL-OUT SHIFT REGISTER

Example 4.19 explains the implementation of the 74595 8-bit serial-in, serial- or parallel-out shift register by using the PIC16F1847-Based PLC. The logic dia-gram of the 74HC595 8-bit serial-in, serial- or parallel-out shift register designed by using rising edge–triggered D flip-flops is shown in Figure 4.96. Table 4.41 depicts the function table of the 74HC595 8-bit serial-in, serial- or parallel-out

TABLE 4.40
The Function Table of the 74194 4-Bit Bidirectional Universal Shift Register Shown in Figure 4.92

Operating Mode	Inputs							Parallel Data Outputs			
	$\overline{\text{MR}}$	CP	S1	S0	DSR	DSL	Pn	Q0	Q1	Q2	Q3
Reset	L	×	×	×	×	×	×	L	L	L	L
Hold	H	↑	l	l	×	×	×	q_0	q_1	q_2	q_3
Shift Right	H	↑	l	h	l	×	×	L	q_0	q_1	q_2
	H	↑	l	h	h	×	×	H	q_0	q_1	q_2
Shift Left	H	↑	h	l	×	l	×	q_1	q_2	q_3	L
	H	↑	h	l	×	h	×	q_1	q_2	q_3	H
Parallel Load	H	↑	h	h	×	×	p_n	p_0	p_1	p_2	p_3

$\overline{\text{MR}}$: Asynchronous Master Reset (Active LOW) Input
CP: Clock (Active HIGH Going Edge) Input
S1,S0: Mode Select Inputs
DSR: Serial (Shift Right) Data Input
DSL: Serial (Shift Left) Data Input
Pn (P0, P1, P2, P3): Parallel Data Inputs
Q0, Q1, Q2, Q3: Parallel Data Outputs
L = LOW Voltage Level, H = HIGH Voltage Level, × = Don't Care
l = LOW voltage level one set-up time prior to the LOW to HIGH clock transition
h = HIGH voltage level one set-up time prior to the LOW to HIGH clock transition
p_n (q_n) = Lower case letters indicate the state of the referenced input (or output) one set-up time prior to the LOW to HIGH clock transition
↑ = LOW to HIGH clock transition

shift register shown in Figure 4.96. The 74HC595 is an 8-bit serial-in, serial- or parallel-out shift register with a latch register and 3-state outputs. Both the shift and latch register have separate clocks. The device features a serial data input (SDI) and a serial data output (SDO) to enable cascading, and an asynchronous reset $\overline{\text{MR}}$ input. A LOW on $\overline{\text{MR}}$ will reset the shift register. Data is shifted on the LOW–HIGH transitions of the SHCP input. The data in the shift register is transferred to the latch register on a LOW–HIGH transition of the LTCP input. If both clocks are connected together, the shift register will always be one clock pulse ahead of the latch register. Data in the latch register appears at the output whenever the output enable input $\left(\overline{\text{OE}}\right)$ is LOW. A HIGH on $\overline{\text{OE}}$ causes the outputs to assume a high-impedance OFF state. Operation of the $\overline{\text{OE}}$ input does not affect the state of the registers.

Figure 4.97 shows the logic diagram of the 74595 8-bit serial-in, serial- or parallel-out shift register implemented by using the PIC16F1847-Based PLC. It is important to note that in this logic diagram, 3-state buffers are represented by

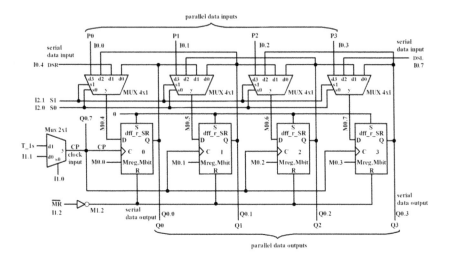

FIGURE 4.93 The logic diagram of the 74194 4-bit bidirectional universal shift register implemented by using the PIC16F1847-Based PLC.

```
user_program_1  macro
;--- PLC codes to be allocated in the "user_program_1" macro start from here ---
;__Example_Bsc_4.18
;
;___ 74194 4-bit bidirectional universal shift register _____

                      ;s0,i1,i0,rego,bito              ;rung 1
        mux_2_1       I1.0,T_1s,I1.1,Q0.7

        inv_out       I1.2,M1.2                        ;rung 2

                      ;s1,s0,i3,i2,i1,i0,rego,bito      ;rung 3
        mux_4_1       I2.1,I2.0,I0.0,Q0.1,I0.4,Q0.0,M0.4

                      ;s1,s0,i3,i2,i1,i0,rego,bito      ;rung 4
        mux_4_1       I2.1,I2.0,I0.1,Q0.2,Q0.0,Q0.1,M0.5

                      ;s1,s0,i3,i2,i1,i0,rego,bito      ;rung 5
        mux_4_1       I2.1,I2.0,I0.2,Q0.3,Q0.1,Q0.2,M0.6

                      ;s1,s0,i3,i2,i1,i0,rego,bito      ;rung 6
        mux_4_1       I2.1,I2.0, ,I0.3,I0.7 Q0.2,Q0.3,M0.7

        ld            Q0.7                             ;rung 7
                      ;Mreg,Mbit,D,Q,S,R
        dff_r_SR      M0.3,M0.7,Q0.3,LOGIC0,M1.2

        ld            Q0.7                             ;rung 8
                      ;Mreg,Mbit,D,Q,S,R
        dff_r_SR      M0.2,M0.6,Q0.2,LOGIC0,M1.2

        ld            Q0.7                             ;rung 9
                      ;Mreg,Mbit,D,Q,S,R
        dff_r_SR      M0.1,M0.5,Q0.1,LOGIC0,M1.2

        ld            Q0.7                             ;rung 10
                      ;Mreg,Mbit,D,Q,S,R
        dff_r_SR      M0.0,M0.4,Q0.0,LOGIC0,M1.2
;
;--- PLC codes to be allocated in the "user_program_1" macro end here ----------
        endm
```

FIGURE 4.94 The user program of Example 4.18 (74194 4-bit bidirectional universal shift register).

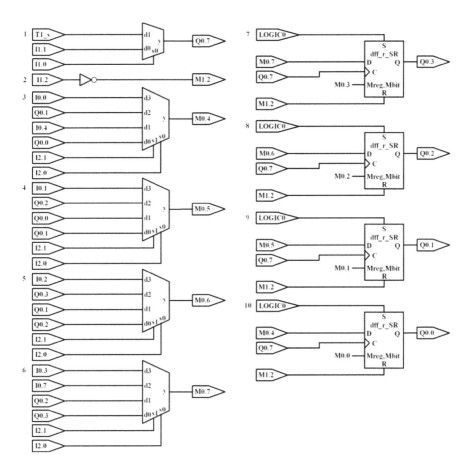

FIGURE 4.95 Schematic diagram of the user program of Example 4.18 (74194 4-bit bidirectional universal shift register).

2-input AND gates, since it is not possible to implement 3-state buffers by using the PIC16F1847-Based PLC. The D flip-flops of the shift register are implemented by using "dff_r_SR" macros, while the D flip-flops of the latch register are implemented by using "dff_r" macros. 2-input AND gates are implemented by using "and" macros.

The user program of Example 4.19 (74595 8-bit serial-in, serial- or parallel-out shift register) is shown in Figure 4.98. The schematic diagram of the user program of Example 4.19 is depicted in Figure 4.99. When the project file of the PIC16F1847-Based PLC is open in the MPLAB X IDE, from the file "PICPLC _PIC16F1847_user_Bsc.inc", if you uncomment Example 4.19 and run the project by pressing the symbol ▷ from the toolbar, then the PIC16F1847 microcontroller within the CPU board of the PIC16F1847-Based PLC will be programmed. After loading the program file to the PIC16F1847 microcontroller, switch the 4PDT to "RUN" and the power switch to the "ON" position. Next you can test the operation of this example.

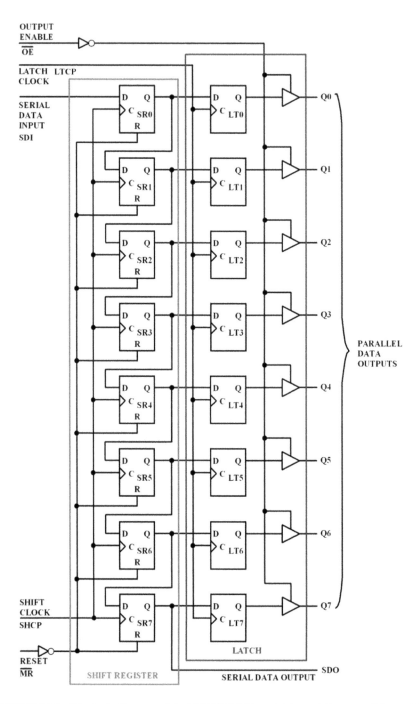

FIGURE 4.96 The logic diagram of the 74HC595 8-bit serial-in, serial- or parallel-out shift register with latched 3-state outputs.

TABLE 4.41

The Function Table of the 74HC595 8-Bit Serial-In, Serial- or Parallel-Out Shift Register with Latched 3-State Outputs Shown in Figure 4.96

Control Inputs				Input	Outputs		Function
SHCP	LTCP	$\overline{OE}$	$\overline{MR}$	SDI	SDO	Qn	
×	×	L	L	×	L	NC	a LOW level on $\overline{MR}$ only affects the shift register
×	↑	L	L	×	L	L	empty shift register loaded into latch register
×	×	H	L	×	L	Z	shift register clear; parallel outputs in high-impedance OFF state
↑	×	L	H	H	Q_{SR6}	NC	logic HIGH level shifted into shift register stage 0 (SR0). Contents of all shift register stages shifted through, e.g., previous state of stage 6 (internal Q_{SR6}) appears on the serial output SDO (Q_{SR7}).
×	↑	L	H	×	NC	QnS	contents of shift register stages (internal QnS) are transferred to the latch register and parallel output stages
↑	↑	L	H	×	Q_{SR6}	QnS	contents of shift register shifted through; previous contents of the shift register is transferred to the storage register and the parallel output stages

SHCP: Shift Register Clock Input (Active HIGH Going Edge)

LTCP: Latch Clock Input (Active HIGH Going Edge)

$\overline{OE}$: Output Enable Input (Active LOW)

$\overline{MR}$: Master Reset Input (Active LOW)

SDI: Serial Data Input

SDO: Serial Data Output

Qn = Q0, Q1, Q2, Q3, Q4, Q5, Q6, Q7: Parallel Data Outputs

L = LOW Voltage Level, H = HIGH Voltage Level, × = Don't Care

↑ = LOW to HIGH clock transition

NC = No Change

Z = high-impedance OFF state

4.15.20   EXAMPLE 4.20: 4-BIT JOHNSON COUNTER

Example 4.20 explains the implementation of a 4-bit Johnson counter by using the PIC16F1847-Based PLC. There are two very common types of shift register counters, namely the Johnson counter and the Ring counter. They are basically shift registers with the serial outputs connected back to the serial inputs in order to produce

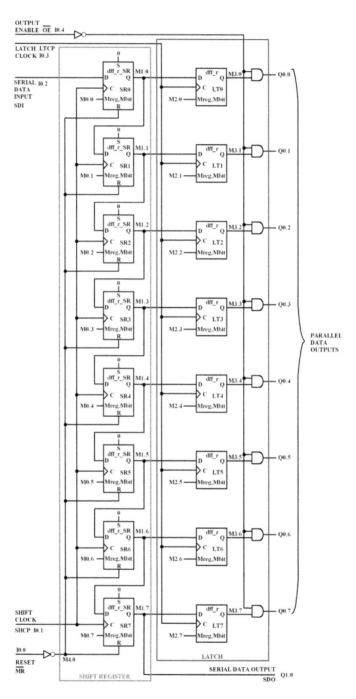

FIGURE 4.97 The logic diagram of the 74595 8-bit serial-in, serial- or parallel-out shift register implemented by using the PIC16F1847-Based PLC.

```
user_program_1  macro
;--- PLC codes to be allocated in the "user_program_1" macro start from here ---
;__Example_Bsc_4.19
;___ 74595 8-bit serial-input/serial or parallel-output shift register _____

        inv_out         I0.0,M4.0                       ;rung 1

        ld_not          I0.4                            ;rung 2
        and             M3.0
        out             Q0.0

        ld_not          I0.4                            ;rung 3
        and             M3.1
        out             Q0.1

        ld_not          I0.4                            ;rung 4
        and             M3.2
        out             Q0.2

        ld_not          I0.4                            ;rung 5
        and             M3.3
        out             Q0.3

        ld_not          I0.4                            ;rung 6
        and             M3.4
        out             Q0.4

        ld_not          I0.4                            ;rung 7
        and             M3.5
        out             Q0.5

        ld_not          I0.4                            ;rung 8
        and             M3.6
        out             Q0.6

        ld_not          I0.4                            ;rung 9
        and             M3.7
        out             Q0.7

        ld              I0.3                            ;rung 10
                        ;Mreg,Mbit,D,Q
        dff_r           M2.7,M1.7,M3.7

        ld              I0.3                            ;rung 11
                        ;Mreg,Mbit,D,Q
        dff_r           M2.6,M1.6,M3.6

        ld              I0.3                            ;rung 12
                        ;Mreg,Mbit,D,Q
        dff_r           M2.5,M1.5,M3.5

        ld              I0.3                            ;rung 13
                        ;Mreg,Mbit,D,Q
        dff_r           M2.4,M1.4,M3.4
```

FIGURE 4.98   (*1 of* 2) The user program of Example 4.19 (74595 8-bit serial-in, serial- or parallel-out shift register).

particular sequences. These registers are classified as counters because they exhibit a specified sequence of states. An example Ring counter is considered in the next section. Johnson counters are a variation of standard ring counters, with the inverted output of the last stage fed back to the input of the first stage. They are also known as twisted ring counters. An n-stage Johnson counter yields a count sequence of length $2n$, so it may be considered to be a mod-$2n$ counter. The logic diagram of the 4-bit Johnson counter designed by using rising edge–triggered D flip-flops is shown in Figure 4.100. Table 4.42 depicts the truth table of the 4-bit Johnson counter shown in Figure 4.68.

```
        ld              I0.3                    ;rung 14
                        ;Mreg,Mbit,D,Q
        dff_r           M2.3,M1.3,M3.3

        ld              I0.3                    ;rung 15
                        ;Mreg,Mbit,D,Q
        dff_r           M2.2,M1.2,M3.2

        ld              I0.3                    ;rung 16
                        ;Mreg,Mbit,D,Q
        dff_r           M2.1,M1.1,M3.1

        ld              I0.3                    ;rung 17
                        ;Mreg,Mbit,D,Q
        dff_r           M2.0,M1.0,M3.0

        ld              I0.1                    ;rung 18
                        ;Mreg,Mbit,D,Q,S,R
        dff_r_SR        M0.7,M1.6,M1.7,LOGIC0,M4.0

        ld              I0.1                    ;rung 19
                        ;Mreg,Mbit,D,Q,S,R
        dff_r_SR        M0.6,M1.5,M1.6,LOGIC0,M4.0

        ld              I0.1                    ;rung 20
                        ;Mreg,Mbit,D,Q,S,R
        dff_r_SR        M0.5,M1.4,M1.5,LOGIC0,M4.0

        ld              I0.1                    ;rung 21
                        ;Mreg,Mbit,D,Q,S,R
        dff_r_SR        M0.4,M1.3,M1.4,LOGIC0,M4.0

        ld              I0.1                    ;rung 22
                        ;Mreg,Mbit,D,Q,S,R
        dff_r_SR        M0.3,M1.2,M1.3,LOGIC0,M4.0

        ld              I0.1                    ;rung 23
                        ;Mreg,Mbit,D,Q,S,R
        dff_r_SR        M0.2,M1.1,M1.2,LOGIC0,M4.0

        ld              I0.1                    ;rung 24
                        ;Mreg,Mbit,D,Q,S,R
        dff_r_SR        M0.1,M1.0,M1.1,LOGIC0,M4.0

        ld              I0.1                    ;rung 25
                        ;Mreg,Mbit,D,Q,S,R
        dff_r_SR        M0.0,I0.2,M1.0,LOGIC0,M4.0

        in_out          M1.7,Q1.0               ;rung 26
;--- PLC codes to be allocated in the "user_program_1" macro end here ----------
        endm
```

FIGURE 4.98 Continued

Figure 4.101 shows the logic diagram of the 4-bit Johnson counter implemented by using the PIC16F1847-Based PLC. D flip-flops are implemented by
using "dff_r" macros. The CP clock signal can be applied through I0.1 or it can
be fixed by using a T_1s reference timing signal. The choice between these two
clock sources is made by using a 2-to-1 multiplexer "Mux 2x1". When the select
input I0.0 of "Mux 2x1" is 0 (or 1, respectively), the CP clock input is I0.1 (or T_1s,
respectively).

The user program of Example 4.20 (4-bit Johnson counter) is shown in Figure 4.102.
The schematic diagram of the user program of Example 4.20 is depicted in Figure
4.103. When the project file of the PIC16F1847-Based PLC is open in the MPLAB
X IDE, from the file "PICPLC_PIC16F1847_user_Bsc.inc", if you uncomment

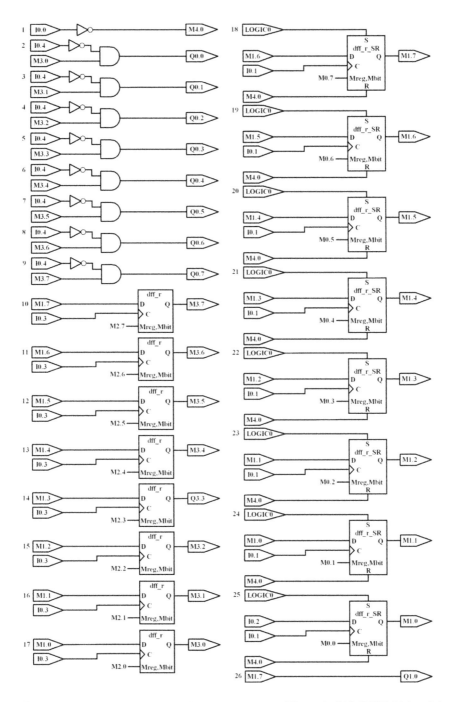

FIGURE 4.99 Schematic diagram of the user program of Example 4.19 (74595 8-bit serial-in, serial- or parallel-out shift register).

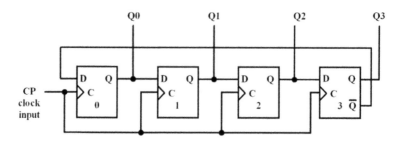

FIGURE 4.100 The logic diagram of the 4-bit Johnson counter.

TABLE 4.42

Truth Table for the 4-Bit Johnson Counter Shown in Figure 4.100

CP clock inputs	Parallel Data Outputs			
	Q0 (Q0.0)	Q1 (Q0.1)	Q2 (Q0.2)	Q3 (Q0.3)
	0	0	0	0
1st ↑	1	0	0	0
2nd ↑	1	1	0	0
3rd ↑	1	1	1	0
4th ↑	1	1	1	1
5th ↑	0	1	1	1
6th ↑	0	0	1	1
7th ↑	0	0	0	1
8th ↑	0	0	0	0
9th ↑	1	0	0	0
.	.	.	.	.

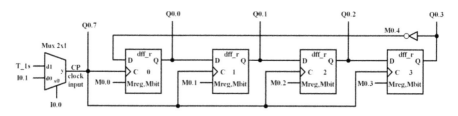

FIGURE 4.101 The logic diagram of the 4-bit Johnson counter implemented by using the PIC16F1847-Based PLC.

Example 4.20 and run the project by pressing the symbol ▷ from the toolbar, then the PIC16F1847 microcontroller within the CPU board of the PIC16F1847-Based PLC will be programmed. After loading the program file to the PIC16F1847 microcontroller, switch the 4PDT to "RUN" and the power switch to the "ON" position. Next you can test the operation of this example.

```
user_program_1   macro
;--- PLC codes to be allocated in the "user_program_1" macro start from here ---
;__Example_Bsc_4.20
;
;_____ 4-bit Johnson counter _____
;
                        ;s0,i1,i0,rego,bito              ;rung 1
        mux_2_1         I0.0,T_1s,I0.1,Q0.7

        inv_out         Q0.3,M0.4                        ;rung 2

        ld              Q0.7                             ;rung 3
                        ;Mreg,Mbit,D,Q
        dff_r           M0.3,Q0.2,Q0.3

        ld              Q0.7                             ;rung 4
                        ;Mreg,Mbit,D,Q
        dff_r           M0.2,Q0.1,Q0.2

        ld              Q0.7                             ;rung 5
                        ;Mreg,Mbit,D,Q
        dff_r           M0.1,Q0.0,Q0.1

        ld              Q0.7                             ;rung 6
                        ;Mreg,Mbit,D,Q
        dff_r           M0.0,M0.4,Q0.0
;
;--- PLC codes to be allocated in the "user_program_1" macro end here ----------
        endm
```

FIGURE 4.102   The user program of Example 4.20 (4-bit Johnson counter).

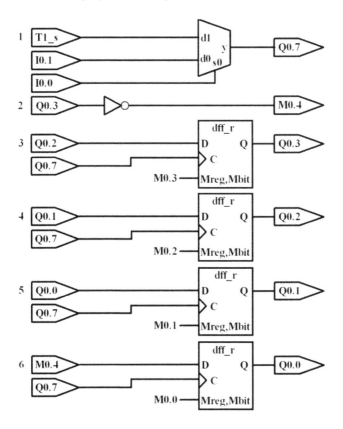

FIGURE 4.103   Schematic diagram of the user program of Example 4.20 (4-bit Johnson counter).

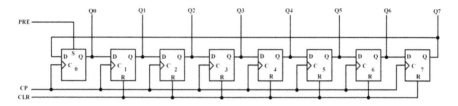

FIGURE 4.104 The logic diagram of the 8-bit ring counter.

TABLE 4.43
**Truth Table for the 8-Bit Ring Counter Shown in Figure 4.104,
Assuming That Initially, Data Q0Q1Q2Q3Q4Q5Q6Q7 = 10000000 Is
Stored in the Register**

CP clock inputs	Parallel Data Outputs							
	Q0 (Q0.0)	Q1 (Q0.1)	Q2 (Q0.2)	Q3 (Q0.3)	Q4 (Q0.4)	Q5 (Q0.5)	Q6 (Q0.6)	Q7 (Q0.7)
...	1	0	0	0	0	0	0	0
1st ↑	0	1	0	0	0	0	0	0
2nd ↑	0	0	1	0	0	0	0	0
3rd ↑	0	0	0	1	0	0	0	0
4th ↑	0	0	0	0	1	0	0	0
5th ↑	0	0	0	0	0	1	0	0
6th ↑	0	0	0	0	0	0	1	0
7th ↑	0	0	0	0	0	0	0	1
8th ↑	1	0	0	0	0	0	0	0
9th ↑	0	1	0	0	0	0	0	0
.	.	.	.	.	.	.	.	.

4.15.21 EXAMPLE 4.21: 8-BIT RING COUNTER

Example 4.21 explains the implementation of an 8-bit ring counter by using the PIC16F1847-Based PLC. A ring counter is basically a circulating shift register in which the output of the most significant stage is fed back to the input of the least significant stage. The logic diagram of the 8-bit ring counter designed by using rising edge–triggered D flip-flops is shown in Figure 4.104. When CLR signal is LOW, the output of each stage is shifted into the next stage on the positive-going edge (rising edge) of a clock pulse. If the CLR signal is HIGH, all flip-flops are reset to 0 except for the first flip-flop FF0. If the PRE signal is HIGH, the first flip-flop FF0 is preset to 1. Different bit patterns can be initially loaded into the counter by using the PRE, CLR, and CP inputs. Table 4.43 shows the truth table for the 8-bit ring counter shown in Figure 4.104, assuming that initially, data Q0Q1Q2Q3Q4Q5Q6Q7 = 10000000 is stored in the register. Likewise, Table 4.44 shows the truth table for the 8-bit ring

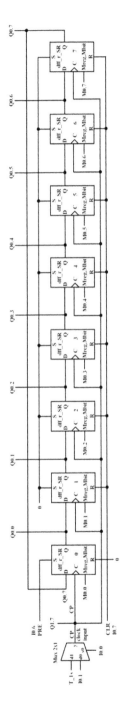

FIGURE 4.105 The logic diagram of the 8-bit ring counter (wrong implementation) implemented by using the PIC16F1847-Based PLC.

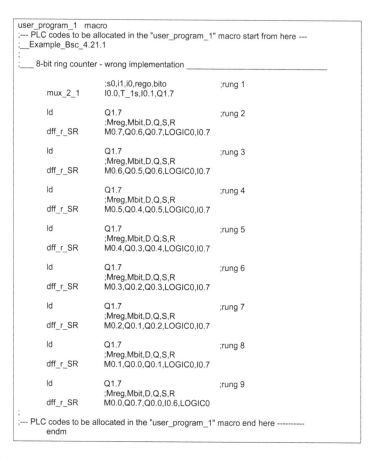

FIGURE 4.106 The user program of Example 4.21.1 (8-bit ring counter—wrong implementation).

counter shown in Figure 4.104, assuming that initially, data Q0Q1Q2Q3Q4Q5Q6Q7 = 11100000 is stored in the register.

Figure 4.105 shows the logic diagram of the 8-bit ring counter implemented by using the PIC16F1847-Based PLC. D flip-flops are implemented by using "dff_r_SR" macros. The CP clock signal can be applied through I0.1 or it can be fixed by using a T_1s reference timing signal. The choice between these two clock sources is made by using a 2-to-1 multiplexer "Mux 2x1". When the select input I0.0 of "Mux 2x1" is 0 (or 1, respectively), the CP clock input is I0.1 (or T_1s, respectively).

The user program of Example 4.21.1 (8-bit ring counter—wrong implementation) is shown in Figure 4.106. The schematic diagram of the user program of Example 4.21.1 is depicted in Figure 4.107. When the project file of the PIC16F1847-Based PLC is open in the MPLAB X IDE, from the file "PICPLC_PIC16F1847_user_Bsc .inc", if you uncomment Example 4.21.1 and run the project by pressing the symbol ▷ from the toolbar, then the PIC16F1847 microcontroller within the CPU board of the PIC16F1847-Based PLC will be programmed. After loading the program file to

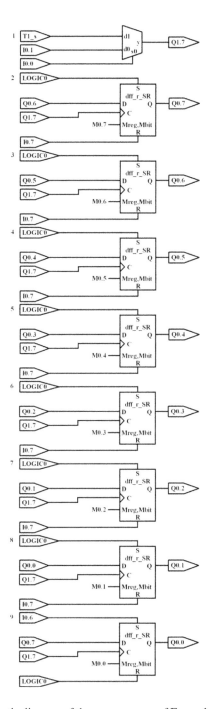

FIGURE 4.107 Schematic diagram of the user program of Example 4.21.1 (8-bit ring counter-wrong implementation).

TABLE 4.44

Truth Table for the 8-Bit Ring Counter Shown in Figure 4.104, Assuming That Initially, Data Q0Q1Q2Q3Q4Q5Q6Q7 = 11100000 Is Stored in the Register

CP clock inputs	Parallel Data Outputs							
	Q0 (Q0.0)	Q1 (Q0.1)	Q2 (Q0.2)	Q3 (Q0.3)	Q4 (Q0.4)	Q5 (Q0.5)	Q6 (Q0.6)	Q7 (Q0.7)
...	1	1	1	0	0	0	0	0
1st ↑	0	1	1	1	0	0	0	0
2nd ↑	0	0	1	1	1	0	0	0
3rd ↑	0	0	0	1	1	1	0	0
4th ↑	0	0	0	0	1	1	1	0
5th ↑	0	0	0	0	0	1	1	1
6th ↑	1	0	0	0	0	0	1	1
7th ↑	1	1	0	0	0	0	0	1
8th ↑	1	1	1	0	0	0	0	0
9th ↑	0	1	1	1	0	0	0	0
.	.	.	.	.	.	.	.	.

TABLE 4.45

Truth Table for the 8-Bit Ring Counter Implemented as Shown in Figure 4.107, Assuming That Initially, Data Q0Q1Q2Q3Q4Q5Q6Q7 = 10000000 Is Stored in the Register

CP clock inputs	Parallel Data Outputs							
	Q0 (Q0.0)	Q1 (Q0.1)	Q2 (Q0.2)	Q3 (Q0.3)	Q4 (Q0.4)	Q5 (Q0.5)	Q6 (Q0.6)	Q7 (Q0.7)
...	1	0	0	0	0	0	0	0
1st ↑	0	1	0	0	0	0	0	0
2nd ↑	0	0	1	0	0	0	0	0
3rd ↑	0	0	0	1	0	0	0	0
4th ↑	0	0	0	0	1	0	0	0
5th ↑	0	0	0	0	0	1	0	0
6th ↑	0	0	0	0	0	0	1	0
7th ↑	1	0	0	0	0	0	0	1
8th ↑	0	1	0	0	0	0	0	0
9th ↑	0	0	1	0	0	0	0	0
10th ↑	0	0	0	1	0	0	0	0
11th ↑	0	0	0	0	1	0	0	0
12th ↑	0	0	0	0	0	1	0	0
13th ↑	0	0	0	0	0	0	1	0
14th ↑	1	0	0	0	0	0	0	1
.	.	.	.	.	.	.	.	.

TABLE 4.46

Truth Table for the 8-Bit Ring Counter Implemented as Shown in Figure 4.107, Assuming That Initially, Data Q0Q1Q2Q3Q4Q5Q6Q7 = 11100000 Is Stored in the Register

CP clock inputs	Parallel Data Outputs							
	Q0 (Q0.0)	Q1 (Q0.1)	Q2 (Q0.2)	Q3 (Q0.3)	Q4 (Q0.4)	Q5 (Q0.5)	Q6 (Q0.6)	Q7 (Q0.7)
...	1	1	1	0	0	0	0	0
1st ↑	0	1	1	1	0	0	0	0
2nd ↑	0	0	1	1	1	0	0	0
3rd ↑	0	0	0	1	1	1	0	0
4th ↑	0	0	0	0	1	1	1	0
5th ↑	1	0	0	0	0	1	1	1
6th ↑	1	1	0	0	0	0	1	1
7th ↑	1	1	1	0	0	0	0	1
8th ↑	0	1	1	1	0	0	0	0
9th ↑	0	0	1	1	1	0	0	0
10th ↑	0	0	0	1	1	1	0	0
11th ↑	0	0	0	0	1	1	1	0
12th ↑	1	0	0	0	0	1	1	1
13th ↑	1	1	0	0	0	0	1	1
14th ↑	1	1	1	0	0	0	0	1
15th ↑	0	1	1	1	0	0	0	0
.	.	.	.	.	.	.	.	.

the PIC16F1847 microcontroller, switch the 4PDT to "RUN" and the power switch to the "ON" position. Next you can test the operation of Example 4.21.1. Rungs from 3 to 10 represent the implementation of flip-flops FF7, FF6, FF5, FF4, FF3, FF2, FF1, and FF0, respectively. In this implementation, the reverse order of rungs is important for a proper operation in order to avoid the avalanche effect, but not enough to make the implementation correct. It can be observed that in this implementation there is a problem. You can test the operation of Example 4.21.1 based on the initial bit patterns shown in Tables 4.45 and 4.46. These tables show the behaviors of the 8-bit ring counter implemented as shown in Figure 4.107. Of course, if you compare these two tables with Tables 4.43 and 4.44, you can see that the results obtained here are not correct.

This incorrect operation of the 8-bit ring counter is due to the fact that the output of the last stage, i.e., Q0.7, is available to the input of the first stage at the same PLC scan time. Therefore, we end up with this incorrect operation. To solve this problem, it is necessary to introduce a delay of one PLC scan time for the output of the last stage, i.e., Q0.7, before it can be made available to the input of the first stage. Figure 4.108 shows the logic diagram of the 8-bit ring counter (correct

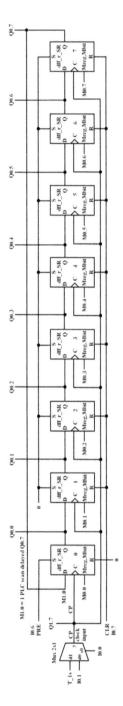

FIGURE 4.108 The logic diagram of the 8-bit ring counter (correct implementation) implemented by using the PIC16F1847-Based PLC.

```
user_program_1  macro
;--- PLC codes to be allocated in the "user_program_1" macro start from here ---
;__Example_Bsc_4.21.2
;
;
;___ 8-bit ring counter - correct implementation _____

                        ;s0,i1,i0,rego,bito           ;rung 1
        mux_2_1         I0.0,T_1s,I0.1,Q1.7

        in_out          Q0.7,M1.0                    ;rung 2

        ld              Q1.7                         ;rung 3
                        ;Mreg,Mbit,D,Q,S,R
        dff_r_SR        M0.7,Q0.6,Q0.7,LOGIC0,I0.7

        ld              Q1.7                         ;rung 4
                        ;Mreg,Mbit,D,Q,S,R
        dff_r_SR        M0.6,Q0.5,Q0.6,LOGIC0,I0.7

        ld              Q1.7                         ;rung 5
                        ;Mreg,Mbit,D,Q,S,R
        dff_r_SR        M0.5,Q0.4,Q0.5,LOGIC0,I0.7

        ld              Q1.7                         ;rung 6
                        ;Mreg,Mbit,D,Q,S,R
        dff_r_SR        M0.4,Q0.3,Q0.4,LOGIC0,I0.7

        ld              Q1.7                         ;rung 7
                        ;Mreg,Mbit,D,Q,S,R
        dff_r_SR        M0.3,Q0.2,Q0.3,LOGIC0,I0.7

        ld              Q1.7                         ;rung 8
                        ;Mreg,Mbit,D,Q,S,R
        dff_r_SR        M0.2,Q0.1,Q0.2,LOGIC0,I0.7

        ld              Q1.7                         ;rung 9
                        ;Mreg,Mbit,D,Q,S,R
        dff_r_SR        M0.1,Q0.0,Q0.1,LOGIC0,I0.7

        ld              Q1.7                         ;rung 10
                        ;Mreg,Mbit,D,Q,S,R
        dff_r_SR        M0.0,M1.0,Q0.0,I0.6,LOGIC0
;
;--- PLC codes to be allocated in the "user_program_1" macro end here ----------
        endm
```

FIGURE 4.109 The user program of Example 4.21.2 (8-bit ring counter—correct implementation).

implementation) implemented by using the PIC16F1847-Based PLC, where memory bit M1.0 represents the signal of the last stage output Q0.7 delayed by one PLC scan time. The user program of Example 4.21.2 (8-bit ring counter—correct implementation) is shown in Figure 4.109. The schematic diagram of the user program of Example 4.21.2 is depicted in Figure 4.110. When the project file of the PIC16F1847-Based PLC is open in the MPLAB X IDE, from the file "PICPLC_PIC16F1847_user_Bsc.inc", if you uncomment Example 4.21.2 and run the project by pressing the symbol ▷ from the toolbar, then the PIC16F1847 microcontroller within the CPU board of the PIC16F1847-Based PLC will be programmed. After loading the program file to the PIC16F1847 microcontroller, switch the 4PDT to "RUN" and the power switch to the "ON" position. Next you can test the operation of Example 4.21.2. This time the operation of the 8-bit ring counter is correct, as described in Tables 4.43 and 4.44.

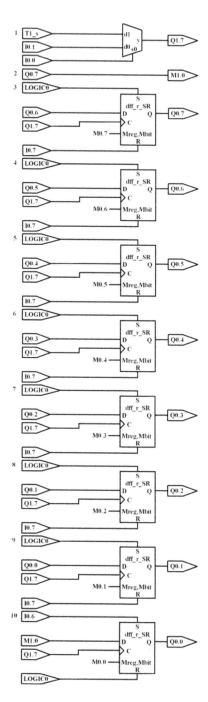

FIGURE 4.110 Schematic diagram of the user program of Example 4.21.2 (8-bit ring counter—correct implementation).

5 Timer Macros

INTRODUCTION

In this chapter, the following timer macros are described:

TON_8 (8-bit on-delay timer),
TON_16 (16-bit on-delay timer),
RTO_8 (8-bit retentive on-delay timer),
RTO_16 (16-bit retentive on-delay timer),
TOF_8 (8-bit off-delay timer),
TOF_16 (16-bit off-delay timer),
TP_8 (8-bit pulse timer),
TP_16 (16-bit pulse timer),
TEP_8 (8-bit extended pulse timer),
TEP_16 (16-bit extended pulse timer),
TOS_8 (8-bit oscillator timer),
TOS_16 (16-bit oscillator timer).

Timers can be used in a wide range of applications where a time delay function is required based on an input signal. The definition of the 8-bit variables to be used for timer macros, namely TV_L, TV_L+1, ..., TV_L+79, TV_H, TV_H+1, ..., TV_H+79, and T_Q0, T_Q1, ..., T_Q9, and their allocation in SRAM data memory are shown in Figures 5.1 and 5.2 , respectively. All 8-bit variables defined for timers must be cleared at the beginning of the PLC operation for a proper operation. Therefore, all variables of timer macros are cleared within the macro "initialize" as shown in Figure 5.3. The status bits, which will be explained in the next sections, of all timers are defined as shown in Figure 5.4. The individual bits (1-bit variables) of 8-bit SRAM registers T_Q0, T_Q1, ..., T_Q9 are shown in Table 5.1.

At any time at most 80 different timers can be used irrespective of the timer type. A unique timer number from 0 to 79 can be assigned to only one of the timer macros TON_8, RTO_8, TOF_8, TP_8, TEP_8, TOS_8, TON_16, RTO_16, TOF_16, TP_16, TEP_16, or TOS_16. All timer macros share 80 8-bit variables, namely TV_L, TV_L+1, ..., TV_L+79. These variables are used to hold current timing values (TV) for 8-bit timers and likewise the low byte of current timing values (TVL) for 16-bit timers. In addition,16-bit timer macros, namely TON_16, RTO_16, TOF_16, TP_16, TEP_16, and TOS_16, share 80 8-bit variables, namely TV_H, TV_H+1, ..., TV_H+79. These variables are used to hold the high byte of current timing values (TVH) for 16-bit timers. Since there is one current timing value for each timer, do not assign the same number to more than one timer.

The file "PICPLC_PIC16F1847_macros_Bsc.inc", which is downloadable from this book's webpage under the downloads section, contains macros defined for the

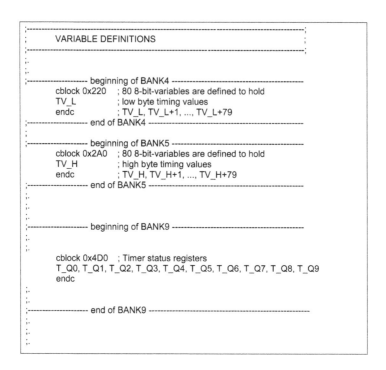

FIGURE 5.1   The definition of 8-bit variables to be used for the timer macros.

PIC16F1847-Based PLC explained in this book (*Hardware and Basic Concepts*). Let us now consider the timer macros. In the following, first of all a general description is given for the considered timer function, and then its 8-bit and 16-bit implementations for the PIC16F1847-Based PLC are provided.

5.1 ON-DELAY TIMER (TON)

The on-delay timer can be used to delay setting an output true (ON) for a fixed period of time after an input signal becomes true (ON). The symbol and timing diagram of the on-delay timer (TON) are both shown in Figure 5.5. As the input signal **IN** goes true (ON), the timing function is started and therefore the elapsed time **ET** starts to increase. When the elapsed time **ET** reaches the time specified by the preset time input **PT**, the output **Q** goes true (ON) and the elapsed time is held. The output **Q** remains true (ON) until the input signal **IN** goes false (OFF). If the input signal **IN** is not true (ON) longer than the delay time specified in **PT**, the output **Q** remains false (OFF). The following two sections explain the implementations of 8-bit and 16-bit on-delay timers, respectively, for the PIC16F1847-Based PLC.

5.2 MACRO "TON_8" (8-BIT ON-DELAY TIMER)

The macro "TON_8" defines 80 8-bit on-delay timers selected with the num = 0, 1, ..., 79. The symbol of the macro "TON_8" is depicted in Table 5.2. The macro

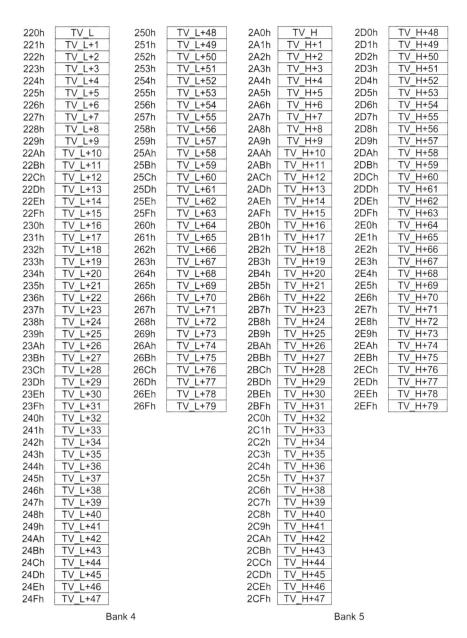

Bank 4   Bank 5

FIGURE 5.2 (*1 of 2*) Allocation of 8-bit variables in SRAM data memory to be used for the timer macros.

"TON_8" and its flowchart are shown in Figures 5.6 and 5.7, respectively. Note that the CLK input has a rising edge symbol, but within the macro "TON_8", there is no rising edge detection mechanism for the signal provided through this input. Instead, it is assumed that the signal connected here is one of the rising edge signals available in the PIC16F1847-Based PLC, i.e., one of the re_T2ms (2 ms), re_T10ms (10

4D0h	T_Q0
4D1h	T_Q1
4D2h	T_Q2
4D3h	T_Q3
4D4h	T_Q4
4D5h	T_Q5
4D6h	T_Q6
4D7h	T_Q7
4D8h	T_Q8
4D9h	T_Q9

Bank 9

FIGURE 5.2 Continued

ms), re_T100ms (100 ms), or re_T1s (1 s). IN (input signal, through "W"), Q (output signal = timer status bit), and CLK (re_Treg, re_Tbit) are all defined as Boolean variables. The time constant "tcnst" is an integer constant (here, for 8-bit resolution, it is chosen from any number in the range 1–255) and is used to define preset time PT, which is obtained by the formula $PT = tcnst \times CLK$, where CLK should be used as the period of the rising edge signal of one of the four reference timing signals—ticks.

```
initialize macro
          local    L1,L2
;·
;·
;·
;·
;·
;·
;-------- clear BANKs from 0 to 12 -------------
          clrf     FSR0L          ; initialize the pointer to the first location
          movlw    b'00100000'    ; of linear memory, i.e., 0x2000
          movwf    FSR0H          ;
          clrw                    ; W = 0
L2        movwi    FSR0++         ; clear INDF0 register
          btfss    FSR0H,2        ; all done?      if not goto L2
          bra      L2             ; if 0x2400 is reached then skip
;-----------------------------
;·
;·
;·
;·
;·
;·
          endm
```

FIGURE 5.3 The initialization of all variables of timer macros within the macro "initialize".

#define	TQ0	T_Q0,0	#define	TQ40	T_Q5,0
#define	TQ1	T_Q0,1	#define	TQ41	T_Q5,1
#define	TQ2	T_Q0,2	#define	TQ42	T_Q5,2
#define	TQ3	T_Q0,3	#define	TQ43	T_Q5,3
#define	TQ4	T_Q0,4	#define	TQ44	T_Q5,4
#define	TQ5	T_Q0,5	#define	TQ45	T_Q5,5
#define	TQ6	T_Q0,6	#define	TQ46	T_Q5,6
#define	TQ7	T_Q0,7	#define	TQ47	T_Q5,7
#define	TQ8	T_Q1,0	#define	TQ48	T_Q6,0
#define	TQ9	T_Q1,1	#define	TQ49	T_Q6,1
#define	TQ10	T_Q1,2	#define	TQ50	T_Q6,2
#define	TQ11	T_Q1,3	#define	TQ51	T_Q6,3
#define	TQ12	T_Q1,4	#define	TQ52	T_Q6,4
#define	TQ13	T_Q1,5	#define	TQ53	T_Q6,5
#define	TQ14	T_Q1,6	#define	TQ54	T_Q6,6
#define	TQ15	T_Q1,7	#define	TQ55	T_Q6,7
#define	TQ16	T_Q2,0	#define	TQ56	T_Q7,0
#define	TQ17	T_Q2,1	#define	TQ57	T_Q7,1
#define	TQ18	T_Q2,2	#define	TQ58	T_Q7,2
#define	TQ19	T_Q2,3	#define	TQ59	T_Q7,3
#define	TQ20	T_Q2,4	#define	TQ60	T_Q7,4
#define	TQ21	T_Q2,5	#define	TQ61	T_Q7,5
#define	TQ22	T_Q2,6	#define	TQ62	T_Q7,6
#define	TQ23	T_Q2,7	#define	TQ63	T_Q7,7
#define	TQ24	T_Q3,0	#define	TQ64	T_Q8,0
#define	TQ25	T_Q3,1	#define	TQ65	T_Q8,1
#define	TQ26	T_Q3,2	#define	TQ66	T_Q8,2
#define	TQ27	T_Q3,3	#define	TQ67	T_Q8,3
#define	TQ28	T_Q3,4	#define	TQ68	T_Q8,4
#define	TQ29	T_Q3,5	#define	TQ69	T_Q8,5
#define	TQ30	T_Q3,6	#define	TQ70	T_Q8,6
#define	TQ31	T_Q3,7	#define	TQ71	T_Q8,7
#define	TQ32	T_Q4,0	#define	TQ72	T_Q9,0
#define	TQ33	T_Q4,1	#define	TQ73	T_Q9,1
#define	TQ34	T_Q4,2	#define	TQ74	T_Q9,2
#define	TQ35	T_Q4,3	#define	TQ75	T_Q9,3
#define	TQ36	T_Q4,4	#define	TQ76	T_Q9,4
#define	TQ37	T_Q4,5	#define	TQ77	T_Q9,5
#define	TQ38	T_Q4,6	#define	TQ78	T_Q9,6
#define	TQ39	T_Q4,7	#define	TQ79	T_Q9,7

FIGURE 5.4 The definition of status bits (timer outputs) for timer macros.

TABLE 5.1

Individual Bits of 8-Bit SRAM Registers T_Q0, T_Q1, ..., T_Q9

Address	Name	Bit 7	Bit 6	Bit 5	Bit 4	Bit 3	Bit 2	Bit 1	Bit 0
4D0h	T_Q0	TQ7	TQ6	TQ5	TQ4	TQ3	TQ2	TQ1	TQ0
4D1h	T_Q1	TQ15	TQ14	TQ13	TQ12	TQ11	TQ10	TQ9	TQ8
4D2h	T_Q2	TQ23	TQ22	TQ21	TQ20	TQ19	TQ18	TQ17	TQ16
4D3h	T_Q3	TQ31	TQ30	TQ29	TQ28	TQ27	TQ26	TQ25	TQ24
4D4h	T_Q4	TQ39	TQ38	TQ37	TQ36	TQ35	TQ34	TQ33	TQ32
4D5h	T_Q5	TQ47	TQ46	TQ45	TQ44	TQ43	TQ42	TQ41	TQ40
4D6h	T_Q6	TQ55	TQ54	TQ53	TQ52	TQ51	TQ50	TQ49	TQ48
4D7h	T_Q7	TQ63	TQ62	TQ61	TQ60	TQ59	TQ58	TQ57	TQ56
4D8h	T_Q8	TQ71	TQ70	TQ69	TQ68	TQ67	TQ66	TQ65	TQ64
4D9h	T_Q9	TQ79	TQ78	TQ77	TQ76	TQ75	TQ74	TQ73	TQ72

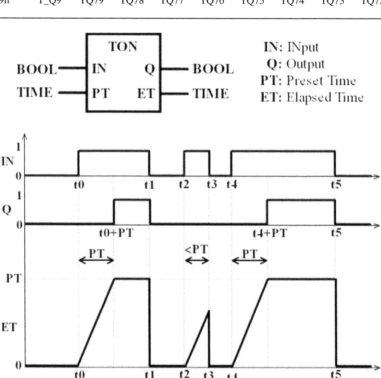

FIGURE 5.5 The symbol and timing diagram of the on-delay timer (TON).

The on-delay timer outputs are represented by the status bits TQ0, TQ1, ..., TQ79, defined by T_Q0+num/8,num−(8*[num/8]), as shown in Figure 5.4. An 8-bit integer variable "TV_L+num" is used to count the rising edge signals connected to the CLK input. The count value of "TV_L+num" defines the elapsed time ET as follows: ET = CLK×count value of "TV_L+num". Let us now briefly consider how the macro

TABLE 5.2
The Symbol of the Macro "TON_8"

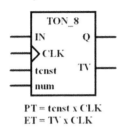

IN (input signal, through "W") = 0 or 1

CLK (rising edge signal of one of the four reference timing signals, re_Treg, re_Tbit) = re_T2ms (2 ms), re_T10ms (10 ms), re_T100ms (100 ms), or re_T1s (1 s)

tcnst (8-bit time constant value) = 1, 2, ..., 255

num (the unique timer number) = 0, 1, ..., 79

Q (timer output – timer status bit) = one of the following: TQ0, TQ1, ..., TQ79, defined by T_Q0+num/8,num–(8*[num/8]) (num = 0, 1, ..., 79)

TV (8-bit timing count value hold in an 8-bit register) = TV_L+num (num = 0, 1, ..., 79)

"TON_8" works. First of all, the preset time PT is defined by means of the period of a rising edge signal "CLK = re_Treg, re_Tbit" and a time constant "tcnst". If the input signal IN, taken into the macro by means of "W", is false (OFF), then the timer output Q, i.e., the status bit T_Q0+num/8,num–(8*[num/8]), is forced to be false (OFF) and the counter "TV_L+num" is loaded with "00h". If the input signal IN is true (ON) and the timer output Q, i.e., the status bit T_Q0+num/8,num–(8*[num/8]), is false (OFF), then with each rising edge signal "CLK = re_Treg, re_Tbit", the related counter "TV_L+num" is incremented by 1. In this case, when the count value of "TV_L+num" is equal to the number "tcnst", then the state change from OFF to ON is issued for the timer output Q, i.e., the status bit T_Q0+num/8,num–(8*[num/8]). If the input signal IN and the timer output Q, i.e., the status bit T_Q0+num/8,num–(8*[num/8]), are both true (ON), then no action is taken and the elapsed time ET is held. In this macro a previously defined 8-bit variable "Temp_1" is also utilized.

In order to explain how an on-delay timer is set up by using the macro "TON_8", let us consider Figure 5.8, where an on-delay timer is obtained with the macro "TON_8".

In this example program we have:

IN = I0.0
num = 57 (decimal)
CLK (re_Treg,re_Tbit) = re_T1s (the rising edge signal with 1-second period)
tcnst = 50 (decimal)

Therefore, we obtain the following from the macro "TON_8":

$$PT = tcnst * CLK \left(\text{the period of rising edge signal re\_T1s} \right)$$

$$= 50 * 1 \text{ second} = 50 \text{ seconds}$$

TV_Ln = TV_L register number

```
TON_8 macro  num, re_Treg,re_Tbit, tcnst
          local    L1,L2
          local    R_n,b_n,TV_Ln
;-------------------------------------------------------------------------
  if num < 80                          ;if num < 80 then carry on, else do not compile.
;-------------------------------------------------------------------------
  if (tcnst>0)&&(tcnst<256)    ;if 0<tcnst<256 then carry on, else do not compile.
;-------------------------------------------------------------------------
R_n     set     T_Q0+num/8           ;T_Q0 Register number
b_n     set     num-(8*(num/8))      ;T_Q0 bit number
TV_Ln set     TV_L+num             ;TV_L register number
;-------------------------------------------------------------------------
          movwf  Temp_1
          btfsc   Temp_1,0
          goto    L2
          banksel TV_Ln
          clrf     TV_Ln
          banksel R_n
          bcf     R_n,b_n
          goto    L1
L2
          banksel R_n
          btfsc   R_n,b_n
          goto    L1
          btfss   re_Treg,re_Tbit
          goto    L1
          banksel TV_Ln
          incf    TV_Ln,F
          movf    TV_Ln,W
          banksel R_n
          xorlw   tcnst
          btfsc   STATUS,Z
          bsf     R_n,b_n
L1
;-------------------------------------------------------------------------
  else
  error "Make sure that 0 < tcnst < 256 !"
  endif
;-------------------------------------------------------------------------
  else
  error "The timer number must be one of 0, 1, ..., 79."
  endif
;-------------------------------------------------------------------------
          endm
```

FIGURE 5.6   The macro "TON_8".

$TV\_Ln = TV\_L + num = TV\_L + 57 = 220h + 57d = 220h + 39h = 259h$ (This is the 8-bit register in Bank4 of SRAM where the timing count value of this timer will be held.)

Note that in the following, the division $\dfrac{num}{8}$ is an integer division.

$R\_n = T\_Q0$ Register number

$$R\_n = T\_Q0 + num/8 = T\_Q0 + \frac{num}{8} = 4D0h + \frac{57}{8} = 4D0h + 7d = 4D7h$$

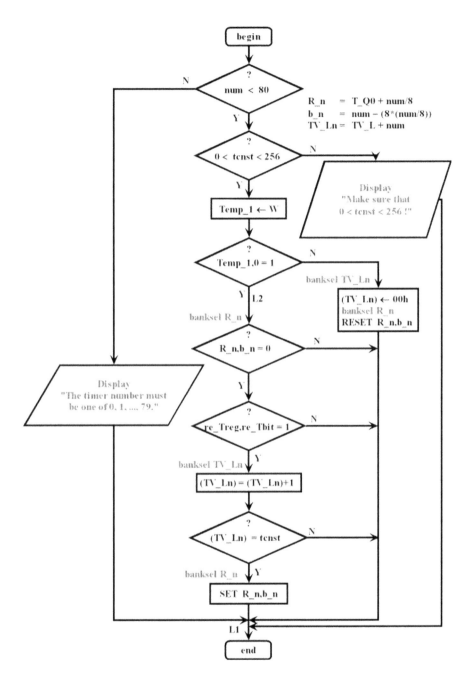

FIGURE 5.7 The flowchart of the macro "TON_8".

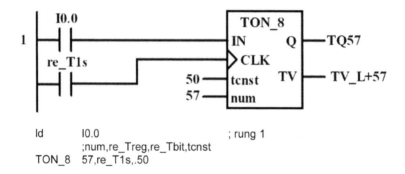

Id   I0.0   ; rung 1
            ;num,re_Treg,re_Tbit,tcnst
TON_8   57,re_T1s,.50

FIGURE 5.8 Example on-delay timer obtained with the macro "TON_8": a user program and its ladder diagram.

$$b\_n = T\_Q0 \text{ bit number}$$

$$b\_n = num - \left(8*\left(num/8\right)\right) = num - \left(8*\left(\frac{num}{8}\right)\right)$$

$$= 57 - \left(8*\left(\frac{57}{8}\right)\right) = 57 - \left(8*7\right) = 57 - 56 = 1$$

timer output (timer status bit) = R_n,b_n = 4D7h,1 = TQ57 (This is the 1-bit variable in Bank9 of SRAM where the state of the timer output will be held.)

5.3 MACRO "TON_16" (16-BIT ON-DELAY TIMER)

The macro "TON_16" defines 80 16-bit on-delay timers selected with the num = 0, 1, ..., 79. The symbol of the macro "TON_16" is depicted in Table 5.3. The macro "TON_16" and its flowchart are shown in Figures 5.9 and 5.10, respectively. Note that the CLK input has a rising edge symbol, but within the macro "TON_16", there

---

TABLE 5.3
The Symbol of the Macro "TON_16"

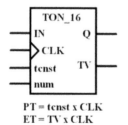

IN (input signal, through "W") = 0 or 1

CLK (rising edge signal of one of the four reference timing signals, re_Treg, re_Tbit) = re_T2ms (2 ms), re_T10ms (10 ms), re_T100ms (100 ms), or re_T1s (1 s)

tcnst (16-bit time constant value) = 1, 2, ..., 65535

num (the unique timer number) = 0, 1, ..., 79

Q (timer output – timer status bit) = one of the following: TQ0, TQ1, ..., TQ79, defined by T_Q0+num/8,num–(8*[num/8]) (num = 0, 1, ..., 79)

TV (16-bit timing count value hold in two 8-bit registers) = TV_H+num and TV_L+num (num = 0, 1, ..., 79)

```
TON_16 macro  num, re_Treg,re_Tbit, tcnst
          local    L1,L2,L3
          local    R_n,b_n,TV_Ln,TV_Hn
;-----------------------------------------------------------------------------
  if num < 80                          ;if num < 80 then carry on, else do not compile.
;-----------------------------------------------------------------------------
  if (tcnst>0)&&(tcnst<65536)     ;if 0<tcnst<65536 then carry on, else do not compile.
;-----------------------------------------------------------------------------
R_n      set     T_Q0+num/8            ;T_Q0 Register number
b_n      set     num-(8*(num/8))       ;T_Q0 bit number
TV_Ln set       TV_L+num              ;TV_L register number
TV_Hn set       TV_H+num              ;TV_H register number
;-----------------------------------------------------------------------------
          movwf  Temp_1
          btfsc   Temp_1,0
          goto    L3
          banksel TV_Ln
          clrf     TV_Ln
          banksel TV_Hn
          clrf     TV_Hn
          banksel R_n
          bcf      R_n,b_n
          goto    L1
L3
          banksel R_n
          btfsc   R_n,b_n
          goto    L1
          btfss   re_Treg,re_Tbit
          goto    L1
          banksel TV_Ln
          incfsz  TV_Ln,F
          goto    L2
          banksel TV_Hn
          incf     TV_Hn,F
          banksel TV_Ln
L2        movf    TV_Ln,W
          xorlw   low tcnst
          btfss   STATUS,Z
          goto    L1
          banksel TV_Hn
          movf    TV_Hn,W
          xorlw   high tcnst
          btfss   STATUS,Z
          goto    L1
          banksel R_n
          bsf      R_n,b_n
L1
;-----------------------------------------------------------------------------
```

FIGURE 5.9 *(1 of 2)* The macro "TON_16".

```
;-----------------------------------------------------------
  else
  error "Make sure that 0 < tcnst < 65536 !"
  endif
;-----------------------------------------------------------
  else
  error "The timer number must be one of 0, 1, ..., 79."
  endif
;-----------------------------------------------------------
          endm
```

FIGURE 5.9 Continued

is no rising edge detection mechanism for the signal provided through this input. Instead, it is assumed that the signal connected here is one of the rising edge signals available in the PIC16F1847-Based PLC, i.e., one of re_T2ms (2 ms), re_T10ms (10 ms), re_T100ms (100 ms), or re_T1s (1 s). IN (input signal, through "W"), Q (output signal = timer status bit), and CLK (re_Treg, re_Tbit) are all defined as Boolean variables. The time constant "tcnst" is an integer constant (here, for 16-bit resolution, it is chosen from any number in the range 1–65,535) and is used to define preset time PT, which is obtained by the formula PT = tcnst×CLK, where CLK should be used as the period of the rising edge signal of one of the four reference timing signals—ticks. The on-delay timer outputs are represented by the status bits TQ0, TQ1, ..., TQ79, defined by T_Q0+num/8,num–(8*[num/8]), as shown in Figure 5.4. A 16-bit integer variable TV consisting of two 8-bit variables "TV_H+num and TV_L+num" is used to count the rising edge signals connected to the CLK input. "TV_H+num" holds the high byte of the TV, while "TV_L+num" holds the low byte of the TV. The count value of the 16-bit integer variable "TV_H+num and TV_L+num" defines the elapsed time ET as follows: ET = CLK×count value of "TV_H+num and TV_L+num". Let us now briefly consider how the macro "TON_16" works. First of all, the preset time PT is defined by means of the period of a rising edge signal "CLK = re_Treg, re_Tbit" and a time constant "tcnst". If the input signal IN, taken into the macro by means of "W", is false (OFF), then the timer output Q, i.e., the status bit T_Q0+num/8,num–(8*[num/8]), is forced to be false (OFF) and the counter "TV_H+num and TV_L+num" is loaded with "0000h". If the input signal IN is true (ON) and the timer output Q, i.e., the status bit T_Q0+num/8,num–(8*[num/8]), is false (OFF), then with each rising edge signal "CLK = re_Treg, re_Tbit", the related counter "TV_H+num and TV_L+num" is incremented by 1. In this case, when the count value of "TV_H+num and TV_L+num" is equal to the number "tcnst", then the state change from OFF to ON is issued for the timer output Q, i.e., the status bit T_Q0+num/8,num–(8*[num/8]). If the input signal IN and the timer output Q, i.e., the status bit T_Q0+num/8,num–(8*[num/8]) are both true (ON), then no action is taken and the elapsed time ET is held. In this macro a previously defined 8-bit variable "Temp_1" is also utilized.

In order to explain how an on-delay timer is set up by using the macro "TON_16", let us consider Figure 5.11, where an on-delay timer is obtained with the macro "TON_16".

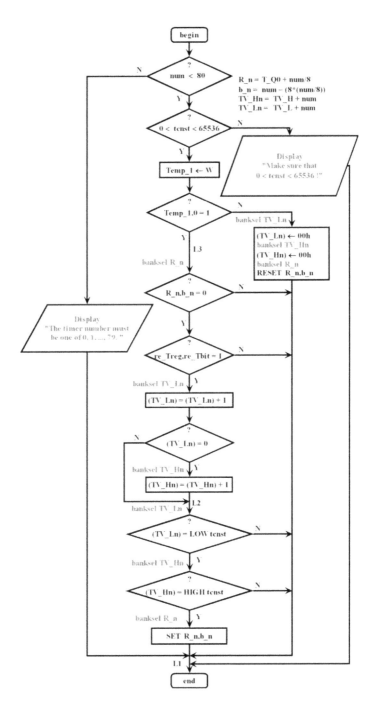

FIGURE 5.10 The flowchart of the macro "TON_16".

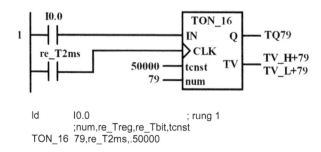

```
Id        I0.0                            ; rung 1
          ;num,re_Treg,re_Tbit,tcnst
TON_16  79,re_T2ms,.50000
```

FIGURE 5.11 Example on-delay timer obtained with the macro "TON_16": a user program and its ladder diagram.

In this example program we have:

IN = I0.0
num = 79 (decimal)
CLK(re_Treg,re_Tbit) = re_T2ms (the rising edge signal with 2-millisecond period)
tcnst = 50000 (decimal)

Therefore, we obtain the following from the macro "TON_16":

$$PT = tcnst * CLK \left(\text{the period of rising edge signal } re\_T2ms \right)$$

$$= 50000 * 2 \text{ ms} = 100 \text{ seconds}$$

TV_Ln = TV_L register number
TV_Ln = TV_L+num = TV_L+79 = 220h + 79d = 220h + 4Fh = 26Fh (This is the 8-bit register in Bank4 of SRAM where the low byte of the timing count value of this timer will be held.)

TV_Hn = TV_H register number
TV_Hn = TV_H+num = TV_H+79 = 2A0h + 79d = 2A0h + 4Fh = 2EFh (This is the 8-bit register in Bank5 of SRAM where the high byte of the timing count value of this timer will be held.)

Note that in the following, the division $\dfrac{num}{8}$ is an integer division.

R_n = T_Q0 Register number

$$R\_n = T\_Q0 + num/8 = T\_Q0 + \frac{num}{8} = 4D0h + \frac{79}{8} = 4D0h + 9d = 4D9h$$

$$b\_n = T\_Q0 \text{ bit number}$$

$$b\_n = \text{num} - \left(8*\left(\text{num}/8\right)\right) = \text{num} - \left(8*\left(\frac{\text{num}}{8}\right)\right)$$

$$= 79 - \left(8*\left(\frac{79}{8}\right)\right) = 79 - \left(8*9\right) = 79 - 72 = 7$$

timer output (timer status bit) = R_n,b_n = 4D9h,7 = TQ79 (This is the 1-bit variable in Bank9 of SRAM where the state of the timer output will be held.)

5.4   RETENTIVE ON-DELAY TIMER (RTO)

The symbol and timing diagram of the retentive on-delay timer (RTO) are both shown in Figure 5.12. The retentive on-delay timer accumulates timing values within a period set by parameter **PT**. When the signal state at input **IN** changes from OFF

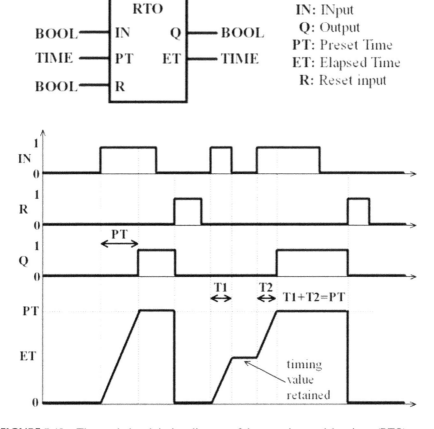

FIGURE 5.12   The symbol and timing diagram of the retentive on-delay timer (RTO).

to ON (positive-going edge—rising edge), the retentive on-delay timer executes and the duration **PT** starts. When the input **IN** has signal state ON, the duration **PT** is running and the timing values are accumulated and recorded. When the input **IN** has signal state OFF, the retentive on-delay timer retains the accumulated timing value. When enabled again (the input **IN** has signal state ON) it starts from the accumulated timing value. When the duration **PT** expires, the output **Q** has the signal state ON and the elapsed time **ET** is held. The output **Q** remains set to ON, even when the signal state at the input **IN** changes from ON to OFF. This timing operation is very useful for keeping track of elapsed times. The **R** input resets the elapsed time **ET** and **Q** regardless of the signal state at the input **IN**. The following two sections explain the implementations of 8-bit and 16-bit retentive on-delay timers, respectively, for the PIC16F1847-Based PLC.

5.5 MACRO "RTO_8" (8-BIT RETENTIVE ON-DELAY TIMER)

The macro "RTO_8" defines 80 8-bit retentive on-delay timers selected with the num = 0, 1, ..., 79. The symbol of the macro "RTO_8" is depicted in Table 5.4. The macro "RTO_8" and its flowchart are shown in Figures 5.13 and 5.14, respectively. Note that the CLK input has a rising edge symbol, but within the macro "RTO_8", there is no rising edge detection mechanism for the signal provided through this input. Instead, it is assumed that the signal connected here is one of the rising edge signals available in the PIC16F1847-Based PLC, i.e., one of re_T2ms (2 ms), re_T10ms (10 ms), re_T100ms (100 ms), or re_T1s (1 s). IN (input signal, through "W"), Q (output signal = timer status bit), CLK (re_Treg, re_Tbit), and R (reset input, R_reg,R_bit) are all defined as Boolean variables. The time constant "tcnst" is an integer constant (here, for 8-bit resolution, it is chosen from any number in the range 1–255) and is used to define preset time PT, which is obtained by the formula PT = tcnst×CLK, where CLK should be used as the period of the rising edge signal of one of the four reference timing signals—ticks. The retentive on-delay timer outputs are represented by the status bits TQ0, TQ1, ..., TQ79, defined by T_Q0+num/8,num−(8*[num/8]), as shown in Figure 5.4. An 8-bit

---

TABLE 5.4
The Symbol of the Macro "RTO_8"

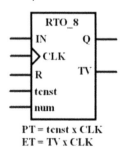

PT = tcnst x CLK
ET = TV x CLK

IN (input signal, through "W") = 0 or 1
CLK (rising edge signal of one of the four reference timing signals, re_Treg, re_Tbit) = re_T2ms (2 ms), re_T10ms (10 ms), re_T100ms (100 ms), or re_T1s (1 s)
R (reset input, R_reg,R_bit) = 0, 1
tcnst (8-bit time constant value) = 1, 2, ..., 255
num (the unique timer number) = 0, 1, ..., 79
Q (timer output – timer status bit) = one of the following: TQ0, TQ1, ..., TQ79, defined by T_Q0+num/8,num−(8*[num/8]) (num = 0, 1, ..., 79)
TV (8-bit timing count value hold in an 8-bit register) = TV_L+num (num = 0, 1, ..., 79)

```
RTO_8 macro    num, R_reg,R_bit, re_Treg,re_Tbit, tcnst
        local     L1,L2
        local     R_n,b_n,TV_Ln
;-----------------------------------------------------------------------
  if num < 80                     ;if num < 80 then carry on, else do not compile.
;-----------------------------------------------------------------------
  if (tcnst>0)&&(tcnst<256)       ;if 0<tcnst<256 then carry on, else do not compile.
;-----------------------------------------------------------------------
R_n     set     T_Q0+num/8              ;T_Q0 Register number
b_n     set     num-(8*(num/8))         ;T_Q0 bit number
TV_Ln   set     TV_L+num                ;TV_L register number
;-----------------------------------------------------------------------
        movwf  Temp_1
        banksel R_reg
        btfss   R_reg,R_bit
        goto    L2
        banksel TV_Ln
        clrf    TV_Ln
        banksel R_n
        bcf     R_n,b_n
        goto    L1
L2      btfss   Temp_1,0
        goto    L1
        banksel R_n
        btfsc   R_n,b_n
        goto    L1
        btfss   re_Treg,re_Tbit
        goto    L1
        banksel TV_Ln
        incf    TV_Ln,F
        movf    TV_Ln,W
        banksel R_n
        xorlw   tcnst
        btfsc   STATUS,Z
        bsf     R_n,b_n
L1
;-----------------------------------------------------------------------
  else
  error "Make sure that 0 < tcnst < 256 !"
  endif
;-----------------------------------------------------------------------
  else
  error "The timer number must be one of 0, 1, ..., 79."
  endif
;-----------------------------------------------------------------------
        endm
```

FIGURE 5.13   The macro "RTO_8".

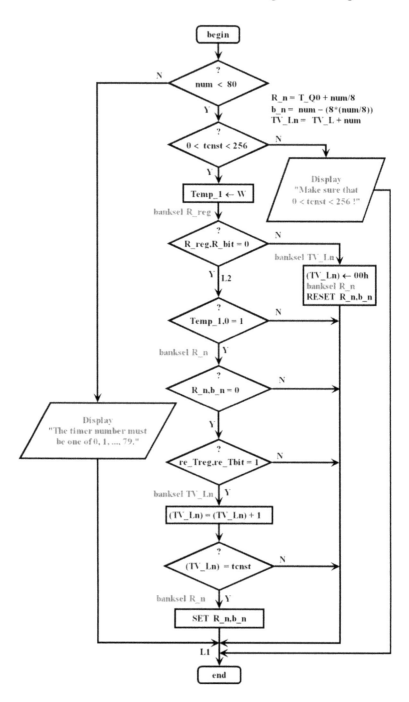

FIGURE 5.14 The flowchart of the macro "RTO_8".

integer variable "TV_L+num" is used to count the rising edge signals connected to the CLK input. The count value of "TV_L+num" defines the elapsed time ET as follows: ET = CLK×count value of "TV_L+num". Let us now briefly consider how the macro "RTO_8" works. First of all, the preset time PT is defined by means of the period of a rising edge signal "CLK = re_Treg, re_Tbit" and a time constant "tcnst". If the reset input signal R (R_reg,R_bit) is true (ON), then regardless of the signal state at the input IN, the timer output Q, i.e., the status bit T_Q0+num/8,num−(8*[num/8]), is forced to be false (OFF) and the counter "TV_L+num" is loaded with "00h". When R (R_reg,R_bit) is false (OFF), then the following operations are carried out. If the input signal IN is true (ON) and the timer output Q, i.e., the status bit T_Q0+num/8,num−(8*[num/8]), is false (OFF), then with each rising edge signal "CLK = re_Treg, re_Tbit", the related counter "TV_L+num" is incremented by 1. In this case, when the count value of "TV_L+num" is equal to the number "tcnst", then the state change from OFF to ON is issued for the timer output Q, i.e., the status bit T_Q0+num/8,num−(8*[num/8]). If the input signal IN is false (OFF) or if the input signal IN and the timer output Q, i.e., the status bit T_Q0+num/8,num−(8*[num/8]), are both true (ON), then no action is taken and the elapsed time ET is held. In this macro a previously defined 8-bit variable "Temp_1" is also utilized.

In order to explain how a retentive on-delay timer is set up by using the macro "RTO_8", let us consider Figure 5.15, where a retentive on-delay timer is obtained with the macro "RTO_8".

In this example program we have:

IN = I0.0
num = 5 (decimal)
R = I0.1
CLK(re_Treg,re_Tbit) = re_T100ms (the rising edge signal with 100-millisecond period)

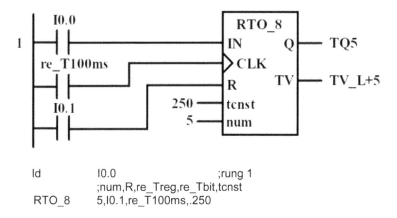

```
ld          I0.0                    ;rung 1
            ;num,R,re_Treg,re_Tbit,tcnst
RTO_8       5,I0.1,re_T100ms,.250
```

FIGURE 5.15 Example retentive on-delay timer obtained with the macro "RTO_8": a user program and its ladder diagram.

tcnst = 250 (decimal)

Therefore we obtain the following from the macro "RTO_8":

$$PT = tcnst * CLK \left(\text{the period of rising edge signal re\_T100ms}\right)$$

$$= 250 * 100 \text{ millisecond} = 25 \text{ seconds}$$

TV_Ln = TV_L register number
TV_Ln = TV_L+num = TV_L+5 = 220h+5d = 220h+5h = 225h (This is the
8-bit register in Bank4 of SRAM where the timing count value of this timer
will be held.)

Note that in the following, the division $\dfrac{num}{8}$ is an integer division.

R_n = T_Q0 **R**egister **n**umber

$$R\_n = T\_Q0 + num/8 = T\_Q0 + \frac{num}{8} = 4D0h + \frac{5}{8} = 4D0h + 0 = 4D0h$$

b_n = T_Q0 **b**it **n**umber

$$b\_n = num - \left(8 * \left(num/8\right)\right) = num - \left(8 * \left(\frac{num}{8}\right)\right)$$

$$= 5 - \left(8 * \left(\frac{5}{8}\right)\right) = 5 - \left(8 * 0\right) = 5 - 0 = 5$$

timer output (timer status bit) = R_n,b_n = 4D0h,5 = TQ5 (This is the 1-bit variable in Bank9 of SRAM where the state of the timer output will be held.)

5.6   MACRO "RTO_16" (16-BIT RETENTIVE ON-DELAY TIMER)

The macro "RTO_16" defines 80 16-bit retentive on-delay timers selected with the num
= 0, 1, …, 79. The symbol of the macro "RTO_16" is depicted in Table 5.5. The macro
"RTO_16" and its flowchart are shown in Figures 5.16 and 5.17, respectively. Note that
the CLK input has a rising edge symbol, but within the macro "RTO_16", there is no
rising edge detection mechanism for the signal provided through this input. Instead, it
is assumed that the signal connected here is one of the rising edge signals available in
the PIC16F1847-Based PLC, i.e., one of re_T2ms (2 ms), re_T10ms (10 ms), re_T100ms
(100 ms), or re_T1s (1 s). IN (input signal, through "W"), Q (output signal = timer status
bit), CLK (re_Treg, re_Tbit), and R (reset input, R_reg,R_bit) are all defined as Boolean
variables. The time constant "tcnst" is an integer constant (here, for 16-bit resolution, it
is chosen from any number in the range 1–65,535) and is used to define preset time PT,
which is obtained by the formula PT = tcnst×CLK, where CLK should be used as the

TABLE 5.5
The Symbol of the Macro "RTO_16"

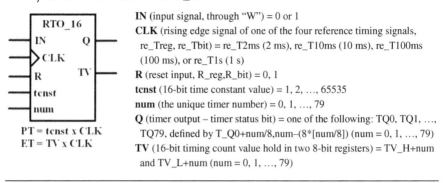

IN (input signal, through "W") = 0 or 1

CLK (rising edge signal of one of the four reference timing signals, re_Treg, re_Tbit) = re_T2ms (2 ms), re_T10ms (10 ms), re_T100ms (100 ms), or re_T1s (1 s)

R (reset input, R_reg,R_bit) = 0, 1

tcnst (16-bit time constant value) = 1, 2, …, 65535

num (the unique timer number) = 0, 1, …, 79

Q (timer output – timer status bit) = one of the following: TQ0, TQ1, …, TQ79, defined by T_Q0+num/8,num–(8*[num/8]) (num = 0, 1, …, 79)

TV (16-bit timing count value hold in two 8-bit registers) = TV_H+num and TV_L+num (num = 0, 1, …, 79)

period of the rising edge signal of one of the four reference timing signals—ticks. The retentive on-delay timer outputs are represented by the status bits TQ0, TQ1, …, TQ79, defined by T_Q0+num/8,num–(8*[num/8]), as shown in Figure 5.4. A 16-bit integer variable TV consisting of two 8-bit variables "TV_H+num and TV_L+num" is used to count the rising edge signals connected to the CLK input. "TV_H+num" holds the high byte of the TV, while "TV_L+num" holds the low byte of the TV. The count value of the 16-bit integer variable "TV_H+num and TV_L+num" defines the elapsed time ET as follows: ET = CLK×count value of "TV_H+num and TV_L+num". Let us now briefly consider how the macro "RTO_16" works. First of all, the preset time PT is defined by means of the period of a rising edge signal "CLK = re_Treg, re_Tbit" and a time constant "tcnst". If the reset input signal R (R_reg,R_bit) is true (ON), then regardless of the signal state at the input IN, the timer output Q, i.e., the status bit T_Q0+num/8,num–(8*[num/8]), is forced to be false (OFF) and the counter "TV_H+num and TV_L+num" is loaded with "0000h". When R (R_reg,R_bit) is false (OFF), then the following operations are carried out. If the input signal IN is true (ON) and the timer output Q, i.e., the status bit T_Q0+num/8,num–(8*[num/8]), is false (OFF), then with each rising edge signal "CLK = re_Treg, re_Tbit", the related counter "TV_H+num and TV_L+num" is incremented by 1. In this case, when the count value of "TV_H+num and TV_L+num" is equal to the number "tcnst", then the state change from OFF to ON is issued for the timer output Q, i.e., the status bit T_Q0+num/8,num–(8*[num/8]). If the input signal IN is false (OFF) or if the input signal IN and the timer output Q, i.e., the status bit T_Q0+num/8,num–(8*[num/8]), are both true (ON), then no action is taken and the elapsed time ET is held. In this macro a previously defined 8-bit variable "Temp_1" is also utilized.

In order to explain how a retentive on-delay timer is set up by using the macro "RTO_16", let us consider Figure 5.18, where a retentive on-delay timer is obtained with the macro "RTO_16".

In this example program we have:

IN = I0.0
num = 63 (decimal)

```
RTO_16 macro  num, R_reg,R_bit, re_Treg,re_Tbit, tcnst
          local    L1,L2,L3
          local    R_n,b_n,TV_Ln,TV_Hn
;-------------------------------------------------------------------------------
  if num < 80                              ;if num < 80 then carry on, else do not compile.
  ;-----------------------------------------------------------------------------
  if (tcnst>0)&&(tcnst<65536)   ;if 0<tcnst<65536 then carry on, else do not compile.
;-------------------------------------------------------------------------------
R_n     set      T_Q0+num/8              ;T_Q0 Register number
b_n     set      num-(8*(num/8))         ;T_Q0 bit number
TV_Ln  set      TV_L+num                ;TV_L register number
TV_Hn  set      TV_H+num                ;TV_H register number
;-------------------------------------------------------------------------------
          movwf  Temp_1
          banksel R_reg
          btfss    R_reg,R_bit
          goto     L3
          banksel R_n
          bcf      R_n,b_n
          banksel TV_Ln
          clrf     TV_Ln
          banksel TV_Hn
          clrf     TV_Hn
          goto     L1
L3        btfss    Temp_1,0
          goto     L1
          banksel R_n
          btfsc    R_n,b_n
          goto     L1
          btfss    re_Treg,re_Tbit
          goto     L1
          banksel TV_Ln
          incfsz   TV_Ln,F
          goto     L2
          banksel TV_Hn
          incf     TV_Hn,F
L2
          banksel TV_Ln
          movf     TV_Ln,W
          xorlw    low tcnst
          btfss    STATUS,Z
          goto     L1
          banksel TV_Hn
          movf     TV_Hn,W
          xorlw    high tcnst
          btfss    STATUS,Z
          goto     L1
          banksel R_n
          bsf      R_n,b_n
L1
;-------------------------------------------------------------------------------
```

FIGURE 5.16   (*1 of 2*) The macro "RTO_16".

```
;-------------------------------------------------------------------
  else
  error "Make sure that 0 < tcnst < 65536 !"
  endif
;-------------------------------------------------------------------
  else
  error "The timer number must be one of 0, 1, ..., 79."
  endif
;-------------------------------------------------------------------
        endm
```

FIGURE 5.16 Continued

$R = I0.1$

$CLK(re\_Treg, re\_Tbit) = re\_T10ms$ (the rising edge signal with 10-millisecond period)

$tcnst = 35000$ (decimal)

Therefore, we obtain the following from the macro "RTO_16":

$$PT = tcnst * CLK \left(\text{the period of rising edge signal } re\_T10ms \right)$$

$$= 35000 * 10 \text{ ms} = 350 \text{ seconds}$$

$TV\_Ln = TV\_L$ register number

$TV\_Ln = TV\_L + num = TV\_L + 63 = 220h + 63d = 220h + 3Fh = 25Fh$ (This is the 8-bit register in Bank4 of SRAM where the low byte of the timing count value of this timer will be held.)

$TV\_Hn = TV\_H$ register number

$TV\_Hn = TV\_H + num = TV\_H + 63 = 2A0h + 63d = 2A0h + 3Fh = 2DFh$ (This is the 8-bit register in Bank5 of SRAM where the high byte of the timing count value of this timer will be held.)

Note that in the following, the division $\dfrac{num}{8}$ is an integer division.

$R\_n = T\_Q0$ **R**egister **n**umber

$$R\_n = T\_Q0 + num/8 = T\_Q0 + \frac{num}{8} = 4D0h + \frac{63}{8} = 4D0h + 7d = 4D7h$$

$b\_n = T\_Q0$ **b**it **n**umber

$$b\_n = num - \left(8 * \left(num/8 \right) \right) = num - \left(8 * \left(\frac{num}{8} \right) \right)$$

$$= 63 - \left(8 * \left(\frac{63}{8} \right) \right) = 63 - \left(8 * 7 \right) = 63 - 56 = 7$$

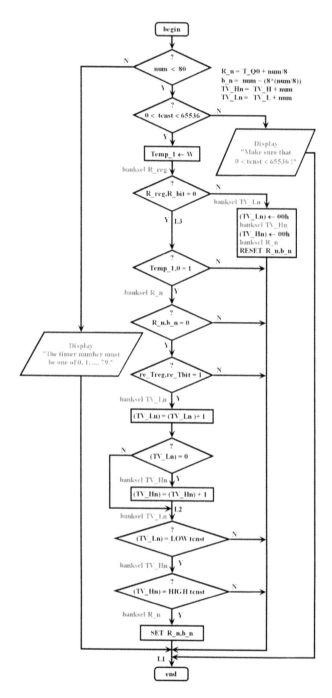

FIGURE 5.17   The flowchart of the macro "RTO_16".

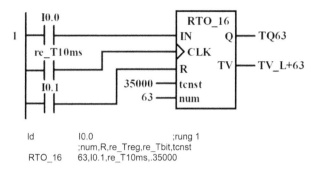

```
ld       I0.0                    ;rung 1
         ;num,R,re_Treg,re_Tbit,tcnst
RTO_16   63,I0.1,re_T10ms,.35000
```

FIGURE 5.18 Example retentive on-delay timer obtained with the macro "RTO_16": a user program and its ladder diagram.

timer output (timer status bit) = R_n,b_n = 4D7h,7 = TQ63 (This is the 1-bit variable in Bank9 of SRAM where the state of the timer output will be held.)

5.7 OFF-DELAY TIMER (TOF)

The off-delay timer can be used to delay setting an output false (OFF) for a fixed period of time after an input signal goes false (OFF), i.e., the output is held ON for a given period longer than the input. The symbol and timing diagram of the off-delay timer (TOF) are both shown in Figure 5.19. As the input signal **IN** goes true (ON), the output **Q** follows and remains true (ON) until the input signal **IN** is false (OFF) for the period specified in preset time input **PT**. As the input signal **IN** goes false (OFF), the elapsed time **ET** starts to increase. It continues to increase until it reaches the preset time input **PT**, at which point the output **Q** is set false (OFF) and the elapsed time is held. If the input signal **IN** is only false (OFF) for a period shorter than the input **PT**, the output **Q** remains true (ON). The following two sections explain the implementations of 8-bit and 16-bit off-delay timers, respectively, for the PIC16F1847-Based PLC.

5.8 MACRO "TOF_8" (8-BIT OFF-DELAY TIMER)

The macro "TOF_8" defines 80 8-bit off-delay timers selected with the num = 0, 1, …, 79. The symbol of the macro "TOF_8" is depicted in Table 5.6. The macro "TOF_8" and its flowchart are shown in Figures 5.20 and 5.21, respectively. Note that the CLK input has a rising edge symbol, but within the macro "TOF_8", there is no rising edge detection mechanism for the signal provided through this input. Instead, it is assumed that the signal connected here is one of the rising edge signals available in the PIC16F1847-Based PLC, i.e., one of the following: re_T2ms (2 ms), re_T10ms (10 ms), re_T100ms (100 ms), or re_T1s (1 s). IN (input signal, through "W"), Q (output signal = timer status bit), and CLK (re_Treg, re_Tbit) are all defined

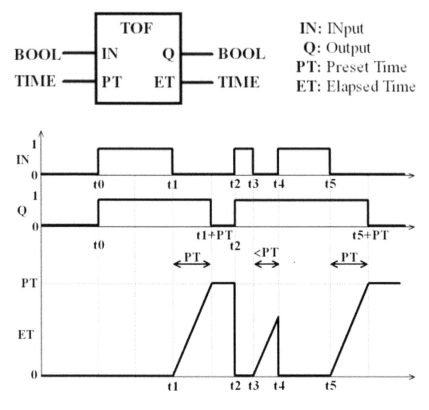

FIGURE 5.19 The symbol and timing diagram of the off-delay timer (TOF).

TABLE 5.6
The Symbol of the Macro "TOF_8"

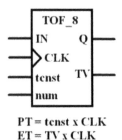

IN (input signal, through "W") = 0 or 1
CLK (rising edge signal of one of the four reference timing signals, re_Treg, re_Tbit) = re_T2ms (2 ms), re_T10ms (10 ms), re_T100ms (100 ms), or re_T1s (1 s)
tcnst (8-bit time constant value) = 1, 2, ..., 255
num (the unique timer number) = 0, 1, ..., 79
Q (timer output – timer status bit) = one of the following: TQ0, TQ1, ..., TQ79, defined by T_Q0+num/8,num–(8*[num/8]) (num = 0, 1, ..., 79)
TV (8-bit timing count value hold in an 8-bit register) = TV_L+num (num = 0, 1, ..., 79)

PT = tcnst x CLK
ET = TV x CLK

---

as Boolean variables. The time constant "tcnst" is an integer constant (here, for 8-bit resolution, it is chosen from any number in the range 1–255) and is used to define preset time PT, which is obtained by the formula PT = tcnst×CLK, where CLK should be used as the period of the rising edge signal of one of the four reference timing signals—ticks. The off-delay timer outputs are represented by the status bits

```
TOF_8 macro   num, re_Treg,re_Tbit, tcnst
        local    L1,L2
        local    R_n,b_n,TV_Ln
;----------------------------------------------------------------------------
  if num < 80                   ;if num < 80 then carry on, else do not compile.
  ;--------------------------------------------------------------------------
  if (tcnst>0)&&(tcnst<256)     ;if 0<tcnst<256 then carry on, else do not compile.
  ;--------------------------------------------------------------------------
R_n   set     T_Q0+num/8            ;T_Q0 Register number
b_n   set     num-(8*(num/8))       ;T_Q0 bit number
TV_Ln set     TV_L+num              ;TV_L register number
;--------------------------------------------------------------------------
        movwf  Temp_1
        btfss  Temp_1,0
        goto   L2
        banksel TV_Ln
        clrf   TV_Ln
        banksel R_n
        bsf    R_n,b_n
        goto   L1
L2
        banksel R_n
        btfss  R_n,b_n
        goto   L1
        btfss  re_Treg,re_Tbit
        goto   L1
        banksel TV_Ln
        incf   TV_Ln,F
        movf   TV_Ln,W
        banksel R_n
        xorlw  tcnst
        btfsc  STATUS,Z
        bcf    R_n,b_n
L1
;--------------------------------------------------------------------------
  else
  error "Make sure that 0 < tcnst < 256 !"
  endif
  ;--------------------------------------------------------------------------
  else
  error "The timer number must be one of 0, 1, ..., 79."
  endif
;--------------------------------------------------------------------------
        endm
```

FIGURE 5.20   The macro "TOF_8".

TQ0, TQ1, ..., TQ79, defined by T_Q0+num/8,num–(8*[num/8]), as shown in Figure 5.4. An 8-bit integer variable "TV_L+num" is used to count the rising edge signals connected to the CLK input. The count value of "TV_L+num" defines the elapsed time ET as follows: $ET = CLK \times$ count value of "TV_L+num". Let us now briefly consider how the macro "TOF_8" works. First of all, the preset time PT is defined

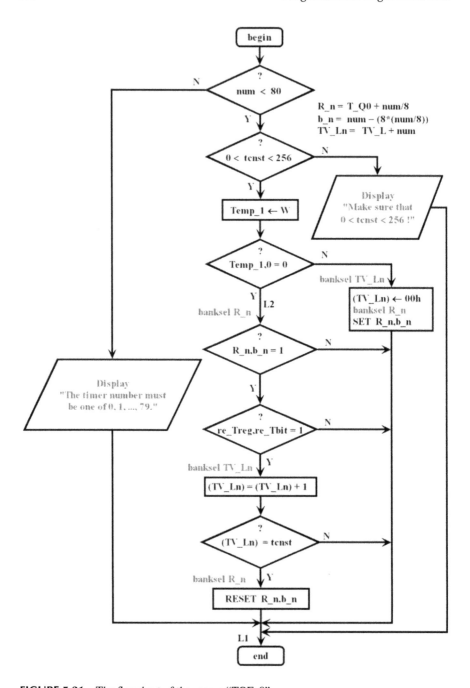

FIGURE 5.21   The flowchart of the macro "TOF_8".

by means of the period of a rising edge signal "CLK = re_Treg, re_Tbit" and a time constant "tcnst". If the input signal IN, taken into the macro by means of "W", is true (ON), then the timer output Q, i.e., the status bit T_Q0+num/8,num−(8*[num/8]), is forced to be true (ON) and the counter "TV_L+num" is loaded with "00h". When IN is ON and the timer output Q, i.e., the status bit T_Q0+num/8,num−(8*[num/8]), is ON, if IN goes false (OFF), then with each rising edge of the reference timing signal "CLK = re_Treg, re_Tbit" the related counter "TV_L+num" is incremented by 1. In this case, when the count value of "TV_L+num" is equal to the number "tcnst", then state change from ON to OFF is issued for the output signal (timer status bit) T_Q0+num/8,num−(8*[num/8]). In this macro a previously defined 8-bit variable "Temp_1" is also utilized.

In order to explain how an off-delay timer is set up by using the macro "TOF_8", let us consider Figure 5.22, where an off-delay timer is obtained with the macro "TOF_8".

In this example program we have:

IN = I0.0
num = 38 (decimal)
CLK(re_Treg,re_Tbit) = re_T1s (the rising edge signal with 1-second period)
tcnst = 150 (decimal)

Therefore, we obtain the following from the macro "TOF_8":

$$PT = tcnst * CLK \ (\text{the period of rising edge signal re\_T1s})$$

$$= 150 * 1 \ second = 150 \ seconds$$

TV_Ln = TV_L register number
TV_Ln = TV_L+num = TV_L+38 = 220h + 38d = 220h + 26h = 246h (This is the 8-bit register in Bank4 of SRAM where the timing count value of this timer will be held.)

```
Id       I0.0                          ; rung 1
         ;num,re_Treg,re_Tbit,tcnst
TOF_8    38,re_T1s,.150
```

FIGURE 5.22 Example off-delay timer obtained with the macro "TOF_8": a user program and its ladder diagram.

Note that in the following, the division $\dfrac{\text{num}}{8}$ is an integer division.

R_n = T_Q0 **Register number**

$$R\_n = T\_Q0 + num/8 = T\_Q0 + \frac{num}{8} = 4D0h + \frac{38}{8} = 4D0h + 4d = 4D4h$$

b_n = T_Q0 **bit number**

$$b\_n = num - \left(8*\left(num/8\right)\right) = num - \left(8*\left(\frac{num}{8}\right)\right)$$

$$= 38 - \left(8*\left(\frac{38}{8}\right)\right) = 38 - \left(8*4\right) = 38 - 32 = 6$$

timer output (timer status bit) = R_n,b_n = 4D4h,6 = TQ38 (This is the 1-bit variable in Bank9 of SRAM where the state of the timer output will be held.)

5.9 MACRO "TOF_16" (16-BIT OFF-DELAY TIMER)

The macro "TOF_16" defines 80 16-bit off-delay timers selected with the num = 0, 1, ..., 79. The symbol of the macro "TOF_16" is depicted in Table 5.7. The macro "TOF_16" and its flowchart are shown in Figures 5.23 and 5.24, respectively. Note that the CLK input has a rising edge symbol, but within the macro "TOF_16", there is no rising edge detection mechanism for the signal provided through this input. Instead, it is assumed that the signal connected here is one of the rising edge signals available in the PIC16F1847-Based PLC, i.e., one of the following: re_T2ms (2 ms), re_T10ms (10 ms), re_T100ms (100 ms), or re_T1s (1 s). IN (input signal, through "W"), Q (output signal = timer status bit), and CLK (re_Treg, re_Tbit) are all defined as Boolean variables. The time constant "tcnst"

TABLE 5.7
The Symbol of the Macro "TOF_16"

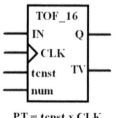

PT = tcnst x CLK
ET = TV x CLK

IN (input signal, through "W") = 0 or 1
CLK (rising edge signal of one of the four reference timing signals, re_Treg, re_Tbit) = re_T2ms (2 ms), re_T10ms (10 ms), re_T100ms (100 ms), or re_T1s (1 s)
tcnst (16-bit time constant value) = 1, 2, ..., 65535
num (the unique timer number) = 0, 1, ..., 79
Q (timer output – timer status bit) = one of the following: TQ0, TQ1, ..., TQ79, defined by T_Q0+num/8,num–(8*[num/8]) (num = 0, 1, ..., 79)
TV (16-bit timing count value hold in two 8-bit registers) = TV_H+num and TV_L+num (num = 0, 1, ..., 79)

```
TOF_16  macro  num, re_Treg,re_Tbit, tcnst
        local    L1,L2,L3
        local    R_n,b_n,TV_Ln,TV_Hn
;---------------------------------------------------------------------------
 if num < 80                    ;if num < 80 then carry on, else do not compile.
;---------------------------------------------------------------------------
 if (tcnst>0)&&(tcnst<65536)    ;if 0<tcnst<65536 then carry on, else do not compile.
;---------------------------------------------------------------------------
R_n   set    T_Q0+num/8          ;T_Q0 Register number
b_n   set    num-(8*(num/8))     ;T_Q0 bit number
TV_Ln set    TV_L+num            ;TV_L register number
TV_Hn set    TV_H+num            ;TV_H register number
;---------------------------------------------------------------------------
        movwf  Temp_1
        btfss  Temp_1,0
        goto   L3
        banksel TV_Ln
        clrf   TV_Ln
        banksel TV_Hn
        clrf   TV_Hn
        banksel R_n
        bsf    R_n,b_n
        goto   L1
L3
        banksel R_n
        btfss  R_n,b_n
        goto   L1
        btfss  re_Treg,re_Tbit
        goto   L1
        banksel TV_Ln
        incfsz TV_Ln,F
        goto   L2
        banksel TV_Hn
        incf   TV_Hn,F
L2
        banksel TV_Ln
        movf   TV_Ln,W
        xorlw  low tcnst
        btfss  STATUS,Z
        goto   L1
        banksel TV_Hn
        movf   TV_Hn,W
        xorlw  high tcnst
        btfss  STATUS,Z
        goto   L1
        banksel R_n
        bcf    R_n,b_n
L1
;---------------------------------------------------------------------------
```

FIGURE 5.23 (*1 of 2*) The macro "TOF_16".

```
;-----------------------------------------------------------------
    else
    error "Make sure that 0 < tcnst < 65536 !"
    endif
;-----------------------------------------------------------------
    else
    error "The timer number must be one of 0, 1, ..., 79."
    endif
;-----------------------------------------------------------------
        endm
```

FIGURE 5.23 Continued

is an integer constant (here, for 16-bit resolution, it is chosen from any number in the range 1–65,535) and is used to define preset time PT, which is obtained by the formula PT = tcnst × CLK, where CLK should be used as the period of the rising edge signal of one of the four reference timing signals—ticks. The off-delay timer outputs are represented by the status bits TQ0, TQ1, ..., TQ79, defined by T_Q0+num/8,num–(8*[num/8]), as shown in Figure 5.4. A 16-bit integer variable TV consisting of two 8-bit variables "TV_H+num and TV_L+num" is used to count the rising edge signals connected to the CLK input. "TV_H+num" holds the high byte of the TV, while "TV_L+num" holds the low byte of the TV. The count value of the 16-bit integer variable "TV_H+num and TV_L+num" defines the elapsed time ET as follows: ET = CLK × count value of "TV_H+num and TV_L+num". Let us now briefly consider how the macro "TOF_16" works. First of all, the preset time PT is defined by means of the period of a rising edge signal "CLK = re_Treg, re_Tbit" and a time constant "tcnst". If the input signal IN, taken into the macro by means of "W", is true (ON), then the timer output Q, i.e., the status bit T_Q0+num/8,num–(8*[num/8]), is forced to be true (ON) and the counter "TV_H+num and TV_L+num" is loaded with "0000h". When IN is ON and the timer output Q, i.e., the status bit T_Q0+num/8,num–(8*[num/8]), is ON, if IN goes false (OFF), then with each rising edge of the reference timing signal "CLK = re_Treg, re_Tbit" the related counter "TV_H+num and TV_L+num" is incremented by 1. In this case, when the count value of "TV_H+num and TV_L+num" is equal to the number "tcnst", then state change from ON to OFF is issued for the output signal (timer status bit) T_Q0+num/8,num–(8*[num/8]). In this macro a previously defined 8-bit variable "Temp_1" is also utilized.

In order to explain how an off-delay timer is set up by using the macro "TOF_16", let us consider Figure 5.25, where an off-delay timer is obtained with the macro "TOF_16".

In this example program we have:

IN = I0.0
num = 11 (decimal)
CLK(re_Treg,re_Tbit) = re_T2ms (the rising edge signal with 2-millisecond period)
tcnst = 25000 (decimal)

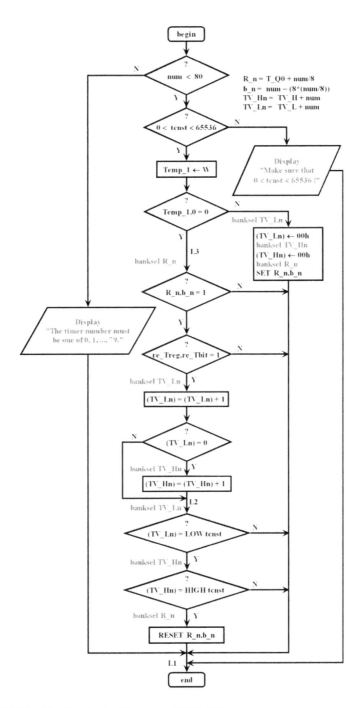

FIGURE 5.24 The flowchart of the macro "TOF_16".

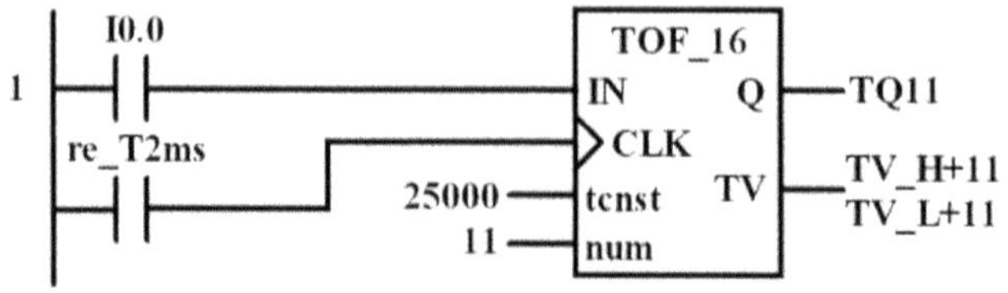

```
Id        I0.0                              ; rung 1
          ;num,re_Treg,re_Tbit,tcnst
TOF_16  11,re_T2ms,.25000
```

FIGURE 5.25 Example off-delay timer obtained with the macro "TOF_16": a user program and its ladder diagram.

Therefore, we obtain the following from the macro "TOF_16":

$$PT = tcnst * CLK \text{ (the period of rising edge signal re\_T2ms)}$$

$$= 25000 * 2 \text{ ms} = 50 \text{ seconds}$$

TV_Ln = TV_L register number
TV_Ln = TV_L+num = TV_L+11 = 220h + 11d = 220h + Bh = 22Bh (This is the 8-bit register in Bank4 of SRAM where the low byte of the timing count value of this timer will be held.)

TV_Hn = TV_H register number
TV_Hn = TV_H+num = TV_H+11 = 2A0h + 11d = 2A0h + Bh = 2ABh (This is the 8-bit register in Bank5 of SRAM where the high byte of the timing count value of this timer will be held.)

Note that in the following, the division $\dfrac{num}{8}$ is an integer division.

R_n = T_Q0 **Register number**

$$R\_n = T\_Q0 + num/8 = T\_Q0 + \frac{num}{8} = 4D0h + \frac{11}{8} = 4D0h + 1d = 4D1h$$

b_n = T_Q0 **bit number**

$$b\_n = num - \left(8*(num/8)\right) = num - \left(8*\left(\frac{num}{8}\right)\right)$$

$$= 11 - \left(8*\left(\frac{11}{8}\right)\right) = 11 - (8*1) = 11 - 8 = 3$$

timer output (timer status bit) = R_n,b_n = 4D1h,3 = TQ11 (This is the 1-bit variable in Bank9 of SRAM where the state of the timer output will be held.)

5.10   PULSE TIMER (TP)

The pulse timer can be used to generate output pulses of a given time duration. The symbol and timing diagram of the pulse timer (TP) are both shown in Figure 5.26. As the input signal **IN** goes true (ON) (t0; t2; t4), the output **Q** follows and remains true (ON) for the pulse duration as specified by the preset time input **PT**. While the pulse output **Q** is true (ON), the elapsed time **ET** is increased (between t0 and t0+PT; between t2 and t2+PT; between t4 and t4+PT). On the termination of the pulse, the elapsed time **ET** is reset. The output **Q** will remain true (ON) until the pulse time has elapsed, irrespective of the state of the input signal **IN**. The following two sections explain the implementations of 8-bit and 16-bit pulse timers, respectively, for the PIC16F1847-Based PLC.

5.11   MACRO "TP_8" (8-BIT PULSE TIMER)

The macro "TP_8" defines 80 8-bit pulse timers selected with the num = 0, 1, ..., 79. The symbol of the macro "TP_8" is depicted in Table 5.8. The macro "TP_8" and its flowchart are shown in Figures 5.27 and 5.28, respectively. Note that the CLK input has a rising edge symbol, but within the macro "TP_8", there is no rising edge detection mechanism for the signal provided through this input. Instead, it is

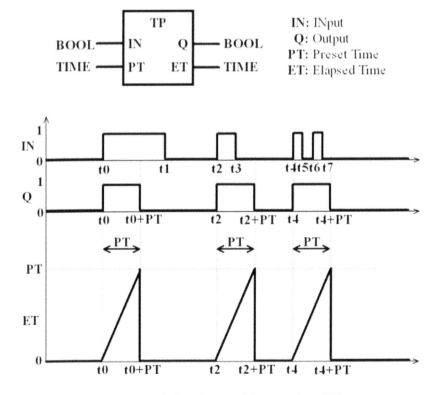

FIGURE 5.26   The symbol and timing diagram of the pulse timer (TP).

TABLE 5.8
The Symbol of the Macro "TP_8"

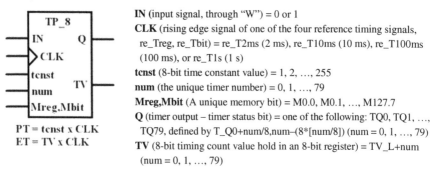

IN (input signal, through "W") = 0 or 1

CLK (rising edge signal of one of the four reference timing signals,
re_Treg, re_Tbit) = re_T2ms (2 ms), re_T10ms (10 ms), re_T100ms
(100 ms), or re_T1s (1 s)

tcnst (8-bit time constant value) = 1, 2, ..., 255

num (the unique timer number) = 0, 1, ..., 79

Mreg,Mbit (A unique memory bit) = M0.0, M0.1, ..., M127.7

Q (timer output – timer status bit) = one of the following: TQ0, TQ1, ...,
TQ79, defined by T_Q0+num/8,num–(8*[num/8]) (num = 0, 1, ..., 79)

TV (8-bit timing count value hold in an 8-bit register) = TV_L+num
(num = 0, 1, ..., 79)

assumed that the signal connected here is one of the rising edge signals available in the PIC16F1847-Based PLC, i.e., one of the following: re_T2ms (2 ms), re_T10ms (10 ms), re_T100ms (100 ms), or re_T1s (1 s). IN (input signal, through "W"), Q (output signal = timer status bit), and CLK (re_Treg, re_Tbit) are all defined as Boolean variables. A unique memory bit (Mreg,Mbit) = M0.0, M0.1, ..., M127.7 is used to detect the rising edge of the input signal IN. The time constant "tcnst" is an integer constant (here, for 8-bit resolution, it is chosen from any number in the range 1–255) and is used to define preset time PT, which is obtained by the formula PT = tcnst×CLK, where CLK should be used as the period of the rising edge signal of one of the four reference timing signals—ticks. The pulse timer outputs are represented by the status bits TQ0, TQ1, ..., TQ79, defined by T_Q0+num/8,num–(8*[num/8]), as shown in Figure 5.4. An 8-bit integer variable "TV_L+num" is used to count the rising edge signals connected to the CLK input. The count value of "TV_L+num" defines the elapsed time ET as follows: ET = CLK×count value of "TV_L+num". Let us now briefly consider how the macro "TP_8" works. First of all, the preset time PT is defined by means of the period of a rising edge signal "CLK = re_Treg, re_Tbit" and a time constant "tcnst". As soon as the rising edge of the input signal IN is detected, by means of Mreg,Mbit, the output signal T_Q0+num/8,num–(8*[num/8]) is forced to be true (ON). After the output becomes true, i.e., T_Q0+num/8,num–(8*[num/8]) = 1, the related counter "TV_L+num" is incremented by 1 with each rising edge signal of one of the four reference timing signals "CLK = re_Treg, re_Tbit". When the count value of "TV_L+num" is equal to the number "tcnst", then state change from ON to OFF is issued for the output signal (timer status bit) T_Q0+num/8,num–(8*[num/8]) and at the same time the counter "TV_L+num" is cleared. In this macro a previously defined 8-bit variable "Temp_1" is also utilized.

In order to explain how a pulse timer is set up by using the macro "TP_8", let us consider Figure 5.29, where a pulse timer is obtained with the macro "TP_8".

In this example program we have:

IN = I0.0
num = 22 (decimal)

```
TP_8    macro   num, Mreg,Mbit, re_Treg,re_Tbit, tcnst
        local   L0,L1,L2,L3
        local   R_n,b_n,TV_Ln
;------------------------------------------------------------------
  if num < 80                   ;if num < 80 then carry on, else do not compile.
;------------------------------------------------------------------
  if (tcnst>0)&&(tcnst<256)     ;if 0<tcnst<256 then carry on, else do not compile.
;------------------------------------------------------------------
R_n     set     T_Q0+num/8              ;T_Q0 Register number
b_n     set     num-(8*(num/8))         ;T_Q0 bit number
TV_Ln   set     TV_L+num                ;TV_L register number
;------------------------------------------------------------------
        movwf   Temp_1
        banksel Mreg
        btfss   Temp_1,0
        bsf     Mreg,Mbit
        btfss   Temp_1,0
        goto    L3
        btfss   Mreg,Mbit
        goto    L3
        bcf     Mreg,Mbit
        banksel R_n
        bsf     R_n,b_n
L3
        banksel R_n
        btfsc   R_n,b_n
        goto    L2
        btfss   Temp_1,0
        goto    L1
L2      btfss   re_Treg,re_Tbit
        goto    L1
        banksel TV_Ln
        incf    TV_Ln,F
        movf    TV_Ln,W
        xorlw   tcnst
        btfss   STATUS,Z
        goto    L1
        banksel R_n
        bcf     R_n,b_n
L1
        banksel R_n
        btfsc   R_n,b_n
        goto    L0
        banksel TV_Ln
        clrf    TV_Ln
L0
;------------------------------------------------------------------
```

FIGURE 5.27   (*1 of 2*) The macro "TP_8".

```
;-----------------------------------------------------------
    else
    error "Make sure that 0 < tcnst < 256 !"
    endif
;-----------------------------------------------------------
    else
    error "The timer number must be one of 0, 1, ..., 79."
    endif
;-----------------------------------------------------------
            endm
```

FIGURE 5.27   Continued

Mreg,Mbit = M0.0
CLK(re_Treg,re_Tbit) = re_T100ms (the rising edge signal with 100-millisec-
    ond period)
tcnst = 120 (decimal)

Therefore, we obtain the following from the macro "TP_8":

$$PT = tcnst * CLK \left(\text{the period of rising edge signal re\_T100ms} \right)$$

$$= 120 * 100 \text{ ms} = 12 \text{ seconds}$$

TV_Ln = TV_L register number
TV_Ln = TV_L+num = TV_L+22 = 220h + 22d = 220h + 16h = 236h (This is
    the 8-bit register in Bank4 of SRAM where the timing count value of this
    timer will be held.)

Note that in the following, the division $\dfrac{num}{8}$ is an integer division.

R_n = T_Q0 **Register number**

$$R\_n = T\_Q0 + num/8 = T\_Q0 + \frac{num}{8} = 4D0h + \frac{22}{8} = 4D0h + 2d = 4D2h$$

b_n = T_Q0 **bit number**

$$b\_n = num - \left(8 * \left(num/8 \right) \right) = num - \left(8 * \left(\frac{num}{8} \right) \right)$$

$$= 22 - \left(8 * \left(\frac{22}{8} \right) \right) = 22 - \left(8 * 2 \right) = 22 - 16 = 6$$

timer output (timer status bit) = R_n,b_n = 4D2h,6 = TQ22 (This is the 1-bit variable in Bank9 of SRAM where the state of the timer output will be held.)

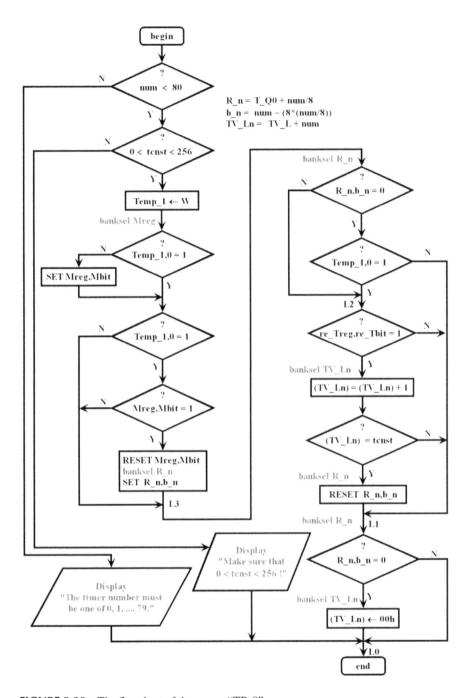

FIGURE 5.28 The flowchart of the macro "TP_8".

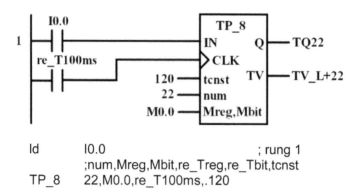

```
ld       I0.0                               ; rung 1
         ;num,Mreg,Mbit,re_Treg,re_Tbit,tcnst
TP_8     22,M0.0,re_T100ms,.120
```

FIGURE 5.29 Example pulse timer obtained with the macro "TP_8": a user program and its ladder diagram.

5.12   MACRO "TP_16" (16-BIT PULSE TIMER)

The macro "TP_16" defines 80 16-bit pulse timers selected with the num = 0, 1, ..., 79. The symbol of the macro "TP_16" is depicted in Table 5.9. The macro "TP_16" and its flowchart are shown in Figures 5.30 and 5.31, respectively. Note that the CLK input has a rising edge symbol, but within the macro "TP_16", there is no rising edge detection mechanism for the signal provided through this input. Instead, it is assumed that the signal connected here is one of the rising edge signals available in the PIC16F1847-Based PLC, i.e., one of the following: re_T2ms (2 ms), re_T10ms (10 ms), re_T100ms (100 ms), or re_T1s (1 s). IN (input signal, through "W"), Q (output signal = timer status bit), and CLK (re_Treg, re_Tbit) are all defined as Boolean variables. A unique memory bit (Mreg,Mbit) = M0.0, M0.1, ..., M127.7 is used to detect the rising edge of the input signal IN. The time constant "tcnst" is an integer constant (here, for 16-bit resolution, it is chosen from any number in the range 1–65,535) and is used to define preset time PT, which is obtained by the formula PT = tcnst × CLK, where CLK should be used as the period of the rising edge signal of one of the four

---

TABLE 5.9
The Symbol of the Macro "TP_16"

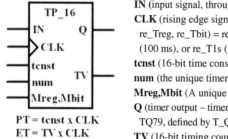

IN (input signal, through "W") = 0 or 1

CLK (rising edge signal of one of the four reference timing signals, re_Treg, re_Tbit) = re_T2ms (2 ms), re_T10ms (10 ms), re_T100ms (100 ms), or re_T1s (1 s)

tcnst (16-bit time constant value) = 1, 2, ..., 65535

num (the unique timer number) = 0, 1, ..., 79

Mreg,Mbit (A unique memory bit) = M0.0, M0.1, ..., M127.7

Q (timer output – timer status bit) = one of the following: TQ0, TQ1, ..., TQ79, defined by T_Q0+num/8,num–(8*[num/8]) (num = 0, 1, ..., 79)

TV (16-bit timing count value hold in two 8-bit registers) = TV_H+num and TV_L+num (num = 0, 1, ..., 79)

```
TP_16 macro    num, Mreg,Mbit, re_Treg,re_Tbit, tcnst
        local    L0,L1,L2,L3,L4
        local    R_n,b_n,TV_Ln,TV_Hn
;-------------------------------------------------------------------------
 if num < 80                    ;if num < 80 then carry on, else do not compile.
;-------------------------------------------------------------------------
 if (tcnst>0)&&(tcnst<65536)    ;if 0<tcnst<65536 then carry on, else do not compile.
;-------------------------------------------------------------------------
R_n     set    T_Q0+num/8          ;T_Q0 Register number
b_n     set    num-(8*(num/8))     ;T_Q0 bit number
TV_Ln   set    TV_L+num            ;TV_L register number
TV_Hn   set    TV_H+num            ;TV_H register number
;-------------------------------------------------------------------------
        movwf  Temp_1
        banksel Mreg
        btfss  Temp_1,0
        bsf    Mreg,Mbit
        btfss  Temp_1,0
        goto   L4
        btfss  Mreg,Mbit
        goto   L4
        bcf    Mreg,Mbit
        banksel R_n
        bsf    R_n,b_n
L4
        banksel R_n
        btfsc  R_n,b_n
        goto   L3
        btfss  Temp_1,0
        goto   L1
L3      btfss  re_Treg,re_Tbit
        goto   L1
        banksel TV_Ln
        incfsz TV_Ln,F
        goto   L2
        banksel TV_Hn
        incf   TV_Hn,F
L2
        banksel TV_Ln
        movf   TV_Ln,W
        xorlw  low tcnst
        btfss  STATUS,Z
        goto   L1
        banksel TV_Hn
        movf   TV_Hn,W
        xorlw  high tcnst
        btfss  STATUS,Z
        goto   L1
        banksel R_n
        bcf    R_n,b_n
```

FIGURE 5.30   *(1 of 2)* The macro "TP_16".

```
L1
        banksel R_n
        btfsc    R_n,b_n
        goto     L0
        banksel TV_Ln
        clrf     TV_Ln
        banksel TV_Hn
        clrf     TV_Hn
LO
;------------------------------------------------------------------
 else
 error "Make sure that 0 < tcnst < 65536 !"
 endif
;------------------------------------------------------------------
 else
 error "The timer number must be one of 0, 1, ..., 79."
 endif
;------------------------------------------------------------------
        endm
```

FIGURE 5.30   Continued

reference timing signals—ticks. The pulse timer outputs are represented by the status bits TQ0, TQ1, ..., TQ79, defined by T_Q0+num/8,num–(8*[num/8]), as shown
in Figure 5.4. A 16-bit integer variable TV consisting of two 8-bit variables "TV_
H+num and TV_L+num" is used to count the rising edge signals connected to the
CLK input. "TV_H+num" holds the high byte of the TV, while "TV_L+num" holds
the low byte of the TV. The count value of the 16-bit integer variable "TV_H+num
and TV_L+num" defines the elapsed time ET as follows: $ET = CLK \times$ count value of
"TV_H+num and TV_L+num". Let us now briefly consider how the macro "TP_16"
works. First of all, the preset time PT is defined by means of the period of a rising
edge signal "CLK = re_Treg, re_Tbit" and a time constant "tcnst". As soon as the rising edge of the input signal IN is detected, by means of Mreg,Mbit, the output signal
T_Q0+num/8,num–(8*[num/8]) is forced to be true (ON). After the output becomes
true, i.e., T_Q0+num/8,num–(8*[num/8]) = 1, the related counter "TV_H+num and
TV_L+num" is incremented by 1 with each rising edge signal of one of the four reference timing signals "CLK = re_Treg, re_Tbit". When the count value of "TV_H+num
and TV_L+num" is equal to the number "tcnst", then state change from ON to OFF
is issued for the output signal (timer status bit) T_Q0+num/8,num–(8*[num/8]) and at
the same time the counter "TV_H+num and "TV_L+num" is cleared. In this macro a
previously defined 8-bit variable "Temp_1" is also utilized.

In order to explain how a pulse timer is set up by using the macro "TP_16", let us
consider Figure 5.32, where a pulse timer is obtained with the macro "TP_16".

In this example program we have:

IN = I0.0
num = 45 (decimal)
Mreg,Mbit = M0.0

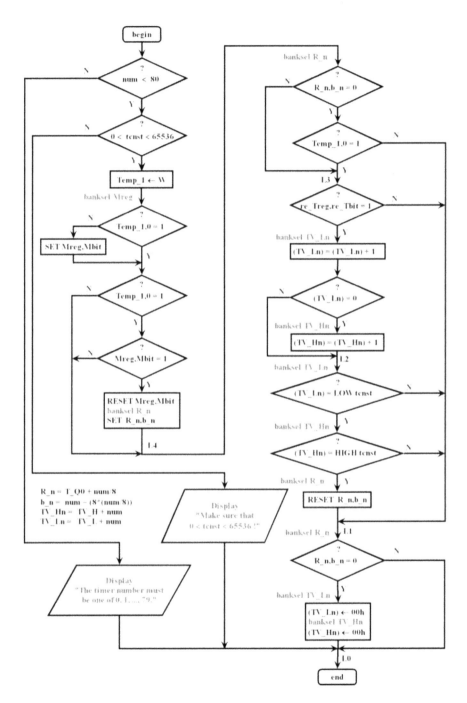

FIGURE 5.31 The flowchart of the macro "TP_16".

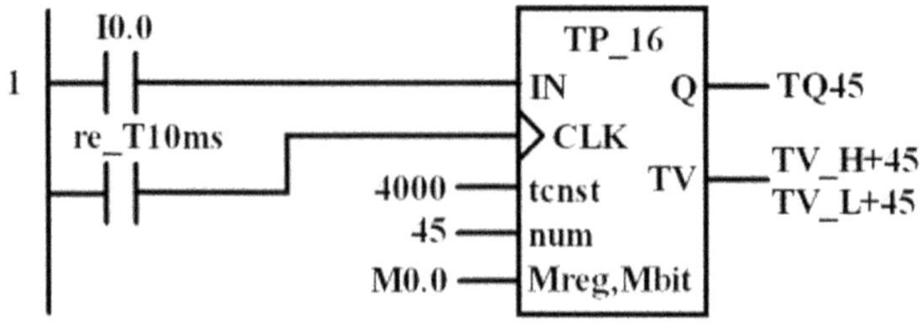

```
Id        I0.0                                      ; rung 1
          ;num,Mreg,Mbit,re_Treg,re_Tbit,tcnst
TP_16     45,M0.0,re_T10ms,.4000
```

FIGURE 5.32 Example pulse timer obtained with the macro "TP_16": a user program and its ladder diagram.

CLK(re_Treg,re_Tbit) = re_T10ms (the rising edge signal with 10-millisecond period)
tcnst = 4000 (decimal)

Therefore, we obtain the following from the macro "TP_16":

$$PT = tcnst * CLK \left(\text{the period of rising edge signal re\_T10ms} \right)$$

$$= 4000 * 10 \text{ ms} = 40 \text{ seconds}$$

TV_Ln = TV_L register number
TV_Ln = TV_L+num = TV_L+45 = 220h + 45d = 220h + 2Dh = 24Dh (This is the 8-bit register in Bank4 of SRAM where the low byte of the timing count value of this timer will be held.)
TV_Hn = TV_H register number
TV_Hn = TV_H+num = TV_H+45 = 2A0h + 45d = 2A0h + 2Dh = 2CDh (This is the 8-bit register in Bank5 of SRAM where the high byte of the timing count value of this timer will be held.)

Note that in the following, the division $\dfrac{num}{8}$ is an integer division.

R_n = T_Q0 Register number

$$R\_n = T\_Q0 + num/8 = T\_Q0 + \frac{num}{8} = 4D0h + \frac{45}{8} = 4D0h + 5d = 4D5h$$

b_n = T_Q0 bit number

$$b\_n = num - \left(8 * \left(num/8\right)\right) = num - \left(8 * \left(\frac{num}{8}\right)\right)$$

$$= 45 - \left(8 * \left(\frac{45}{8}\right)\right) = 45 - \left(8 * 5\right) = 45 - 40 = 5$$

timer output (timer status bit) = R_n,b_n = 4D5h,5 = TQ45 (This is the 1-bit variable in Bank9 of SRAM where the state of the timer output will be held.)

5.13   EXTENDED PULSE TIMER (TEP)

The extended pulse timer functions in the same way as the pulse timer, as explained in the previous section, with an additional feature. The symbol and timing diagram of the extended pulse timer (TEP) are both shown in Figure 5.33. As the input signal **IN** goes true (ON) (t0; t2; t4), the output **Q** follows and remains true (ON), for the extended pulse duration as specified by the preset time input PT. While the extended pulse output **Q** is true (ON), the elapsed time **ET** is increased (between t0 and t0 + PT; between t2 and t2 + PT; between t4 and t6; between t6 and t6 + PT). On the termination of the pulse, the elapsed time **ET** is reset. The output **Q** will remain true (ON) as long as the timer is running. The timer will be restarted ("re-triggered") with the preset time value **PT**, and thus the pulse duration will be extended if the signal state at the input **IN** changes from OFF to ON (t6) while the timer is running. The following two sections explain the implementations of 8-bit and 16-bit extended pulse timers, respectively, for the PIC16F1847-Based PLC.

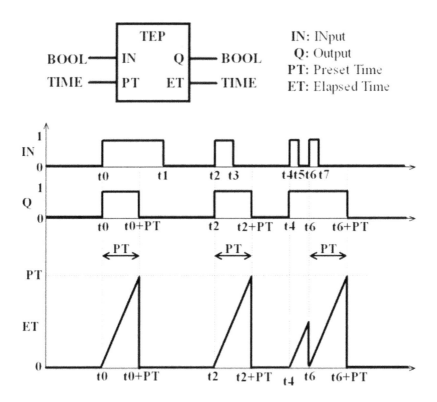

FIGURE 5.33   The symbol and timing diagram of the extended pulse timer (TEP).

5.14   MACRO "TEP_8" (8-BIT EXTENDED PULSE TIMER)

The macro "TEP_8" defines 80 8-bit extended pulse timers selected with the num = 0, 1, ..., 79. The symbol of the macro "TEP_8" is depicted in Table 5.10. The macro "TEP_8" and its flowchart are shown in Figures 5.34 and 5.35, respectively. Note that the CLK input has a rising edge symbol, but within the macro "TEP_8", there is no rising edge detection mechanism for the signal provided through this input. Instead, it is assumed that the signal connected here is one of the rising edge signals available in the PIC16F1847-Based PLC, i.e., one of the following: re_T2ms (2 ms), re_T10ms (10 ms), re_T100ms (100 ms), or re_T1s (1 s). IN (input signal, through "W"), Q (output signal = timer status bit), and CLK (re_Treg, re_Tbit) are all defined as Boolean variables. The time constant "tcnst" is an integer constant (here, for 8-bit resolution, it is chosen from any number in the range 1–255) and is used to define preset time PT, which is obtained by the formula PT = tcnst×CLK, where CLK should be used as the period of the rising edge signal of one of the four reference timing signals—ticks. The extended pulse timer outputs are represented by the status bits TQ0, TQ1, ..., TQ79, defined by T_Q0+num/8,num–(8*[num/8]), as shown in Figure 5.4. A unique memory bit Mreg,Mbit = M0.0, M0.1, ..., M127.7 is used to detect the rising edge of the input signal IN. An 8-bit integer variable "TV_L+num" is used to count the rising edge signals connected to the CLK input. The count value of "TV_L+num" defines the elapsed time ET as follows: ET = CLK×count value of "TV_L+num". Let us now briefly consider how the macro "TEP_8" works. First of all, the preset time PT is defined by means of the period of a rising edge signal "CLK = re_Treg, re_Tbit" and a time constant "tcnst". When the rising edge of the input signal IN is detected, by means of Mreg,Mbit, the output signal T_Q0+num/8,num–(8*[num/8]) is forced to be true (ON) and the related counter "TV_L+num" is cleared. After the output becomes true, i.e., T_Q0+num/8,num–(8*[num/8]) = 1, the related counter "TV_L+num" is incremented by 1 with each rising edge signal of one of the four reference timing signals "CLK = re_Treg, re_Tbit". When the count value of "TV_L+num" is equal to the number "tcnst", then state change from ON to

---

TABLE 5.10
The Symbol of the Macro "TEP_8"

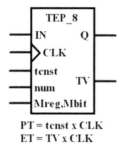

PT = tcnst x CLK
ET = TV x CLK

IN (input signal, through "W") = 0 or 1
CLK (rising edge signal of one of the four reference timing signals, re_Treg, re_Tbit) = re_T2ms (2 ms), re_T10ms (10 ms), re_T100ms (100 ms), or re_T1s (1 s)
tcnst (8-bit time constant value) = 1, 2, ..., 255
num (the unique timer number) = 0, 1, ..., 79
Mreg,Mbit (A unique memory bit) = M0.0, M0.1, ..., M127.7
Q (timer output – timer status bit) = one of the following: TQ0, TQ1, ..., TQ79, defined by T_Q0+num/8,num–(8*[num/8]) (num = 0, 1, ..., 79)
TV (8-bit timing count value hold in an 8-bit register) = TV_L+num (num = 0, 1, ..., 79)

```
TEP_8 macro    num, Mreg,Mbit, re_Treg,re_Tbit, tcnst
        local    L0,L1,L2,L3
        local    R_n,b_n,TV_Ln
;-----------------------------------------------------------------
 if num < 80                    ;if num < 80 then carry on, else do not compile.
;-----------------------------------------------------------------
 if (tcnst>0)&&(tcnst<256)     ;if 0<tcnst<256 then carry on, else do not compile.
;-----------------------------------------------------------------
R_n     set      T_Q0+num/8        ;T_Q0 Register number
b_n     set      num-(8*(num/8))   ;T_Q0 bit number
TV_Ln   set      TV_L+num          ;TV_L register number
;-----------------------------------------------------------------
        movwf  Temp_1
        banksel Mreg
        btfss    Temp_1,0
        bsf      Mreg,Mbit
        btfss    Temp_1,0
        goto     L3
        btfss    Mreg,Mbit
        goto     L3
        bcf      Mreg,Mbit
        banksel TV_Ln
        clrf     TV_Ln
        banksel R_n
        bsf      R_n,b_n
L3
        banksel R_n
        btfsc    R_n,b_n
        goto     L2
        btfss    Temp_1,0
        goto     L1
L2      btfss    re_Treg,re_Tbit
        goto     L1
        banksel TV_Ln
        incf     TV_Ln,F
        movf     TV_Ln,W
        xorlw    tcnst
        btfss    STATUS,Z
        goto     L1
        banksel R_n
        bcf      R_n,b_n
L1
        banksel R_n
        btfsc    R_n,b_n
        goto     L0
        banksel TV_Ln
        clrf     TV_Ln
L0
;-----------------------------------------------------------------
```

FIGURE 5.34 *(1 of 2)* The macro "TEP_8".

```
;-----------------------------------------------------------------
    else
    error "Make sure that 0 < tcnst < 256 !"
    endif
;-----------------------------------------------------------------
    else
    error "The timer number must be one of 0, 1, ..., 79."
    endif
;-----------------------------------------------------------------
         endm
```

FIGURE 5.34 Continued

OFF is issued for the output signal (timer status bit) T_Q0+num/8,num−(8*[num/8]) and at the same time the counter "TV_L+num" is cleared. In this macro a previously defined 8-bit variable "Temp_1" is also utilized.

In order to explain how an extended pulse timer is set up by using the macro "TEP_8", let us consider Figure 5.36, where an extended pulse timer is obtained with the macro "TEP_8".

In this example program we have:

IN = I0.0
num = 0 (decimal)
Mreg,Mbit = M0.0
CLK(re_Treg,re_Tbit) = re_T1s (the rising edge signal with 1-second period)
tcnst = 120 (decimal)

Therefore, we obtain the following from the macro "TEP_8":

$$PT = tcnst * CLK \left(\text{the period of rising edge signal re\_T1s} \right)$$

$$= 120 * 1 \text{ second} = 120 \text{ seconds}$$

TV_Ln = TV_L register number
TV_Ln = TV_L+num = TV_L+0 = 220h+0d = 220h+0 = 220h (This is the 8-bit register in Bank4 of SRAM where the timing count value of this timer will be held.)

Note that in the following, the division $\dfrac{num}{8}$ is an integer division.

R_n = T_Q0 **R**egister **n**umber

$$R\_n = T\_Q0 + num/8 = T\_Q0 + \frac{num}{8} = 4D0h + \frac{0}{8} = 4D0h + 0d = 4D0h$$

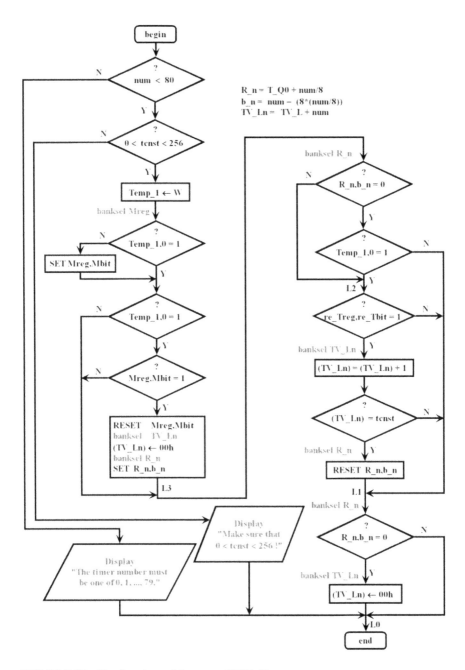

FIGURE 5.35 The flowchart of the macro "TEP_8".

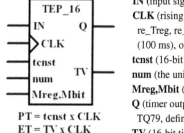

```
Id        I0.0                                    ; rung 1
          ;num,Mreg,Mbit,re_Treg,re_Tbit,tcnst
TEP_8     0,M0.0,re_T1s,.120
```

FIGURE 5.36 Example extended pulse timer obtained with the macro "TEP_8": a user program and its ladder diagram.

$$b\_n = T\_Q0 \text{ bit number}$$

$$b\_n = num - \left(8*(num/8)\right) = num - \left(8*\left(\frac{num}{8}\right)\right)$$

$$= 0 - \left(8*\left(\frac{0}{8}\right)\right) = 0 - (8*0) = 0$$

timer output (timer status bit) = R_n,b_n = 4D0h,0 = TQ0 (This is the 1-bit variable in Bank9 of SRAM where the state of the timer output will be held.)

5.15   MACRO "TEP_16" (16-BIT EXTENDED PULSE TIMER)

The macro "TEP_16" defines 80 16-bit extended pulse timers selected with the num = 0, 1, ..., 79. The symbol of the macro "TEP_16" is depicted in Table 5.11. The

TABLE 5.11
The Symbol of the Macro "TEP_16"

IN (input signal, through "W") = 0 or 1

CLK (rising edge signal of one of the four reference timing signals, re_Treg, re_Tbit) = re_T2ms (2 ms), re_T10ms (10 ms), re_T100ms (100 ms), or re_T1s (1 s)

tcnst (16-bit time constant value) = 1, 2, ..., 65535

num (the unique timer number) = 0, 1, ..., 79

Mreg,Mbit (A unique memory bit) = M0.0, M0.1, ..., M127.7

Q (timer output – timer status bit) = one of the following: TQ0, TQ1, ..., TQ79, defined by T_Q0+num/8,num–(8*[num/8]) (num = 0, 1, ..., 79)

TV (16-bit timing count value hold in two 8-bit registers) = TV_H+num and TV_L+num (num = 0, 1, ..., 79)

```
TEP_16
IN      Q
CLK
tcnst   TV
num
Mreg,Mbit

PT = tcnst x CLK
ET = TV x CLK
```

macro "TEP_16" and its flowchart are shown in Figures 5.37 and 5.38, respectively. Note that the CLK input has a rising edge symbol, but within the macro "TEP_16", there is no rising edge detection mechanism for the signal provided through this input. Instead, it is assumed that the signal connected here is one of the rising edge signals available in the PIC16F1847-Based PLC, i.e., one of the following: re_T2ms (2 ms), re_T10ms (10 ms), re_T100ms (100 ms), or re_T1s (1 s). IN (input signal, through "W"), Q (output signal = timer status bit), and CLK (re_Treg, re_Tbit) are all defined as Boolean variables. The time constant "tcnst" is an integer constant

```
TEP_16 macro  num,Mreg,Mbit,re_Treg,re_Tbit,tcnst
         local    L0,L1,L2,L3,L4
         local    R_n,b_n,TV_Ln,TV_Hn
;-------------------------------------------------------------------
 if num < 80                          ;if num < 80 then carry on, else do not compile.
;-------------------------------------------------------------------
 if (tcnst>0)&&(tcnst<65536)    ;if 0<tcnst<65536 then carry on, else do not compile.
;-------------------------------------------------------------------
R_n    set    T_Q0+num/8             ;T_Q0 Register number
b_n    set    num-(8*(num/8))        ;T_Q0 bit number
TV_Ln  set    TV_L+num               ;TV_L register number
TV_Hn  set    TV_H+num               ;TV_H register number
;-------------------------------------------------------------------
         movwf  Temp_1
         banksel Mreg
         btfss   Temp_1,0
         bsf     Mreg,Mbit
         btfss   Temp_1,0
         goto    L4
         btfss   Mreg,Mbit
         goto    L4
         bcf     Mreg,Mbit
         banksel TV_Ln
         clrf    TV_Ln
         banksel TV_Hn
         clrf    TV_Hn
         banksel R_n
         bsf     R_n,b_n
L4
         banksel R_n
         btfsc   R_n,b_n
         goto    L3
         btfss   Temp_1,0
         goto    L1
L3       btfss   re_Treg,re_Tbit
         goto    L1
         banksel TV_Ln
         incfsz  TV_Ln,F
         goto    L2
         banksel TV_Hn
         incf    TV_Hn,F
```

FIGURE 5.37   *(1 of 2)* The macro "TEP_16".

```
L2
        banksel TV_Ln
        movf    TV_Ln,W
        xorlw   low tcnst
        btfss   STATUS,Z
        goto    L1
        banksel TV_Hn
        movf    TV_Hn,W
        xorlw   high tcnst
        btfss   STATUS,Z
        goto    L1
        banksel R_n
        bcf     R_n,b_n
L1
        banksel R_n
        btfsc   R_n,b_n
        goto    L0
        banksel TV_Ln
        clrf    TV_Ln
        banksel TV_Hn
        clrf    TV_Hn
L0
;------------------------------------------------------------------
  else
  error "Make sure that 0 < tcnst < 65536 !"
  endif
;------------------------------------------------------------------
  else
  error "The timer number must be one of 0, 1, ..., 79."
  endif
;------------------------------------------------------------------
        endm
```

FIGURE 5.37   Continued

(here, for 16-bit resolution, it is chosen from any number in the range 1–65,535) and is used to define preset time PT, which is obtained by the formula $PT = tcnst \times CLK$, where CLK should be used as the period of the rising edge signal of one of the four reference timing signals—ticks. The extended pulse timer outputs are represented by the status bits TQ0, TQ1, ..., TQ79, defined by T_Q0+num/8,num–(8*[num/8]), as shown in Figure 5.4. A unique memory bit Mreg,Mbit = M0.0, M0.1, ..., M127.7 is used to detect the rising edge of the input signal IN. A 16-bit integer variable TV consisting of two 8-bit variables "TV_H+num and TV_L+num" is used to count the rising edge signals connected to the CLK input. "TV_H+num" holds the high byte of the TV, while "TV_L+num" holds the low byte of the TV. The count value of the 16-bit integer variable "TV_H+num and TV_L+num" defines the elapsed time ET as follows: $ET = CLK \times$ count value of "TV_H+num and TV_L+num". Let us now

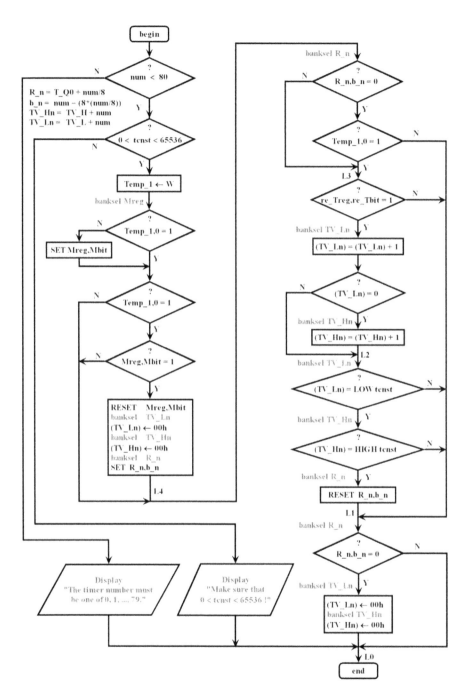

FIGURE 5.38 The flowchart of the macro "TEP_16".

briefly consider how the macro "TEP_16" works. First of all, the preset time PT is defined by means of the period of a rising edge signal "CLK = re_Treg, re_Tbit" and a time constant "tcnst". When the rising edge of the input signal IN is detected, by means of Mreg,Mbit, the output signal T_Q0+num/8,num−(8*[num/8]) is forced to be true (ON) and the related counter "TV_H+num and TV_L+num" is cleared. After the output becomes true, i.e., T_Q0+num/8,num−(8*[num/8]) = 1, the related counter "TV_H+num and TV_L+num" is incremented by 1 with each rising edge signal of one of the four reference timing signals "CLK = re_Treg, re_Tbit". When the count value of "TV_L+num" is equal to the number "tcnst", then state change from ON to OFF is issued for the output signal (timer status bit) T_Q0+num/8,num−(8*[num/8]) and at the same time the counter "TV_H+num and TV_L+num" is cleared. In this macro a previously defined 8-bit variable "Temp_1" is also utilized.

In order to explain how an extended pulse timer is set up by using the macro "TEP_16", let us consider Figure 5.39, where an extended pulse timer is obtained with the macro "TEP_16".

In this example program we have:

IN = I0.0
num = 25 (decimal)
Mreg,Mbit = M0.0
CLK(re_Treg,re_Tbit) = re_T2ms (the rising edge signal with 2-millisecond
    period)
tcnst = 25000 (decimal)

Therefore, we obtain the following from the macro "TEP_16":

$$PT = tcnst * CLK \ (\text{the period of rising edge signal re\_T2ms})$$

$$= 25000 * 2 \text{ ms} = 50 \text{ seconds}$$

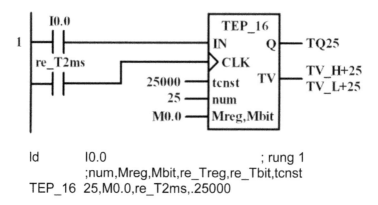

```
Id          I0.0                                   ; rung 1
            ;num,Mreg,Mbit,re_Treg,re_Tbit,tcnst
TEP_16 25,M0.0,re_T2ms,.25000
```

FIGURE 5.39 Example extended pulse timer obtained with the macro "TEP_16": a user program and its ladder diagram.

TV_Ln = TV_L register number

TV_Ln = TV_L+num = TV_L+25 = 220h + 25d = 220h + 19h = 239h (This is the 8-bit register in Bank4 of SRAM where the low byte of the timing count value of this timer will be held.)

TV_Hn = TV_H register number

TV_Hn = TV_H+num = TV_H+25 = 2A0h + 25d = 2A0h + 19h = 2B9h (This is the 8-bit register in Bank5 of SRAM where the high byte of the timing count value of this timer will be held.)

Note that in the following, the division $\dfrac{\text{num}}{8}$ is an integer division.

R_n = T_Q0 **R**egister **n**umber

$$R\_n = T\_Q0 + \text{num}/8 = T\_Q0 + \frac{\text{num}}{8} = 4D0h + \frac{25}{8} = 4D0h + 3d = 4D3h$$

b_n = T_Q0 **b**it **n**umber

$$b\_n = \text{num} - \left(8*\left(\text{num}/8\right)\right) = \text{num} - \left(8*\left(\frac{\text{num}}{8}\right)\right)$$

$$= 25 - \left(8*\left(\frac{25}{8}\right)\right) = 25 - \left(8*3\right) = 25 - 24 = 1$$

timer output (timer status bit) = R_n,b_n = 4D3h,1 = TQ25 (This is the 1-bit variable in Bank9 of SRAM where the state of the timer output will be held.)

5.16   OSCILLATOR TIMER (TOS)

The oscillator timer can be used to generate pulse trains with given durations for true (ON) and false (OFF) times. The symbol and timing diagram of the oscillator timer (TOS) are both shown in Figure 5.40. **PT0** defines the false (OFF) time and **PT1** defines the true (ON) time of the pulse. As the input signal IN goes and remains true (ON), the OFF timing function is started and therefore the elapsed time **ET0** is increased. When the elapsed time **ET0** reaches the time specified by the preset time input **PT0**, the output Q goes true (ON) and **ET0** is cleared. At the same time, as long as the input signal IN remains true (ON), the ON timing function is started and therefore the elapsed time **ET1** is increased. When the elapsed time **ET1** reaches the time specified by the preset time input **PT1**, the output **Q** goes false (OFF – 1) and **ET1** is cleared. Then it is time for next operation for OFF and ON times. This operation will carry on as long as the input signal **IN** remains true (ON) generating the pulse trains based on **PT0** and **PT1**. If the input signal **IN** goes and remains false (OFF), then the output **Q** is forced to be false (OFF). The following two sections explain the implementations of 8-bit and 16-bit oscillator timers, respectively for the PIC16F1847-Based PLC.

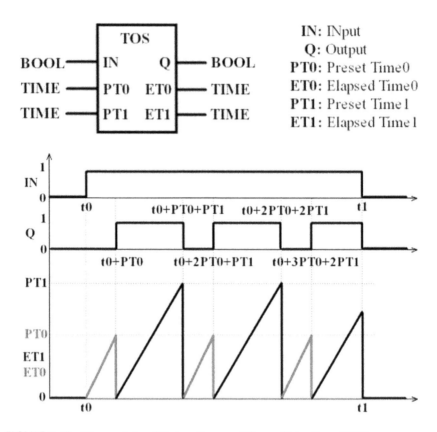

FIGURE 5.40 The symbol and timing diagram of the oscillator timer (TOS).

5.17 MACRO "TOS_8" (8-BIT OSCILLATOR TIMER)

The macro "TOS_8" defines 80 8-bit oscillator timers selected with the num = 0, 1, …, 79. The symbol of the macro "TOS_8" is depicted in Table 5.12. The macro "TOS_8" and its flowchart are shown in Figures 5.41 and 5.42, respectively. Note that the CLK input has a rising edge symbol, but within the macro "TOS_8", there is no rising edge detection mechanism for the signal provided through this input. Instead, it is assumed that the signal connected here is one of the rising edge signals available in the PIC16F1847-Based PLC, i.e., one of the following: re_T2ms (2 ms), re_T10ms (10 ms), re_T100ms (100 ms), or re_T1s (1 s). IN (input signal, through "W"), Q (output signal = timer status bit), and CLK (re_Treg, re_Tbit) are all defined as Boolean variables. The time constant "tcnst0" is an integer constant (here, for 8-bit resolution, it is chosen from any number in the range 1–255) and is used to define preset time PT0, which is obtained by the formula PT0 = tcnst0×CLK, where CLK should be used as the period of the rising edge signal of one of the four reference timing signals—ticks. The time constant "tcnst1" is an integer constant (here, for 8-bit resolution, it is chosen from any number in the range 1–255) and is used to define preset time PT1, which is obtained by the

TABLE 5.12

The Symbol of the Macro "TOS_8"

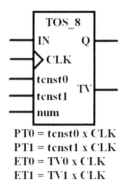

PT0 = tcnst0 x CLK
PT1 = tcnst1 x CLK
ET0 = TV0 x CLK
ET1 = TV1 x CLK

IN (input signal, through "W") = 0 or 1

CLK (rising edge signal of one of the four reference timing signals, re_Treg, re_Tbit) = re_T2ms (2 ms), re_T10ms (10 ms), re_T100ms (100 ms), or re_T1s (1 s)

tcnst0 (8-bit time constant value for Time0) = 1, 2, ..., 255

tcnst1 (8-bit time constant value for Time1) = 1, 2, ..., 255

num (the unique timer number) = 0, 1, ..., 79

Q (timer output – timer status bit) = one of the following: TQ0, TQ1, ..., TQ79, defined by T_Q0+num/8,num–(8*[num/8]) (num = 0, 1, ..., 79)

TV (8-bit timing count value hold in an 8-bit register) = TV_L+num (num = 0, 1, ..., 79)

formula PT1 = tcnst1 × CLK, where CLK should be used as the period of the rising edge signal of one of the four reference timing signals—ticks. The oscillator timer outputs are represented by the status bits TQ0, TQ1, ..., TQ79, defined by T_Q0+num/8,num–(8*[num/8]), as shown in Figure 5.4. An 8-bit integer variable "TV_L+num" is used to count the rising edge signals connected to the CLK input. Note that we use the same counter "TV_L+num" to obtain the time delays for both OFF and ON times, as these durations are mutually exclusive. The count value of "TV_L+num" defines the elapsed time ET0 or ET1 as follows: ET(0 or 1) = CLK × count value of "TV_L+num". Let us now briefly consider how the macro "TOS_8" works. First of all, preset time PT0 (and PT1, respectively) is defined by means of the period of a rising edge signal "CLK = re_Treg, re_Tbit" and a time constant "tcnst0" ("tcnst1", respectively). If the input signal IN, taken into the macro by means of "W", is false (OFF), then the output signal T_Q0+num/8,num–(8*[num/8]) is forced to be false (OFF) and the counter "TV_L+num" is loaded with "00h". If the input signal IN is true (ON) and the output signal Q, i.e., the status bit T_Q0+num/8,num–(8*[num/8]), is false (OFF), then with each rising edge signal "CLK = re_Treg, re_Tbit" the related counter "TV_L+num" is incremented by 1. In this case, when the count value of "TV_L+num" is equal to the number "tcnst0", then "TV_L+num" is cleared and a state change from OFF to ON is issued for the output signal (timer status bit) T_Q0+num/8,num–(8*[num/8]). If both the input signal IN and the output signal Q, i.e., the status bit T_Q0+num/8,num–(8*[num/8]), are true (ON), then with each rising edge signal "CLK = re_Treg, re_Tbit", the related counter "TV_L+num" is incremented by 1. In this case, when the count value of "TV_L+num" is equal to the number "tcnst1", then "TV_L+num" is cleared and a state change from ON to OFF is issued for the output signal (timer status bit) T_Q0+num/8,num–(8*[num/8]). This process will continue as long as the input signal IN remains true (ON). In this macro a previously defined 8-bit variable "Temp_1" is also utilized.

```
TOS_8 macro    num, re_Treg,re_Tbit, tcnst0, tcnst1
        local      L1,L2,L3
        local      R_n,b_n,TV_Ln
;-----------------------------------------------------------------------
  if num < 80                      ;if num < 80 then carry on, else do not compile.
;-----------------------------------------------------------------------
  if ((tcnst0>0)&&(tcnst0<256))&&((tcnst1>0)&&(tcnst1<256))
  ;if (0<tcnst0<256)&(0<tcnst1<256) then carry on, else do not compile.
;-----------------------------------------------------------------------
R_n    set      T_Q0+num/8              ;T_Q0 Register number
b_n    set      num-(8*(num/8))         ;T_Q0 bit number
TV_Ln  set      TV_L+num                ;TV_L register number
;-----------------------------------------------------------------------
        movwf    Temp_1
        btfsc    Temp_1,0
        goto     L3
        banksel TV_Ln
        clrf     TV_Ln
        banksel R_n
        bcf      R_n,b_n
        goto     L1
L3      btfss    re_Treg,re_Tbit
        goto     L1
        banksel TV_Ln
        incf     TV_Ln,F
        banksel R_n
        btfsc    R_n,b_n
        goto     L2
        banksel TV_Ln
        movf     TV_Ln,W
        xorlw    tcnst0
        btfss    STATUS,Z
        goto     L1
        clrf     TV_Ln
        banksel R_n
        bsf      R_n,b_n
        goto     L1
L2
        banksel TV_Ln
        movf     TV_Ln,W
        xorlw    tcnst1
        btfss    STATUS,Z
        goto     L1
        clrf     TV_Ln
        banksel R_n
        bcf      R_n,b_n
L1
;-----------------------------------------------------------------------
```

FIGURE 5.41   (*1 of 2*) The macro "TOS_8".

```
;------------------------------------------------------------------
    else
    error "Make sure that (0 < tcnst0 < 256) AND (0 < tcnst1 < 256) !"
    endif
;------------------------------------------------------------------
  else
  error "The timer number must be one of 0, 1, ..., 79."
  endif
;------------------------------------------------------------------
        endm
```

FIGURE 5.41 Continued

In order to explain how an oscillator timer is set up by using the macro "TOS_8", let us consider Figure 5.43, where an oscillator timer is obtained with the macro "TOS_8".

In this example program we have:

$IN = I0.0$
$num = 50$ (decimal)
$CLK(re\_Treg,re\_Tbit) = re\_T100ms$ (the rising edge signal with 100-millisecond period)
$tcnst0 = 20$ (decimal)
$tcnst1 = 10$ (decimal)

Therefore, we obtain the following from the macro "TOS_8":

$$PT0 = tcnst0 * CLK \left(\text{the period of rising edge signal } re\_T100ms\right)$$

$$= 20 * 100 \text{ ms} = 2 \text{ seconds}$$

$$PT1 = tcnst1 * CLK \left(\text{the period of rising edge signal } re\_T100ms\right)$$

$$= 10 * 100 \text{ ms} = 1 \text{ second}$$

$TV\_Ln = TV\_L$ register number
$TV\_Ln = TV\_L+num = TV\_L+50 = 220h + 50d = 220h + 32h = 252h$ (This is the 8-bit register in Bank4 of SRAM where the timing count value of this timer will be held.)

Note that in the following, the division $\dfrac{num}{8}$ is an integer division.

$R\_n = T\_Q0$ **R**egister **n**umber

$$R\_n = T\_Q0 + num/8 = T\_Q0 + \frac{num}{8} = 4D0h + \frac{50}{8} = 4D0h + 6d = 4D6h$$

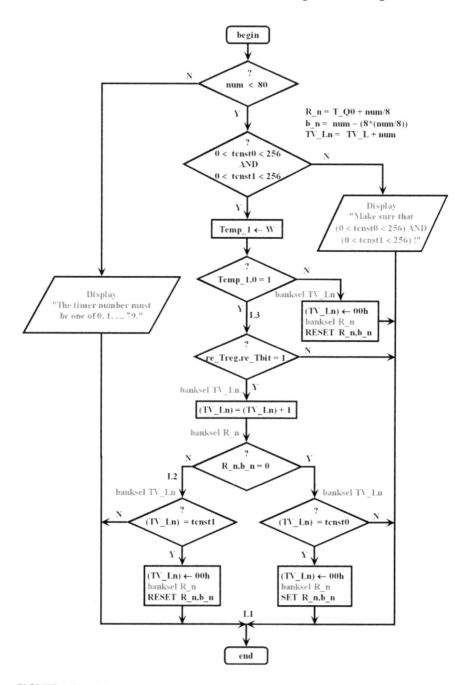

FIGURE 5.42 The flowchart of the macro "TOS_8".

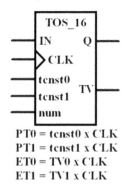

$$Id \qquad I0.0 \qquad\qquad\qquad\qquad ; rung\ 1$$

```
Id        I0.0                              ; rung 1
          ;num,re_Treg,re_Tbit,tcnst0,tcnst1
TOS_8     50,re_T100ms,.20,.10
```

FIGURE 5.43 Example oscillator timer obtained with the macro "TOS_8": a user program and its ladder diagram.

$$b\_n = T\_Q0\ \textbf{bit number}$$

$$b\_n = num - \left(8*\left(num/8\right)\right) = num - \left(8*\left(\frac{num}{8}\right)\right)$$

$$= 50 - \left(8*\left(\frac{50}{8}\right)\right) = 50 - \left(8*6\right) = 50 - 48 = 2$$

timer output (timer status bit) = R_n,b_n = 4D6h,2 = TQ50 (This is the 1-bit variable in Bank9 of SRAM where the state of the timer output will be held.)

5.18   MACRO "TOS_16" (16-BIT OSCILLATOR TIMER)

The macro "TOS_16" defines 80 16-bit oscillator timers selected with the num = 0, 1, …, 79. The symbol of the macro "TOS_16" is depicted in Table 5.13. The macro

---

TABLE 5.13
The Symbol of the Macro "TOS_16"

TOS_16 IN Q CLK tcnst0 TV tcnst1 num PT0 = tcnst0 x CLK PT1 = tcnst1 x CLK ET0 = TV0 x CLK ET1 = TV1 x CLK	**IN** (input signal, through "W") = 0 or 1 **CLK** (rising edge signal of one of the four reference timing signals, re_Treg, re_Tbit) = re_T2ms (2 ms), re_T10ms (10 ms), re_T100ms (100 ms), or re_T1s (1 s) **tcnst0** (16-bit time constant value for Time0) = 1, 2, …, 65535 **tcnst1** (16-bit time constant value for Time1) = 1, 2, …, 65535 **num** (the unique timer number) = 0, 1, …, 79 **Q** (timer output – timer status bit) = one of the following: TQ0, TQ1, …, TQ79, defined by T_Q0+num/8,num–(8*[num/8]) (num = 0, 1, …, 79) **TV** (8-bit timing count value hold in two 8-bit registers) = TV_H+num and TV_L+num (num = 0, 1, …, 79)

```
TOS_16 macro  num, re_Treg, re_Tbit, tcnst0, tcnst1
          local    L1,L2,L3,L4
          local    R_n,b_n,TV_Ln,TV_Hn
;-----------------------------------------------------------------------
  if num < 80                          ;if num < 80 then carry on, else do not compile.
;-----------------------------------------------------------------------
  if ((tcnst0>0)&&(tcnst0<65536))&&((tcnst1>0)&&(tcnst1<65536))
  ;if (0<tcnst0<65536)&(0<tcnst1<65536) then carry on, else do not compile.
;-----------------------------------------------------------------------
R_n    set      T_Q0+num/8              ;T_Q0 Register number
b_n    set      num-(8*(num/8))         ;T_Q0 bit number
TV_Ln  set      TV_L+num                ;TV_L register number
TV_Hn  set      TV_H+num                ;TV_H register number
;-----------------------------------------------------------------------
          movwf  Temp_1
          btfsc  Temp_1,0
          goto   L4
          banksel TV_Ln
          clrf   TV_Ln
          banksel TV_Hn
          clrf   TV_Hn
          banksel R_n
          bcf    R_n,b_n
          goto   L1
L4        btfss  re_Treg,re_Tbit
          goto   L1
          banksel TV_Ln
          incfsz TV_Ln,F
          goto   L3
          banksel TV_Hn
          incf   TV_Hn,F
L3
          banksel R_n
          btfsc  R_n,b_n
          goto   L2
          banksel T V_Ln
          movf   TV_Ln,W
          xorlw  low tcnst0
          btfss  STATUS,Z
          goto   L1
          banksel TV_Hn
          movf   TV_Hn,W
          xorlw  high tcnst0
          btfss  STATUS,Z
```

FIGURE 5.44   (*1 of 2*) The macro "TOS_16".

"TOS_16" and its flowchart are shown in Figures 5.44 and 5.45, respectively. Note that the CLK input has a rising edge symbol, but within the macro "TOS_16", there is no rising edge detection mechanism for the signal provided through this input. Instead, it is assumed that the signal connected here is one of the rising edge signals available in the PIC16F1847-Based PLC, i.e., one of the following: re_T2ms (2 ms), re_T10ms (10

```
        goto    L1
        banksel TV_Ln
        clrf    TV_Ln
        banksel TV_Hn
        clrf    TV_Hn
        banksel R_n
        bsf     R_n,b_n
        goto    L1
L2
        banksel TV_Ln
        movf    TV_Ln,W
        xorlw   low tcnst1
        btfss   STATUS,Z
        goto    L1
        banksel TV_Hn
        movf    TV_Hn,W
        xorlw   high tcnst1
        btfss   STATUS,Z
        goto    L1
        banksel TV_Ln
        clrf    TV_Ln
        banksel TV_Hn
        clrf    TV_Hn
        banksel R_n
        bcf     R_n,b_n
L1
;-----------------------------------------------------------------------
  else
  error "Make sure that (0 < tcnst0 < 65536) AND (0 < tcnst1 < 65536) !"
  endif
;-----------------------------------------------------------------------
  else
  error "The timer number must be one of 0, 1, ..., 79."
  endif
;-----------------------------------------------------------------------
        endm
```

FIGURE 5.44 Continued

ms), re_T100ms (100 ms), or re_T1s (1 s). IN (input signal, through "W"), Q (output signal = timer status bit), and CLK (re_Treg, re_Tbit) are all defined as Boolean variables. The time constant "tcnst0" is an integer constant (here, for 16-bit resolution, it is chosen from any number in the range 1–65,535) and is used to define preset time PT0, which is obtained by the formula PT0 = tcnst0 × CLK, where CLK should be used as the period of the rising edge signal of one of the four reference timing signals—ticks. The time constant "tcnst1" is an integer constant (here, for 16-bit resolution, it is chosen from any number in the range 1–65,535) and is used to define preset time PT1, which is obtained by the formula PT1 = tcnst1 × CLK, where CLK should be used as the period of the rising edge signal of one of the four reference timing signals—ticks. The oscillator timer outputs are represented by the status bits TQ0, TQ1, ..., TQ79, defined

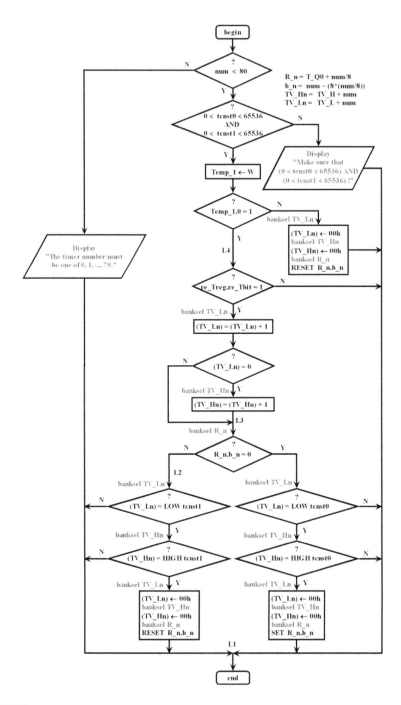

FIGURE 5.45 The flowchart of the macro "TOS_16".

by T_Q0+num/8,num–(8*[num/8]), as shown in Figure 5.4. A 16-bit integer variable TV consisting of two 8-bit variables "TV_H+num and TV_L+num" is used to count the rising edge signals connected to the CLK input. "TV_H+num" holds the high byte of the TV, while "TV_L+num" holds the low byte of the TV. Note that we use the same counter "TV_H+num and TV_L+num" to obtain the time delays for both OFF and ON times, as these durations are mutually exclusive. The count value of "TV_H+num and TV_L+num" defines the elapsed time ET0 or ET1 as follows: ET(0 or 1) = CLK×count value of "TV_H+num and TV_L+num". Let us now briefly consider how the macro "TOS_16" works. First of all, preset time PT0 (and PT1, respectively) is defined by means of the period of a rising edge signal "CLK = re_Treg, re_Tbit" and a time constant "tcnst0" ("tcnst1", respectively). If the input signal IN, taken into the macro by means of "W", is false (OFF), then the output signal TOS8_Q,num (num = 0, 1, …, 79) is forced to be false (OFF) and the counter "TV_H+num and TV_L+num" is loaded with "0000h". If the input signal IN is true (ON) and the output signal Q, i.e., the status bit T_Q0+num/8,num–(8*[num/8]), is false (OFF), then with each rising edge signal "CLK = re_Treg, re_Tbit" the related counter "TV_H+num and TV_L+num" is incremented by 1. In this case, when the count value of "TV_H+num and TV_L+num" is equal to the number "tcnst0", then "TV_H+num and TV_L+num" is cleared and a state change from OFF to ON is issued for the output signal (timer status bit) T_Q0+num/8,num–(8*[num/8]). If both the input signal IN is and the output signal Q, i.e., the status bit T_Q0+num/8,num–(8*[num/8]), are true (ON), then with each rising edge signal "CLK = re_Treg, re_Tbit", the related counter "TV_H+num and TV_L+num" is incremented by 1. In this case, when the count value of "TV_H+num and TV_L+num" is equal to the number "tcnst1", then "TV_H+num and TV_L+num" is cleared and a state change from ON to OFF is issued for the output signal (timer status bit) T_Q0+num/8,num–(8*[num/8]). This process will continue as long as the input signal IN remains true (ON). In this macro a previously defined 8-bit variable "Temp_1" is also utilized.

In order to explain how an oscillator timer is set up by using the macro "TOS_16", let us consider Figure 5.46, where an oscillator timer is obtained with the macro "TOS_16".

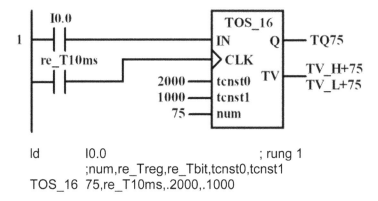

FIGURE 5.46   Example oscillator timer obtained with the macro "TOS_16": a user program and its ladder diagram.

In this example program we have:

IN = I0.0
num = 75 (decimal)
CLK(re_Treg,re_Tbit) = re_T10ms (the rising edge signal with 10-millisecond
    period)
tcnst0 = 2000 (decimal)
tcnst1 = 1000 (decimal)

Therefore, we obtain the following from the macro "TOS_16":

$$PT0 = tcnst0 * CLK \ (\text{the period of rising edge signal re\_T10ms})$$

$$= 2000 * 10 \text{ ms} = 20 \text{ seconds}$$

$$PT1 = tcnst1 * CLK \ (\text{the period of rising edge signal re\_T10ms})$$

$$= 1000 * 10 \text{ ms} = 10 \text{ seconds}$$

TV_Ln = TV_L register number
TV_Ln = TV_L+num = TV_L+75 = 220h + 75d = 220h + 4Bh = 26Bh (This is
    the 8-bit register in Bank4 of SRAM where the low byte of the timing count
    value of this timer will be held.)
TV_Hn = TV_H register number
TV_Hn = TV_H+num = TV_H+75 = 2A0h + 75d = 2A0h + 4Bh = 2EBh (This
    is the 8-bit register in Bank5 of SRAM where the high byte of the timing
    count value of this timer will be held.)

Note that in the following, the division $\dfrac{\text{num}}{8}$ is an integer division.

R_n = T_Q0 **Register number**

$$R\_n = T\_Q0 + num/8 = T\_Q0 + \frac{num}{8} = 4D0h + \frac{75}{8} = 4D0h + 9d = 4D9h$$

b_n = T_Q0 **bit number**

$$b\_n = num - \left(8 * \left(num/8\right)\right) = num - \left(8 * \left(\frac{num}{8}\right)\right)$$

$$= 75 - \left(8 * \left(\frac{75}{8}\right)\right) = 75 - \left(8 * 9\right) = 75 - 72 = 3$$

timer output (timer status bit) = R_n,b_n = 4D9h,3 = TQ75 (This is the 1-bit variable in Bank9 of SRAM where the state of the timer output will be held.)

5.19   EXAMPLES FOR TIMER MACROS

Up to now in this chapter, we have seen timer macros developed for the PIC16F1847-Based PLC. It is now time to consider some examples related to these macros. Before you can run the example programs considered here, you are expected to construct your own PIC16F1847-Based PLC hardware by using the necessary PCB files and by producing your PCBs with their components. For an effective use of examples, all example programs considered in this book are allocated within the file "PICPLC_PIC16F1847 _user_Bsc.inc", which is downloadable from this book's webpage under the downloads section. Initially all example programs are commented out by putting a semicolon ";" in front of each line. When you would like to test one of the example programs, you must uncomment each line of the example program by following the steps shown below:

1. Highlight the block of source lines you want to uncomment by dragging the mouse over these lines with the left mouse button held down. With default coloring in MPLAB X IDE you will now see green characters on a blue background.
2. Release the mouse button.
3. Press Ctrl/Shift/C or press the Alt, S, and M keys in succession, or select "Toggle Comment" from the toolbar "Source" menu. Now a semicolon will be removed from all selected source lines. With default coloring you will see red characters on a white background.

Then, you can run the project by pressing the symbol ▷ from the toolbar. Next, the MPLAB X IDE will produce the "PICPLC_PIC16F1847.X.production.hex" file for the project. Then the MPLAB X IDE will be connected to the PICkit 3 programmer and finally it will program the PIC16F1847 microcontroller within the CPU board of the PIC16F1847-Based PLC. During these steps, make sure that in the CPU board of the PIC16F1847-Based PLC, the 4PDT switch is in the "PROG" position and the power switch is in the "OFF" position. After loading the program file to the PIC16F1847 microcontroller, switch the 4PDT to "RUN" and the power switch to the "ON" position. Finally, you are ready to test the example program. **Warning**: When you finish your study with an example and try to take a look at another example, do not forget to comment the current example program before uncommenting another one. In other words, make sure that only one example program is uncommented and tested at the same time. Otherwise, if you somehow leave more than one example uncommented, the example you are trying to test will probably not function as expected since it may try to access the same resources that are being used and changed by other examples.

Please check the accuracy of each program by cross-referencing it with the related macros.

5.19.1   EXAMPLE 5.1

Example 5.1 shows the usage of the following on-delay timer macros: "TON_8" and "TON_16". The user program of Example 5.1 is shown in Figure 5.47. The

```
user_program_1  macro
;--- PLC codes to be allocated in the "user_program_1" macro start from here ---
;__Example_Bsc_5.1
          ld      I0.0                                      ;rung 1
                  ;num,re_Treg,re_Tbit,tcnst
          TON_8 0,re_T1s,.5      ; time delay = 1 s x 5 = 5 s
          in_out  TQ0,Q0.0                                  ;rung 2

          ld      I0.1                                      ;rung 3
                  ;num,re_Treg,re_Tbit,tcnst
          TON_8 10,re_T1s,.5     ; time delay = 1 s x 5 = 5 s
          in_out  TQ10,Q0.1                                 ;rung 4

          ld      I0.2                                      ;rung 5

          TON_8   20,re_T1s,.5  ; time delay = 1 s x 5 = 5 s
          in_out  TQ20,Q0.2                                 ;rung 6

          ld      I0.3                                      ;rung 7
                  ;num,re_Treg,re_Tbit,tcnst
          TON_8   30,re_T1s,.10; time delay = 1 s x 10 = 10 s
          in_out  TQ30,Q0.3                                 ;rung 8

          ld      I1.2                                      ;rung 9
                          ;s1,s0,R3,R2,R1,R0,OUT
          B_mux_4_1_E  I1.1,I1.0,TV_L+30,TV_L+20,TV_L+10,TV_L,Q1

          ld      I0.4                                      ;rung 10
                  ;num,re_Treg,re_Tbit,tcnst
          TON_16  49,re_T10ms,.500   ; time delay = 10 ms x 500 = 5 s
          in_out  TQ49,Q0.4                                 ;rung 11

          ld      I0.5                                      ;rung 12
                  ;num,re_Treg,re_Tbit,tcnst
          TON_16  59,re_T10ms,.1000   ; time delay = 10 ms x 1000 = 10 s
          in_out  TQ59,Q0.5                                 ;rung 13

          ld      I0.6                                      ;rung 14
                  ;num,re_Treg,re_Tbit,tcnst
          TON_16  69,re_T10ms,.1500   ; time delay = 10 ms x 1500 = 15 s
          in_out  TQ69,Q0.6                                 ;rung 15

          ld      I0.7                                      ;rung 16
                  ;num,re_Treg,re_Tbit,tcnst
          TON_16  79,re_T10ms,.2000   ; time delay = 10 ms x 2000 = 20 s
          in_out  TQ79,Q0.7                                 ;rung 17

          ld      I2.2                                      ;rung 18
                          ;s1,s0,R3,R2,R1,R0,OUT
          B_mux_4_1_E  I2.1,I2.0,TV_L+79,TV_L+69,TV_L+59,TV_L+49,Q2

          ld      I2.2                                      ;rung 19
                          ;s1,s0,R3,R2,R1,R0,OUT
          B_mux_4_1_E  I2.1,I2.0,TV_H+79,TV_H+69,TV_H+59,TV_H+49,Q3
;
;--- PLC codes to be allocated in the "user_program_1" macro end here ----------
          endm
```

FIGURE 5.47 The user program of Example 5.1.

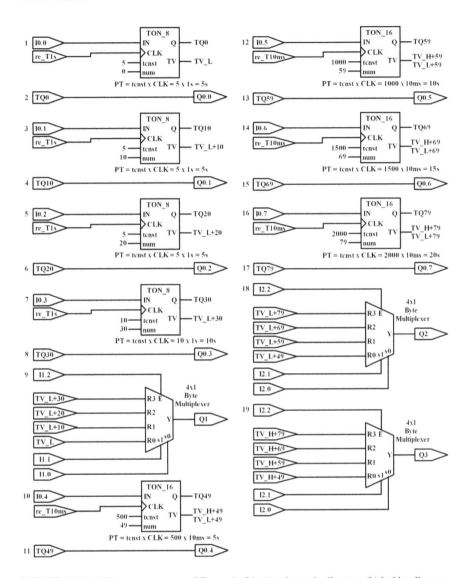

FIGURE 5.48A   The user program of Example 5.1: (a) schematic diagram (b) ladder diagram.

schematic diagram and ladder diagram of Example 5.1 are depicted in Figure 5.48(a) and Figure 5.48(b), respectively. When the project file of the PIC16F1847-Based PLC is open in the MPLAB X IDE, from the file "PICPLC_PIC16F1847 _user_Bsc.inc", if you uncomment Example 5.1 and run the project by pressing the symbol ▷ from the toolbar, then the PIC16F1847 microcontroller within the CPU board of the PIC16F1847-Based PLC will be programmed. After loading the program file to the PIC16F1847 microcontroller, switch the 4PDT to "RUN" and the power switch to the "ON" position. Next you can test the operation of this example.

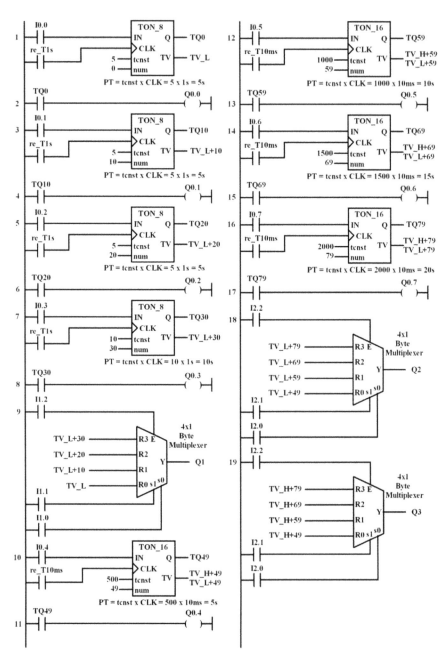

FIGURE 5.48B Continued

In rungs 1 and 2, an on-delay timer "TON_8" is implemented as follows: the input signal IN is taken from I0.0. num = 0 and therefore we choose the first on-delay timer, whose timer status bit (or output Q) is TQ0. The 8-bit integer variable used to count the rising edge signals connected to the CLK input is TV_L+num = TV_L+0. The preset time PT = tcnst×CLK = 5×1 s (re_T1s) = 5 s. As can be seen from the second rung, the timer status bit TQ0 is sent to output Q0.0.

In rungs 3 and 4, an on-delay timer "TON_8" is implemented as follows: the input signal IN is taken from I0.1. num = 10 and therefore the timer status bit (or output Q) of this on-delay timer is TQ10. The 8-bit integer variable used to count the rising edge signals connected to the CLK input is TV_L+num = TV_L+10. The preset time PT = tcnst×CLK = 5×1 s (re_T1s) = 5 s. As can be seen from rung 4, the timer status bit TQ10 is sent to output Q0.1.

In rungs 5 and 6, an on-delay timer "TON_8" is implemented as follows: the input signal IN is taken from I0.2. num = 20 and therefore the timer status bit (or output Q) of this on-delay timer is TQ20. The 8-bit integer variable used to count the rising edge signals connected to the CLK input is TV_L+num = TV_L+20. The preset time PT = tcnst×CLK = 5×1 s (re_T1s) = 5 s. As can be seen from rung 6, the timer status bit TQ20 is sent to output Q0.2.

In rungs 7 and 8, an on-delay timer "TON_8" is implemented as follows: the input signal IN is taken from I0.3. num = 30 and therefore the timer status bit (or output Q) of this on-delay timer is TQ30. The 8-bit integer variable used to count the rising edge signals connected to the CLK input is TV_L+num = TV_L+30. The preset time PT = tcnst×CLK = 10×1 s (re_T1s) = 10 s. As can be seen from rung 8, the timer status bit TQ30 is sent to output Q0.3.

In rung 9, a 4x1 byte multiplexer (see Chapter 4 of *Intermediate Concepts*) is used to observe the current count values of one of the four 8-bit variables considered in the previous rungs from Q1. The following table shows the operation of this 4x1 byte multiplexer.

inputs			output
E	s1	s0	Y
I1.2	I1.1	I1.0	Q1
0	×	×	U
1	0	0	TV_L
1	0	1	TV_L+10
1	1	0	TV_L+20
1	1	1	TV_L+30

×: don't care.
U: The contents of the destination register Y(Q1) remain unchanged.

In rungs 10 and 11, an on-delay timer "TON_16" is implemented as follows: the input signal IN is taken from I0.4. num = 49 and therefore the timer status bit (or output Q) of this on-delay timer is TQ49. The 16-bit integer variable used to count the rising edge signals connected to the CLK input is TV_H+num and TV_L+num = TV_H+49 and TV_L+49. The preset time PT = tcnst×CLK = 500×10 ms (re_T10ms) = 5 s. As can be seen from rung 11, the timer status bit TQ49 is sent to output Q0.4.

In rungs 12 and 13, an on-delay timer "TON_16" is implemented as follows: the input signal IN is taken from I0.5. num = 59 and therefore the timer status bit (or output Q) of this on-delay timer is TQ59. The 16-bit integer variable used to count the rising edge signals connected to the CLK input is TV_H+num and TV_L+num = TV_H+59 and TV_L+59. The preset time PT = tcnst×CLK = 1000×10 ms (re_T10ms) = 10 s. As can be seen from rung 13, the timer status bit TQ59 is sent to output Q0.5.

In rungs 14 and 15, an on-delay timer "TON_16" is implemented as follows: the input signal IN is taken from I0.6. num = 69 and therefore the timer status bit (or output Q) of this on-delay timer is TQ69. The 16-bit integer variable used to count the rising edge signals connected to the CLK input is TV_H+num and TV_L+num = TV_H+69 and TV_L+69. The preset time PT = tcnst×CLK = 1500×10 ms (re_T10ms) = 15 s. As can be seen from rung 15, the timer status bit TQ69 is sent to output Q0.6.

In rungs 16 and 17, an on-delay timer "TON_16" is implemented as follows: the input signal IN is taken from I0.7. num = 79 and therefore the timer status bit (or output Q) of this on-delay timer is TQ79. The 16-bit integer variable used to count the rising edge signals connected to the CLK input is TV_H+num and TV_L+num = TV_H+79 and TV_L+79. The preset time PT = tcnst×CLK = 2000×10 ms (re_T10ms) = 20 s. As can be seen from rung 17, the timer status bit TQ79 is sent to output Q0.7.

In rung 18, a 4x1 byte multiplexer is used to observe the low byte of current count values of one of the four 16-bit variables considered between rungs 10 and 17 from Q2. The following table shows the operation of this 4x1 byte multiplexer.

inputs			output
E	s1	s0	Y
I2.2	I2.1	I2.0	Q2
0	×	×	U
1	0	0	TV_L+49
1	0	1	TV_L+59
1	1	0	TV_L+69
1	1	1	TV_L+79

×: don't care.
U: The contents of the destination
register Y(Q2) remain unchanged.

In rung 19, a 4x1 byte multiplexer (Chapter 4 of *Intermediate Concepts*)is used to observe the high byte of current count values of one of the four 16-bit variables considered between rungs 10 and 17 from Q3. The following table shows the operation of this 4x1 byte multiplexer.

inputs			output
E	s1	s0	Y
I2.2	I2.1	I2.0	Q3
0	×	×	U
1	0	0	TV_H+49
1	0	1	TV_H+59
1	1	0	TV_H+69
1	1	1	TV_H+79

×: don't care.
U: The contents of the destination
register Y(Q3) remain unchanged.

5.19.2 EXAMPLE 5.2

Example 5.2 shows the usage of following retentive on-delay timer macros: "RTO_8"
and "RTO_16". The user program of Example 5.2 is shown in Figure 5.49. The schematic diagram and ladder diagram of Example 5.2 are depicted in Figure 5.50(a)
and Figure 5.50(b), respectively. When the project file of the PIC16F1847-Based
PLC is open in the MPLAB X IDE, from the file "PICPLC_PIC16F1847_user_Bsc
.inc", if you uncomment Example 5.2 and run the project by pressing the symbol ▷
from the toolbar, then the PIC16F1847 microcontroller within the CPU board of the
PIC16F1847-Based PLC will be programmed. After loading the program file to the
PIC16F1847 microcontroller, switch the 4PDT to "RUN" and the power switch to the
"ON" position. Next you can test the operation of this example.

In rungs 1 and 2, a retentive on-delay timer "RTO_8" is implemented as
follows: the input signal IN is taken from I0.0. num = 0 and therefore we
choose the first retentive on-delay timer, whose timer status bit (or output
Q) is TQ0. The reset input R = I3.0. The 8-bit integer variable used to count
the rising edge signals connected to the CLK input is TV_L+num = TV_
L+0. The preset time PT = tcnst×CLK = 5×1 s (re_T1s) = 5 s. As can be
seen from the second rung, the timer status bit TQ0 is sent to output Q0.0.

In rungs 3 and 4, a retentive on-delay timer "RTO_8" is implemented as follows: the input signal IN is taken from I0.1. The reset input R = I3.1. num
= 10 and therefore the timer status bit (or output Q) of this retentive on-
delay timer is TQ10. The 8-bit integer variable used to count the rising edge
signals connected to the CLK input is TV_L+num = TV_L+10. The preset
time PT = tcnst×CLK = 5×1 s (re_T1s) = 5 s. As can be seen from rung 4,
the timer status bit TQ10 is sent to output Q0.1.

In rungs 5 and 6, a retentive on-delay timer "RTO_8" is implemented as follows: the input signal IN is taken from I0.2. The reset input R = I3.2. num
= 20 and therefore the timer status bit (or output Q) of this retentive on-
delay timer is TQ20. The 8-bit integer variable used to count the rising edge
signals connected to the CLK input is TV_L+num = TV_L+20. The preset

```
user_program_1   macro
;--- PLC codes to be allocated in the "user_program_1" macro start from here ---
;__Example_Bsc_5.2
          Id          I0.0                                      ;rung 1
                      ;num,R_reg,R_bit,re_Treg,re_Tbit,tcnst
          RTO_8       0,I3.0,re_T1s,.5 ; time delay = 1 s x 5 = 5 s
          in_out      TQ0,Q0.0                                  ;rung 2

          Id          I0.1                                      ;rung 3
                      ;num,R_reg,R_bit,re_Treg,re_Tbit,tcnst
          RTO_8       10,I3.1,re_T1s,.5 ; time delay = 1 s x 5 = 5 s
          in_out      TQ10,Q0.1                                 ;rung 4

          Id          I0.2                                      ;rung 5
                      ;num,R_reg,R_bit,re_Treg,re_Tbit,tcnst
          RTO_8       20,I3.2,re_T1s,.5 ; time delay = 1 s x 5 = 5 s
          in_out      TQ20,Q0.2                                 ;rung 6

          Id          I0.3                                      ;rung 7
                      ;num,R_reg,R_bit,re_Treg,re_Tbit,tcnst
          RTO_8       30,I3.3,re_T1s,.10 ; time delay = 1 s x 10 = 10 s
          in_out      TQ30,Q0.3                                 ;rung 8

          Id          I1.2                                      ;rung 9
                            ;s1,s0,R3,R2,R1,R0,OUT
          B_mux_4_1_E  I1.1,I1.0,TV_L+30,TV_L+20,TV_L+10,TV_L,Q1

          Id          I0.4                                      ;rung 10
                      ;num,R_reg,R_bit,re_Treg,re_Tbit,tcnst
          RTO_16      49,I3.4,re_T10ms,.500 ; time delay = 10 ms x 500 = 5 s
          in_out      TQ49,Q0.4                                 ;rung 11

          Id          I0.5                                      ;rung 12
                      ;num,R_reg,R_bit,re_Treg,re_Tbit,tcnst
          RTO_16      59,I3.5,re_T10ms,.1000 ; time delay = 10 ms x 1000 = 10 s
          in_out      TQ59,Q0.5                                 ;rung 13

          Id          I0.6                                      ;rung 14
                      ;num,R_reg,R_bit,re_Treg,re_Tbit,tcnst
          RTO_16      69,I3.6,re_T10ms,.1500 ; time delay = 10 ms x 1500 = 15 s
          in_out      TQ69,Q0.6                                 ;rung 15

          Id          I0.7                                      ;rung 16
                      ;num,R_reg,R_bit,re_Treg,re_Tbit,tcnst
          RTO_16      79,I3.7,re_T10ms,.2000 ; time delay = 10 ms x 2000 = 20 s
          in_out      TQ79,Q0.7                                 ;rung 17

          Id          I2.2                                      ;rung 18
                            ;s1,s0,R3,R2,R1,R0,OUT
          B_mux_4_1_E  I2.1,I2.0,TV_L+79,TV_L+69,TV_L+59,TV_L+49,Q2

          Id          I2.2                                      ;rung 19
                            ;s1,s0,R3,R2,R1,R0,OUT
          B_mux_4_1_E  I2.1,I2.0,TV_H+79,TV_H+69,TV_H+59,TV_H+49,Q3
;--- PLC codes to be allocated in the "user_program_1" macro end here ----------
          endm
```

FIGURE 5.49   The user program of Example 5.2.

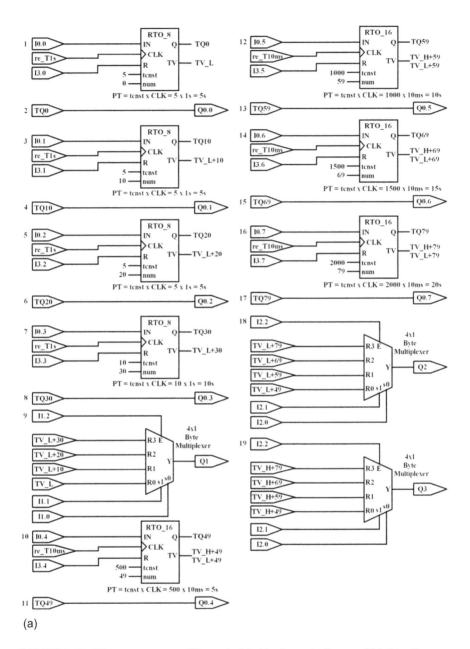

(a)

FIGURE 5.50 The user program of Example 5.2: (a) schematic diagram (b) ladder diagram.

time PT = tcnst×CLK = 5×1 s (re_T1s) = 5 s. As can be seen from rung 6, the timer status bit TQ20 is sent to output Q0.2.

In rungs 7 and 8, a retentive on-delay timer "RTO_8" is implemented as follows: the input signal IN is taken from I0.3. The reset input R = I3.3. num = 30 and therefore the timer status bit (or output Q) of this retentive on-delay timer is TQ30. The 8-bit integer variable used to count the rising edge

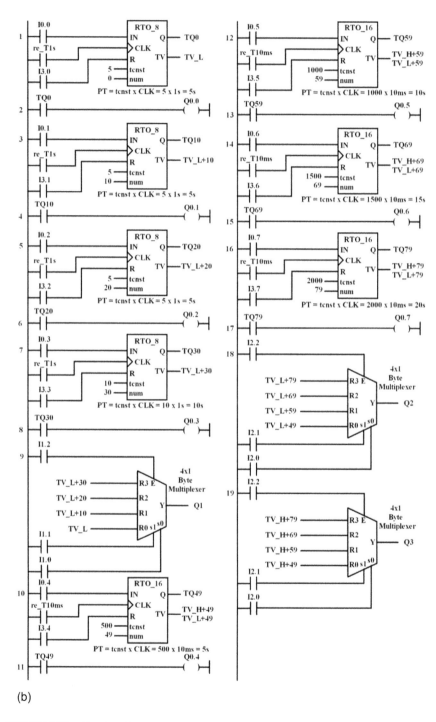

(b)

FIGURE 5.50 Continued

signals connected to the CLK input is TV_L+num = TV_L+30. The preset time PT = tcnst×CLK = 10×1 s (re_T1s) = 10 s. As can be seen from rung 8, the timer status bit TQ30 is sent to output Q0.3.

In rung 9, a 4x1 byte multiplexer is used to observe the current count values of one of the four 8-bit variables considered in the previous rungs from Q1. The following table shows the operation of this 4x1 byte multiplexer.

inputs			output
E	s1	s0	Y
I1.2	I1.1	I1.0	Q1
0	×	×	U
1	0	0	TV_L
1	0	1	TV_L+10
1	1	0	TV_L+20
1	1	1	TV_L+30

×: don't care.

U: The contents of the destination register Y(Q1) remain unchanged.

In rungs 10 and 11, a retentive on-delay timer "RTO_16" is implemented as follows: the input signal IN is taken from I0.4. The reset input R = I3.4. num = 49 and therefore the timer status bit (or output Q) of this retentive on-delay timer is TQ49. The 16-bit integer variable used to count the rising edge signals connected to the CLK input is TV_H+num and TV_L+num = TV_H+49 and TV_L+49. The preset time PT = tcnst×CLK = 500×10 ms (re_T10ms) = 5 s. As can be seen from rung 11, the timer status bit TQ49 is sent to output Q0.4.

In rungs 12 and 13, a retentive on-delay timer "RTO_16" is implemented as follows: the input signal IN is taken from I0.5. The reset input R = I3.5. num = 59 and therefore the timer status bit (or output Q) of this retentive on-delay timer is TQ59. The 16-bit integer variable used to count the rising edge signals connected to the CLK input is TV_H+num and TV_L+num = TV_H+59 and TV_L+59. The preset time PT = tcnst×CLK = 1000×10 ms (re_T10ms) = 10 s. As can be seen from rung 13, the timer status bit TQ59 is sent to output Q0.5.

In rungs 14 and 15, a retentive on-delay timer "RTO_16" is implemented as follows: the input signal IN is taken from I0.6. The reset input R = I3.6. num = 69 and therefore the timer status bit (or output Q) of this retentive on-delay timer is TQ69. The 16-bit integer variable used to count the rising edge signals connected to the CLK input is TV_H+num and TV_L+num = TV_H+69 and TV_L+69. The preset time PT = tcnst×CLK = 1500×10 ms (re_T10ms) = 15 s. As can be seen from rung 15, the timer status bit TQ69 is sent to output Q0.6.

In rungs 16 and 17, a retentive on-delay timer "RTO_16" is implemented as follows: the input signal IN is taken from I0.7. The reset input R = I3.7. num = 79 and therefore the timer status bit (or output Q) of this retentive on-delay timer is TQ79. The 16-bit integer variable used to count the rising edge signals connected to the CLK input is TV_H+num and TV_L+num = TV_H+79 and TV_L+79. The preset time PT = tcnst×CLK = 2000×10 ms (re_T10ms) = 20 s. As can be seen from rung 17, the timer status bit TQ79 is sent to output Q0.7.

In rung 18, a 4x1 byte multiplexer is used to observe the low byte of current count values of one of the four 16-bit variables considered between rungs 10 and 17 from Q2. The following table shows the operation of this 4x1 byte multiplexer.

inputs			output
E	s1	s0	Y
I2.2	I2.1	I2.0	Q2
0	×	×	U
1	0	0	TV_L+49
1	0	1	TV_L+59
1	1	0	TV_L+69
1	1	1	TV_L+79

×: don't care.
U: The contents of the destination register Y(Q2) remain unchanged.

In rung 19, a 4x1 byte multiplexer is used to observe the high byte of current count values of one of the four 16-bit variables considered between rungs 10 and 17 from Q3. The following table shows the operation of this 4x1 byte multiplexer.

inputs			output
E	s1	s0	Y
I2.2	I2.1	I2.0	Q3
0	×	×	U
1	0	0	TV_H+49
1	0	1	TV_H+59
1	1	0	TV_H+69
1	1	1	TV_H+79

×: don't care.
U: The contents of the destination register Y(Q3) remain unchanged.

5.19.3 EXAMPLE 5.3

Example 5.3 shows the usage of following off-delay timer macros: "TOF_8" and "TOF_16". The user program of Example 5.3 is shown in Figure 5.51. The schematic diagram and ladder diagram of Example 5.3 are depicted in Figure 5.52(a) and Figure 5.52(b), respectively. When the project file of the PIC16F1847-Based PLC is open in the MPLAB X IDE, from the file "PICPLC_PIC16F1847_user_Bsc

```
user_program_1   macro
;--- PLC codes to be allocated in the "user_program_1" macro start from here ---
;__Example_Bsc_5.3
         ld       I0.0                                            ;rung 1
                  ;num,re_Treg,re_Tbit,tcnst
         TOF_8  0,re_T1s,.5     ; time delay = 1 s x 5 = 5 s
         in_out  TQ0,Q0.0                                         ;rung 2

         ld       I0.1                                            ;rung 3
                  ;num,re_Treg,re_Tbit,tcnst
         TOF_8  10,re_T1s,.5    ; time delay = 1 s x 5 = 5 s
         in_out  TQ10,Q0.1                                        ;rung 4

         ld       I0.2                                            ;rung 5
                  ;num,re_Treg,re_Tbit,tcnst
         TOF_8  20,re_T1s,.5  ; time delay = 1 s x 5 = 5 s
         in_out  TQ20,Q0.2                                        ;rung 6

         ld       I0.3                                            ;rung 7
                  ;num,re_Treg,re_Tbit,tcnst
         TOF_8  30,re_T1s,.10 ; time delay = 1 s x 10 = 10 s
         in_out  TQ30,Q0.3                                        ;rung 8

         ld       I1.2                                            ;rung 9
                     ;s1,s0,R3,R2,R1,R0,OUT
         B_mux_4_1_E  I1.1,I1.0,TV_L+30,TV_L+20,TV_L+10,TV_L,Q1

         ld       I0.4                                            ;rung 10
                  ;num,re_Treg,re_Tbit,tcnst
         TOF_16  49,re_T10ms,.500   ; time delay = 10 ms x 500 = 5 s
         in_out     TQ49,Q0.4                                     ;rung 11

         ld       I0.5                                            ;rung 12
                  ;num,re_Treg,re_Tbit,tcnst
         TOF_16  59,re_T10ms,.1000   ; time delay = 10 ms x 1000 = 10 s
         in_out     TQ59,Q0.5                                     ;rung 13

         ld       I0.6                                            ;rung 14
                  ;num,re_Treg,re_Tbit,tcnst
         TOF_16  69,re_T10ms,.1500   ; time delay = 10 ms x 1500 = 15 s
         in_out     TQ69,Q0.6                                     ;rung 15

         ld       I0.7                                            ;rung 16
                  ;num,re_Treg,re_Tbit,tcnst
         TOF_16  79,re_T10ms,.2000   ; time delay = 10 ms x 2000 = 20 s
         in_out     TQ79,Q0.7                                     ;rung 17

         ld       I2.2                                            ;rung 18
                     ;s1,s0,R3,R2,R1,R0,OUT
         B_mux_4_1_E  I2.1,I2.0,TV_L+79,TV_L+69,TV_L+59,TV_L+49,Q2

         ld       I2.2                                            ;rung 19
                     ;s1,s0,R3,R2,R1,R0,OUT
         B_mux_4_1_E  I2.1,I2.0,TV_H+79,TV_H+69,TV_H+59,TV_H+49,Q3
;--- PLC codes to be allocated in the "user_program_1" macro end here ----------
         endm
```

FIGURE 5.51   The user program of Example 5.3.

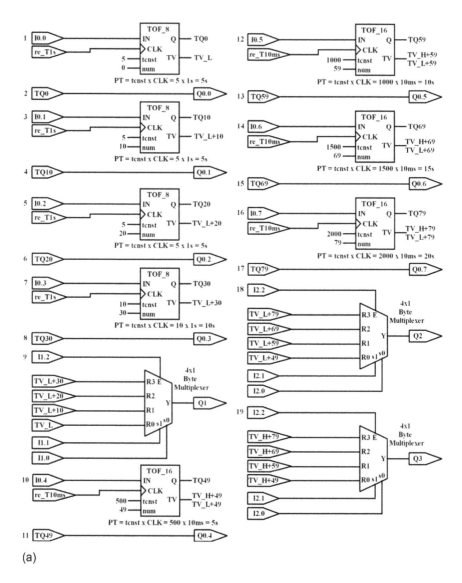

(a)

FIGURE 5.52   The user program of Example 5.3: (a) schematic diagram, (b) ladder diagram.

.inc", if you uncomment Example 5.3 and run the project by pressing the symbol ▷ from the toolbar, then the PIC16F1847 microcontroller within the CPU board of the PIC16F1847-Based PLC will be programmed. After loading the program file to the PIC16F1847 microcontroller, switch the 4PDT to "RUN" and the power switch to the "ON" position. Next you can test the operation of this example.

> **In rungs 1 and 2**, an off-delay timer "TOF_8" is implemented as follows: the input signal IN is taken from I0.0. num = 0 and therefore we choose the first off-delay timer, whose timer status bit (or output Q) is TQ0. The 8-bit integer variable used to count the rising edge signals connected to the CLK input is TV_L+num

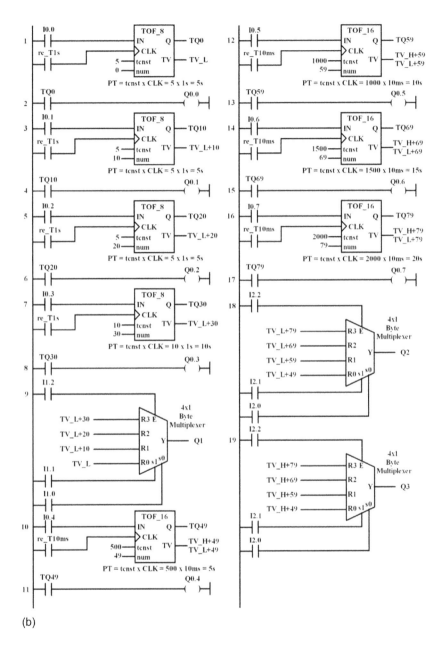

(b)

FIGURE 5.52 Continued

= TV_L+0. The preset time PT = tcnst×CLK = 5×1 s (re_T1s) = 5 s. As can
be seen from the second rung, the timer status bit TQ0 is sent to output Q0.0.

In rungs 3 and 4, an off-delay timer "TOF_8" is implemented as follows: the
input signal IN is taken from I0.1. num = 10 and therefore the timer status bit
(or output Q) of this off-delay timer is TQ10. The 8-bit integer variable used

to count the rising edge signals connected to the CLK input is TV_L+num = TV_L+10. The preset time PT = tcnst×CLK = 5×1 s (re_T1s) = 5 s. As can be seen from rung 4, the timer status bit TQ10 is sent to output Q0.1.

In rungs 5 and 6, an off-delay timer "TOF_8" is implemented as follows: the input signal IN is taken from I0.2. num = 20 and therefore the timer status bit (or output Q) of this off-delay timer is TQ20. The 8-bit integer variable used to count the rising edge signals connected to the CLK input is TV_L+num = TV_L+20. The preset time PT = tcnst×CLK = 5×1 s (re_T1s) = 5 s. As can be seen from rung 6, the timer status bit TQ20 is sent to output Q0.2.

In rungs 7 and 8, an off-delay timer "TOF_8" is implemented as follows: the input signal IN is taken from I0.3. num = 30 and therefore the timer status bit (or output Q) of this off-delay timer is TQ30. The 8-bit integer variable used to count the rising edge signals connected to the CLK input is TV_L+num = TV_L+30. The preset time PT = tcnst×CLK = 10×1 s (re_T1s) = 10 s. As can be seen from rung 8, the timer status bit TQ30 is sent to output Q0.3.

In rung 9, a 4x1 byte multiplexer is used to observe the current count values of one of the four 8-bit variables considered in the previous rungs from Q1. The following table shows the operation of this 4x1 byte multiplexer.

inputs			output
E	s1	s0	Y
I1.2	I1.1	I1.0	Q1
0	×	×	U
1	0	0	TV_L
1	0	1	TV_L+10
1	1	0	TV_L+20
1	1	1	TV_L+30

×: don't care.
U: The contents of the destination
register Y(Q1) remain unchanged.

In rungs 10 and 11, an off-delay timer "TOF_16" is implemented as follows: the input signal IN is taken from I0.4. num = 49 and therefore the timer status bit (or output Q) of this off-delay timer is TQ49. The 16-bit integer variable used to count the rising edge signals connected to the CLK input is TV_H+num and TV_L+num = TV_H+49 and TV_L+49. The preset time PT = tcnst×CLK = 500×10 ms (re_T10ms) = 5 s. As can be seen from rung 11, the timer status bit TQ49 is sent to output Q0.4.

In rungs 12 and 13, an off-delay timer "TOF_16" is implemented as follows: the input signal IN is taken from I0.5. num = 59 and therefore the timer status bit (or output Q) of this off-delay timer is TQ59. The 16-bit integer variable used to count the rising edge signals connected to the CLK input is TV_H+num and TV_L+num = TV_H+59 and TV_L+59. The preset time PT = tcnst×CLK = 1000×10 ms (re_T10ms) = 10 s. As can be seen from rung 13, the timer status bit TQ59 is sent to output Q0.5.

In rungs 14 and 15, an off-delay timer "TOF_16" is implemented as follows: the input signal IN is taken from I0.6. num = 69 and therefore the timer status bit (or output Q) of this off-delay timer is TQ69. The 16-bit integer variable used to count the rising edge signals connected to the CLK input is TV_H+num and TV_L+num = TV_H+69 and TV_L+69. The preset time PT = tcnst×CLK = 1500×10 ms (re_T10ms) = 15 s. As can be seen from rung 15, the timer status bit TQ69 is sent to output Q0.6.

In rungs 16 and 17, an off-delay timer "TOF_16" is implemented as follows: the input signal IN is taken from I0.7. num = 79 and therefore the timer status bit (or output Q) of this off-delay timer is TQ79. The 16-bit integer variable used to count the rising edge signals connected to the CLK input is TV_H+num and TV_L+num = TV_H+79 and TV_L+79. The preset time PT = tcnst×CLK = 2000×10 ms (re_T10ms) = 20 s. As can be seen from rung 17, the timer status bit TQ79 is sent to output Q0.7.

In rung 18, a 4x1 byte multiplexer is used to observe the low byte of current count values of one of the four 16-bit variables considered between rungs 10 and 17 from Q2. The following table shows the operation of this 4x1 byte multiplexer.

inputs			output
E	s1	s0	Y
I2.2	I2.1	I2.0	Q2
0	×	×	U
1	0	0	TV_L+49
1	0	1	TV_L+59
1	1	0	TV_L+69
1	1	1	TV_L+79

×: don't care.
U: The contents of the destination register Y(Q2) remain unchanged.

In rung 19, a 4x1 byte multiplexer is used to observe the high byte of current count values of one of the four 16-bit variables considered between rungs 10 and 17 from Q3. The following table shows the operation of this 4x1 byte multiplexer.

inputs			output
E	s1	s0	Y
I2.2	I2.1	I2.0	Q3
0	×	×	U
1	0	0	TV_H+49
1	0	1	TV_H+59
1	1	0	TV_H+69
1	1	1	TV_H+79

×: don't care.
U: The contents of the destination register Y(Q3) remain unchanged.

5.19.4 EXAMPLE 5.4

Example 5.4 shows the usage of following pulse timer macros: "TP_8" and "TP_16". The user program of Example 5.4 is shown in Figure 5.53. The schematic diagram and ladder diagram of Example 5.4 are depicted in Figure 5.54(a) and Figure 5.54(b), respectively. When the project file of the PIC16F1847-Based PLC is open in the MPLAB X IDE, from the file "PICPLC_PIC16F1847_user _Bsc.inc", if you uncomment Example 5.4 and run the project by pressing the symbol ▷ from the toolbar, then the PIC16F1847 microcontroller within the CPU board of the PIC16F1847-Based PLC will be programmed. After loading the program file to the PIC16F1847 microcontroller, switch the 4PDT to "RUN" and the power switch to the "ON" position. Next you can test the operation of this example.

> **In rungs 1 and 2**, a pulse timer "TP _8" is implemented as follows: the input signal IN is taken from I0.0. The unique memory bit Mreg,Mbit = M0.0. num = 0 and therefore we choose the first pulse timer, whose timer status bit (or output Q) is TQ0. The 8-bit integer variable used to count the rising edge signals connected to the CLK input is TV_L+num = TV_L+0. The preset time PT = tcnst×CLK = 5×1 s (re_T1s) = 5 s. As can be seen from the second rung, the timer status bit TQ0 is sent to output Q0.0.
>
> **In rungs 3 and 4**, a pulse timer "TP _8" is implemented as follows: the input signal IN is taken from I0.1. The unique memory bit Mreg,Mbit = M0.1. num = 10 and therefore the timer status bit (or output Q) of this pulse timer is TQ10. The 8-bit integer variable used to count the rising edge signals connected to the CLK input is TV_L+num = TV_L+10. The preset time PT = tcnst×CLK = 5×1 s (re_T1s) = 5 s. As can be seen from rung 4, the timer status bit TQ10 is sent to output Q0.1.
>
> **In rungs 5 and 6**, a pulse timer "TP _8" is implemented as follows: the input signal IN is taken from I0.2. The unique memory bit Mreg,Mbit = M0.2. num = 20 and therefore the timer status bit (or output Q) of this pulse timer is TQ20. The 8-bit integer variable used to count the rising edge signals connected to the CLK input is TV_L+num = TV_L+20. The preset time PT = tcnst×CLK = 5×1 s (re_T1s) = 5 s. As can be seen from rung 6, the timer status bit TQ20 is sent to output Q0.2.
>
> **In rungs 7 and 8**, a pulse timer "TP _8" is implemented as follows: the input signal IN is taken from I0.3. The unique memory bit Mreg,Mbit = M0.3. num = 30 and therefore the timer status bit (or output Q) of this pulse timer is TQ30. The 8-bit integer variable used to count the rising edge signals connected to the CLK input is TV_L+num = TV_L+30. The preset time PT = tcnst×CLK = 10×1 s (re_T1s) = 10 s. As can be seen from rung 8, the timer status bit TQ30 is sent to output Q0.3.
>
> **In rung 9**, a 4x1 byte multiplexer is used to observe the current count values of one of the four 8-bit variables considered in the previous rungs from Q1. The following table shows the operation of this 4x1 byte multiplexer.

```
user_program_1   macro
;--- PLC codes to be allocated in the "user_program_1" macro start from here ---
;__Example_Bsc_5.4
        ld          I0.0                                              ;rung 1
                    ;num,Mreg,Mbit,re_Treg,re_Tbit,tcnst
        TP_8        0,M0.0,re_T1s,.5 ; time delay = 1 s x 5 = 5 s
        in_out      TQ0,Q0.0                                          ;rung 2

        ld          I0.1                                              ;rung 3
                    ;num,Mreg,Mbit,re_Treg,re_Tbit,tcnst
        TP_8        10,M0.1,re_T1s,.5 ; time delay = 1 s x 5 = 5 s
        in_out      TQ10,Q0.1                                         ;rung 4

        ld          I0.2                                              ;rung 5
                    ;num,Mreg,Mbit,re_Treg,re_Tbit,tcnst
        TP_8        20,M0.2,re_T1s,.5 ; time delay = 1 s x 5 = 5 s
        in_out      TQ20,Q0.2                                         ;rung 6

        ld          I0.3                                              ;rung 7
                    ;num,Mreg,Mbit,re_Treg,re_Tbit,tcnst
        TP_8        30,M0.3,re_T1s,.10 ; time delay = 1 s x 10 = 10 s
        in_out      TQ30,Q0.3                                         ;rung 8

        ld          I1.2                                              ;rung 9
                        ;s1,s0,R3,R2,R1,R0,OUT
        B_mux_4_1_E  I1.1,I1.0,TV_L+30,TV_L+20,TV_L+10,TV_L,Q1

        ld          I0.4                                              ;rung 10
                    ;num,Mreg,Mbit,re_Treg,re_Tbit,tcnst
        TP_16       49,M0.4,re_T10ms,.500 ; time delay = 10 ms x 500 = 5 s
        in_out      TQ49,Q0.4                                         ;rung 11

        ld          I0.5                                              ;rung 12
                    ;num,Mreg,Mbit,re_Treg,re_Tbit,tcnst
        TP_16       59,M0.5,re_T10ms,.1000 ; time delay = 10 ms x 1000 = 10 s
        in_out      TQ59,Q0.5                                         ;rung 13

        ld          I0.6                                              ;rung 14
                    ;num,Mreg,Mbit,re_Treg,re_Tbit,tcnst
        TP_16       69,M0.6,re_T10ms,.1500 ; time delay = 10 ms x 1500 = 15 s
        in_out      TQ69,Q0.6                                         ;rung 15

        ld          I0.7                                              ;rung 16
                    ;num,Mreg,Mbit,re_Treg,re_Tbit,tcnst
        TP_16       79,M0.7,re_T10ms,.2000 ; time delay = 10 ms x 2000 = 20 s
        in_out      TQ79,Q0.7                                         ;rung 17

        ld          I2.2                                              ;rung 18
                        ;s1,s0,R3,R2,R1,R0,OUT
        B_mux_4_1_E  I2.1,I2.0,TV_L+79,TV_L+69,TV_L+59,TV_L+49,Q2

        ld          I2.2                                              ;rung 19
                        ;s1,s0,R3,R2,R1,R0,OUT
        B_mux_4_1_E  I2.1,I2.0,TV_H+79,TV_H+69,TV_H+59,TV_H+49,Q3
;--- PLC codes to be allocated in the "user_program_1" macro end here ----------
        endm
```

FIGURE 5.53   The user program of Example 5.4.

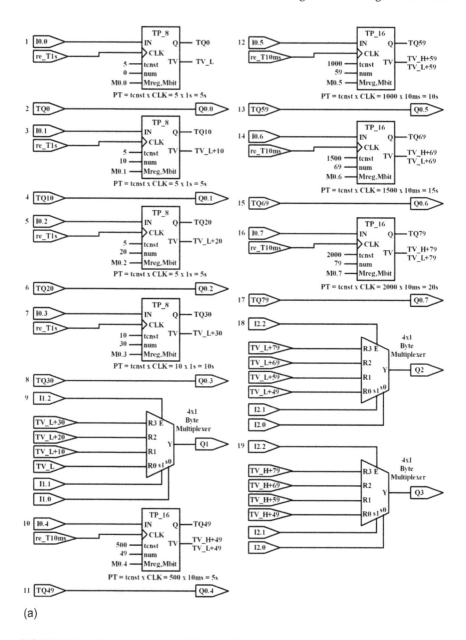

(a)

FIGURE 5.54   The user program of Example 5.4: (a) schematic diagram, (b) ladder diagram.

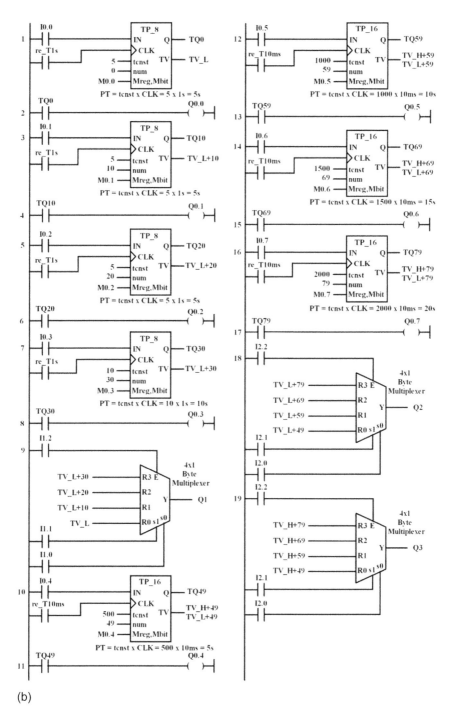

(b)

FIGURE 5.54 Continued

inputs			output
E	s1	s0	Y
I1.2	I1.1	I1.0	Q1
0	×	×	U
1	0	0	TV_L
1	0	1	TV_L+10
1	1	0	TV_L+20
1	1	1	TV_L+30

×: don't care.

U: The contents of the destination
register Y(Q1) remain unchanged.

In rungs 10 and 11, a pulse timer "TP_16" is implemented as follows: the input signal IN is taken from I0.4. The unique memory bit Mreg,Mbit = M0.4. num = 49 and therefore the timer status bit (or output Q) of this pulse timer is TQ49. The 16-bit integer variable used to count the rising edge signals connected to the CLK input is TV_H+num and TV_L+num = TV_H+49 and TV_L+49. The preset time PT = tcnst×CLK = 500×10 ms (re_T10ms) = 5 s. As can be seen from rung 11, the timer status bit TQ49 is sent to output Q0.4.

In rungs 12 and 13, a pulse timer "TP_16" is implemented as follows: the input signal IN is taken from I0.5. The unique memory bit Mreg,Mbit = M0.5. num = 59 and therefore the timer status bit (or output Q) of this pulse timer is TQ59. The 16-bit integer variable used to count the rising edge signals connected to the CLK input is TV_H+num and TV_L+num = TV_H+59 and TV_L+59. The preset time PT = tcnst×CLK = 1000×10 ms (re_T10ms) = 10 s. As can be seen from rung 13, the timer status bit TQ59 is sent to output Q0.5.

In rungs 14 and 15, a pulse timer "TP_16" is implemented as follows: the input signal IN is taken from I0.6. The unique memory bit Mreg,Mbit = M0.6. num = 69 and therefore the timer status bit (or output Q) of this pulse timer is TQ69. The 16-bit integer variable used to count the rising edge signals connected to the CLK input is TV_H+num and TV_L+num = TV_H+69 and TV_L+69. The preset time PT = tcnst×CLK = 1500×10 ms (re_T10ms) = 15 s. As can be seen from rung 15, the timer status bit TQ69 is sent to output Q0.6.

In rungs 16 and 17, a pulse timer "TP_16" is implemented as follows: the input signal IN is taken from I0.7. The unique memory bit Mreg,Mbit = M0.7. num = 79 and therefore the timer status bit (or output Q) of this pulse timer is TQ79. The 16-bit integer variable used to count the rising edge signals connected to the CLK input is TV_H+num and TV_L+num = TV_H+79 and TV_L+79. The preset time PT = tcnst×CLK = 2000×10 ms (re_T10ms) = 20 s. As can be seen from rung 17, the timer status bit TQ79 is sent to output Q0.7.

In rung 18, a 4x1 byte multiplexer is used to observe the low byte of current count values of one of the four 16-bit variables considered between rungs 10 and 17 from Q2. The following table shows the operation of this 4x1 byte multiplexer.

inputs			output
E`	s1	s0	Y
I2.2	I2.1	I2.0	Q2
0	×	×	U
1	0	0	TV_L+49
1	0	1	TV_L+59
1	1	0	TV_L+69
1	1	1	TV_L+79

×: don't care.
U: The contents of the destination
register Y(Q2) remain unchanged.

In rung 19, a 4x1 byte multiplexer is used to observe the high byte of current count values of one of the four 16-bit variables considered between rungs 10 and 17 from Q3. The following table shows the operation of this 4x1 byte multiplexer.

inputs			output
E	s1	s0	Y
I2.2	I2.1	I2.0	Q3
0	×	×	U
1	0	0	TV_H+49
1	0	1	TV_H+59
1	1	0	TV_H+69
1	1	1	TV_H+79

×: don't care.
U: The contents of the destination
register Y(Q3) remain unchanged.

5.19.5 EXAMPLE 5.5

Example 5.5 shows the usage of following extended pulse timer macros: "TEP_8" and "TEP_16". The user program of Example 5.5 is shown in Figure 5.55. The schematic diagram and ladder diagram of Example 5.5 are depicted in Figure 5.56(a) and Figure 5.56(b), respectively. When the project file of the PIC16F1847-Based PLC is open in the MPLAB X IDE, from the file "PICPLC_PIC16F1847_user_Bsc .inc", if you uncomment Example 5.5 and run the project by pressing the symbol ▷ from the toolbar, then the PIC16F1847 microcontroller within the CPU board of the PIC16F1847-Based PLC will be programmed. After loading the program file to the PIC16F1847 microcontroller, switch the 4PDT to "RUN" and the power switch to the "ON" position. Next you can test the operation of this example.

In rungs 1 and 2, an extended pulse timer "TEP_8" is implemented as follows: the input signal IN is taken from I0.0. The unique memory bit Mreg,Mbit = M0.0. num = 0 and therefore we choose the first extended pulse timer,

```
user_program_1   macro
;--- PLC codes to be allocated in the "user_program_1" macro start from here ---
;__Example_Bsc_5.5
        ld              I0.0                                            ;rung 1
                        ;num,Mreg,Mbit,re_Treg,re_Tbit,tcnst
        TEP_8           0,M0.0,re_T1s,.5 ; time delay = 1 s x 5 = 5 s
        in_out          TQ0,Q0.0                                        ;rung 2

        ld              I0.1                                            ;rung 3
                        ;num,Mreg,Mbit,re_Treg,re_Tbit,tcnst
        TEP_8           10,M0.1,re_T1s,.5 ; time delay = 1 s x 5 = 5 s
        in_out          TQ10,Q0.1                                       ;rung 4

        ld              I0.2                                            ;rung 5
                        ;num,Mreg,Mbit,re_Treg,re_Tbit,tcnst
        TEP_8           20,M0.2,re_T1s,.5 ; time delay = 1 s x 5 = 5 s
        in_out          TQ20,Q0.2                                       ;rung 6

        ld              I0.3                                            ;rung 7
                        ;num,Mreg,Mbit,re_Treg,re_Tbit,tcnst
        TEP_8           30,M0.3,re_T1s,.10 ; time delay = 1 s x 10 = 10 s
        in_out          TQ30,Q0.3                                       ;rung 8

        ld              I1.2                                            ;rung 9
                            ;s1,s0,R3,R2,R1,R0,OUT
        B_mux_4_1_E     I1.1,I1.0,TV_L+30,TV_L+20,TV_L+10,TV_L,Q1

        ld              I0.4                                            ;rung 10
                        ;num,Mreg,Mbit,re_Treg,re_Tbit,tcnst
        TEP_16          49,M0.4,re_T10ms,.500 ; time delay = 10 ms x 500 = 5 s
        in_out          TQ49,Q0.4                                       ;rung 11

        ld              I0.5                                            ;rung 12
                        ;num,Mreg,Mbit,re_Treg,re_Tbit,tcnst
        TEP_16          59,M0.5,re_T10ms,.1000 ; time delay = 10 ms x 1000 = 10 s
        in_out          TQ59,Q0.5                                       ;rung 13

        ld              I0.6                                            ;rung 14
                        ;num,Mreg,Mbit,re_Treg,re_Tbit,tcnst
        TEP_16          69,M0.6,re_T10ms,.1500 ; time delay = 10 ms x 1500 = 15 s
        in_out          TQ69,Q0.6                                       ;rung 15

        ld              I0.7                                            ;rung 16
                        ;num,Mreg,Mbit,re_Treg,re_Tbit,tcnst
        TP_16           79,M0.7,re_T10ms,.2000 ; time delay = 10 ms x 2000 = 20 s
        in_out          TQ79,Q0.7                                       ;rung 17

        ld              I2.2                                            ;rung 18
                            ;s1,s0,R3,R2,R1,R0,OUT
        B_mux_4_1_E     I2.1,I2.0,TV_L+79,TV_L+69,TV_L+59,TV_L+49,Q2

        ld              I2.2                                            ;rung 19
                            ;s1,s0,R3,R2,R1,R0,OUT
        B_mux_4_1_E     I2.1,I2.0,TV_H+79,TV_H+69,TV_H+59,TV_H+49,Q3
;--- PLC codes to be allocated in the "user_program_1" macro end here ----------
        endm
```

FIGURE 5.55   The user program of Example 5.5

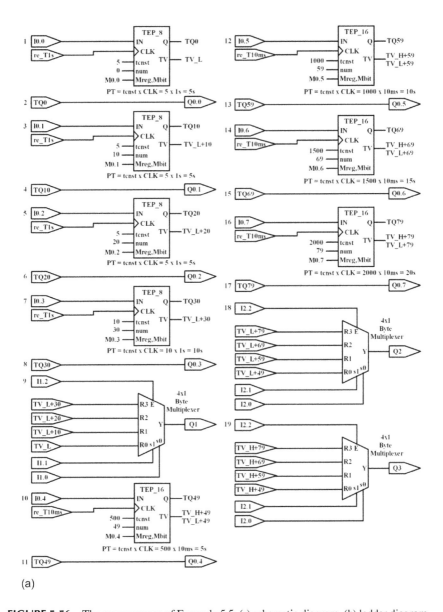

(a)

FIGURE 5.56 The user program of Example 5.5: (a) schematic diagram, (b) ladder diagram.

whose timer status bit (or output Q) is TQ0. The 8-bit integer variable used to count the rising edge signals connected to the CLK input is TV_L+num = TV_L+0. The preset time PT = tcnst × CLK = 5 × 1 s (re_T1s) = 5 s. As can be seen from the second rung, the timer status bit TQ0 is sent to output Q0.0.

In rungs 3 and 4, an extended pulse timer "TEP_8" is implemented as follows: the input signal IN is taken from I0.1. The unique memory bit Mreg,Mbit = M0.1. num = 10 and therefore the timer status bit (or output Q) of this extended pulse timer is TQ10. The 8-bit integer variable used to count the

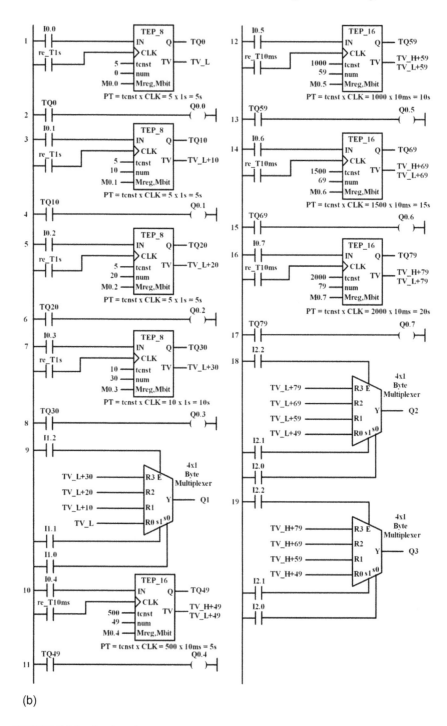

(b)

FIGURE 5.56 Continued

rising edge signals connected to the CLK input is TV_L+num = TV_L+10. The preset time PT = tcnst×CLK = 5×1 s (re_T1s) = 5 s. As can be seen from rung 4, the timer status bit TQ10 is sent to output Q0.1.

In rungs 5 and 6, an extended pulse timer "TEP_8" is implemented as follows: the input signal IN is taken from I0.2. The unique memory bit Mreg,Mbit = M0.2. num = 20 and therefore the timer status bit (or output Q) of this extended pulse timer is TQ20. The 8-bit integer variable used to count the rising edge signals connected to the CLK input is TV_L+num = TV_L+20. The preset time PT = tcnst×CLK = 5×1 s (re_T1s) = 5 s. As can be seen from rung 6, the timer status bit TQ20 is sent to output Q0.2.

In rungs 7 and 8, an extended pulse timer "TEP_8" is implemented as follows: the input signal IN is taken from I0.3. The unique memory bit Mreg,Mbit = M0.3. num = 30 and therefore the timer status bit (or output Q) of this extended pulse timer is TQ30. The 8-bit integer variable used to count the rising edge signals connected to the CLK input is TV_L+num = TV_L+30. The preset time PT = tcnst×CLK = 10×1 s (re_T1s) = 10 s. As can be seen from rung 8, the timer status bit TQ30 is sent to output Q0.3.

In rung 9, a 4x1 byte multiplexer is used to observe the current count values of one of the four 8-bit variables considered in the previous rungs from Q1. The following table shows the operation of this 4x1 byte multiplexer.

	inputs		output
E	s1	s0	Y
I1.2	I1.1	I1.0	Q1
0	×	×	U
1	0	0	TV_L
1	0	1	TV_L+10
1	1	0	TV_L+20
1	1	1	TV_L+30

×: don't care.
U: The contents of the destination register Y(Q1) remain unchanged.

In rungs 10 and 11, an extended pulse timer "TEP_16" is implemented as follows: the input signal IN is taken from I0.4. The unique memory bit Mreg,Mbit = M0.4. num = 49 and therefore the timer status bit (or output Q) of this extended pulse timer is TQ49. The 16-bit integer variable used to count the rising edge signals connected to the CLK input is TV_H+num and TV_L+num = TV_H+49 and TV_L+49. The preset time PT = tcnst×CLK = 500×10 ms (re_T10ms) = 5 s. As can be seen from rung 11, the timer status bit TQ49 is sent to output Q0.4.

In rungs 12 and 13, an extended pulse timer "TEP_16" is implemented as follows: the input signal IN is taken from I0.5. The unique memory bit

Mreg,Mbit = M0.5. num = 59 and therefore the timer status bit (or output Q) of this extended pulse timer is TQ59. The 16-bit integer variable used to count the rising edge signals connected to the CLK input is TV_H+num and TV_L+num = TV_H+59 and TV_L+59. The preset time PT = tcnst×CLK = 1000×10 ms (re_T10ms) = 10 s. As can be seen from rung 13, the timer status bit TQ59 is sent to output Q0.5.

In rungs 14 and 15, an extended pulse timer "TEP_16" is implemented as follows: the input signal IN is taken from I0.6. The unique memory bit Mreg,Mbit = M0.6. num = 69 and therefore the timer status bit (or output Q) of this extended pulse timer is TQ69. The 16-bit integer variable used to count the rising edge signals connected to the CLK input is TV_H+num and TV_L+num = TV_H+69 and TV_L+69. The preset time PT = tcnst×CLK = 1500×10 ms (re_T10ms) = 15 s. As can be seen from rung 15, the timer status bit TQ69 is sent to output Q0.6.

In rungs 16 and 17, an extended pulse timer "TEP_16" is implemented as follows: the input signal IN is taken from I0.7. The unique memory bit Mreg,Mbit = M0.7. num = 79 and therefore the timer status bit (or output Q) of this extended pulse timer is TQ79. The 16-bit integer variable used to count the rising edge signals connected to the CLK input is TV_H+num and TV_L+num = TV_H+79 and TV_L+79. The preset time PT = tcnst×CLK = 2000×10 ms (re_T10ms) = 20 s. As can be seen from rung 17, the timer status bit TQ79 is sent to output Q0.7.

In rung 18, a 4x1 byte multiplexer is used to observe the low byte of current count values of one of the four 16-bit variables considered between rungs 10 and 17 from Q2. The following table shows the operation of this 4x1 byte multiplexer.

inputs			output
E	s1	s0	Y
I2.2	I2.1	I2.0	Q2
0	×	×	U
1	0	0	TV_L+49
1	0	1	TV_L+59
1	1	0	TV_L+69
1	1	1	TV_L+79

×: don't care.
U: The contents of the destination register Y(Q2) remain unchanged.

In rung 19, a 4x1 byte multiplexer is used to observe the high byte of current count values of one of the four 16-bit variables considered between rungs 10 and 17 from Q3. The following table shows the operation of this 4x1 byte multiplexer.

inputs			output
E	s1	s0	Y
I2.2	I2.1	I2.0	Q3
0	×	×	U
1	0	0	TV_H+49
1	0	1	TV_H+59
1	1	0	TV_H+69
1	1	1	TV_H+79

×: don't care.

U: The contents of the destination register Y(Q3) remain unchanged.

5.19.6 EXAMPLE 5.6

Example 5.6 shows the usage of following oscillator timer macros: "TOS_8" and "TOS_16". The user program of Example 5.6 is shown in Figure 5.57. The schematic diagram and ladder diagram of Example 5.6 are depicted in Figure 5.58(a) and Figure 5.58(b), respectively. When the project file of the PIC16F1847-Based PLC is open in the MPLAB X IDE, from the file "PICPLC_PIC16F1847_user_Bsc .inc", if you uncomment Example 5.6 and run the project by pressing the symbol ▷ from the toolbar, then the PIC16F1847 microcontroller within the CPU board of the PIC16F1847-Based PLC will be programmed. After loading the program file to the PIC16F1847 microcontroller, switch the 4PDT to "RUN" and the power switch to the "ON" position. Next you can test the operation of this example.

In rungs 1 and 2, an oscillator timer "TOS_8" is implemented as follows: the input signal IN is taken from I0.0. num = 0 and therefore we choose the first oscillator timer, whose timer status bit (or output Q) is TQ0. The 8-bit integer variable used to count the rising edge signals connected to the CLK input is TV_L+num = TV_L+0. The preset time 0 PT0 = tcnst0×CLK = 5×1 s (re_T1s) = 5 s. The preset time 1 PT1 = tcnst1×CLK = 5×1 s (re_T1s) = 5 s. As can be seen from the second rung, the timer status bit TQ0 is sent to output Q0.0.

In rungs 3 and 4, an oscillator timer "TOS_8" is implemented as follows: the input signal IN is taken from I0.1. num = 10 and therefore the timer status bit (or output Q) of this oscillator timer is TQ10. The 8-bit integer variable used to count the rising edge signals connected to the CLK input is TV_L+num = TV_L+10. The preset time 0 PT0 = tcnst0×CLK = 5×1 s (re_T1s) = 5 s. The preset time 1 PT1 = tcnst1×CLK = 10×1 s (re_T1s) = 10 s. As can be seen from rung 4, the timer status bit TQ10 is sent to output Q0.1.

In rungs 5 and 6, an oscillator timer "TOS_8" is implemented as follows: the input signal IN is taken from I0.2. num = 20 and therefore the timer status bit (or output Q) of this oscillator timer is TQ20. The 8-bit integer

```
user_program_1  macro
;--- PLC codes to be allocated in the "user_program_1" macro start from here ---
;__Example_Bsc_5.6
        ld          I0.0                                      ;rung 1
                    ;num,re_Treg,re_Tbit,tcnst0,tcnst1  ;off time delay = 1 s x 5 = 5 s
        TOS_8       0,re_T1s,.5,.5                            ;on time delay = 1 s x 5 = 5 s
        in_out      TQ0,Q0.0                                  ;rung 2

        ld          I0.1                                      ;rung 3
                    ;num,re_Treg,re_Tbit,tcnst0,tcnst1  ;off time delay = 1 s x 5 = 5 s
        TOS_8       10,re_T1s,.5,.10                          ;on time delay = 1 s x 10 = 10 s
        in_out      TQ10,Q0.1                                 ;rung 4

        ld          I0.2                                      ;rung 5
                    ;num,re_Treg,re_Tbit,tcnst0,tcnst1  ;off time delay = 1 s x 10 = 10 s
        TOS_8       20,re_T1s,.10,.5                          ;on time delay = 1 s x 5 = 5 s
        in_out      TQ20,Q0.2                                 ;rung 6

        ld          I0.3                                      ;rung 7
                    ;num,re_Treg,re_Tbit,tcnst0,tcnst1  ;off time delay = 1 s x 1 = 1 s
        TOS_8       30,re_T1s,.1,.2                           ;on time delay = 1 s x 2 = 2 s
        in_out      TQ30,Q0.3                                 ;rung 8

        ld          I1.2                                      ;rung 9
                    ;s1,s0,R3,R2,R1,R0,OUT
        B_mux_4_1_E I1.1,I1.0,TV_L+30,TV_L+20,TV_L+10,TV_L,Q1

        ld          I0.4                                      ;rung 10
                    ;num,re_Treg,re_Tbit,tcnst0,tcnst1  ;off time delay = 10ms x 50 = 0,5 s
        TOS_16      49,re_T10ms,.50,.100                      ;on time delay = 10ms x 100 = 1 s
        in_out      TQ49,Q0.4                                 ;rung 11

        ld          I0.5                                      ;rung 12
                    ;num,re_Treg,re_Tbit,tcnst0,tcnst1  ;off time delay = 10ms x 1000 = 1 s
        TOS_16      59,re_T10ms,.100,.50                      ;on time delay = 10ms x 500 = 0,5 s
        in_out      TQ59,Q0.5                                 ;rung 13

        ld          I0.6                                      ;rung 14
                    ;num,re_Treg,re_Tbit,tcnst0,tcnst1  ;off time delay = 10ms x 200 = 2 s
        TOS_16      69,re_T10ms,.200,.100                     ;on time delay = 10ms x 100 = 1 s
        in_out      TQ69,Q0.6                                 ;rung 15

        ld          I0.7                                      ;rung 16
                    ;num,re_Treg,re_Tbit,tcnst0,tcnst1  ;off time delay = 10ms x 100 = 1 s
        TOS_16      79,re_T10ms,.100,.200                     ;on time delay = 10ms x 200 = 2 s
        in_out      TQ79,Q0.7                                 ;rung 17

        ld          I2.2                                      ;rung 18
                    ;s1,s0,R3,R2,R1,R0,OUT
        B_mux_4_1_E I2.1,I2.0,TV_L+79,TV_L+69,TV_L+59,TV_L+49,Q2

        ld          I2.2                                      ;rung 19
                    ;s1,s0,R3,R2,R1,R0,OUT
        B_mux_4_1_E I2.1,I2.0,TV_H+79,TV_H+69,TV_H+59,TV_H+49,Q3
;--- PLC codes to be allocated in the "user_program_1" macro end here ----------
        endm
```

FIGURE 5.57   The user program of Example 5.6.

variable used to count the rising edge signals connected to the CLK input is $TV\_L+num = TV\_L+20$. The preset time 0 $PT0 = tcnst0 \times CLK = 10 \times 1$ s (re_T1s) = 10 s. The preset time 1 $PT1 = tcnst1 \times CLK = 5 \times 1$ s (re_T1s) = 5 s. As can be seen from rung 6, the timer status bit TQ20 is sent to output Q0.2.

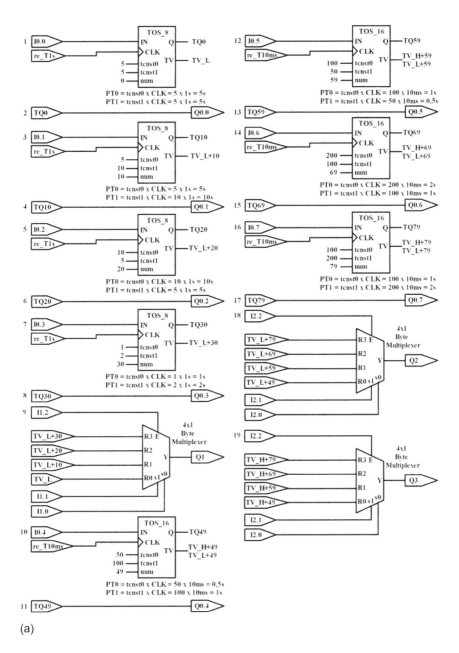

(a)

FIGURE 5.58   The user program of Example 5.6: (a) schematic diagram, (b) ladder diagram.

In rungs 7 and 8, an oscillator timer "TOS_8" is implemented as follows: the input signal IN is taken from I0.3. num = 30 and therefore the timer status bit (or output Q) of this oscillator timer is TQ30. The 8-bit integer variable used to count the rising edge signals connected to the CLK input is TV_L+num = TV_L+30. The preset time 0 PT0 = tcnst0×CLK = 1×1 s

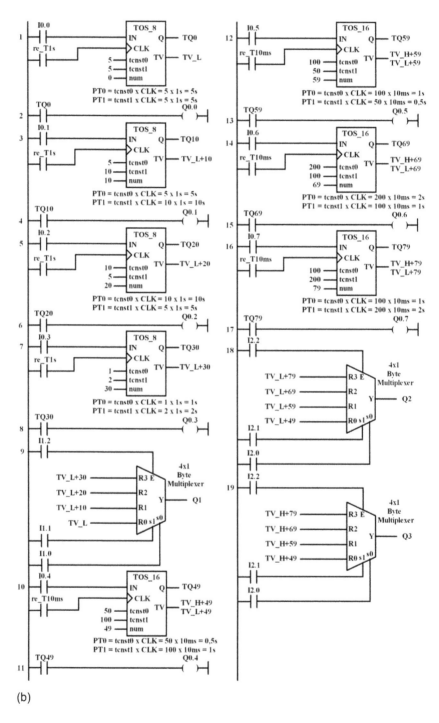

(b)

FIGURE 5.58   Continued

(re_T1s) = 1 s. The preset time 1 PT1 = tcnst1 ×CLK = 2×1 s (re_T1s) = 2 s. As can be seen from rung 8, the timer status bit TQ30 is sent to output Q0.3.

In rung 9, a 4x1 byte multiplexer is used to observe the current count values of one of the four 8-bit variables considered in the previous rungs from Q1. The following table shows the operation of this 4x1 byte multiplexer.

inputs			output
E	s1	s0	Y
I1.2	I1.1	I1.0	Q1
0	×	×	U
1	0	0	TV_L
1	0	1	TV_L+10
1	1	0	TV_L+20
1	1	1	TV_L+30

×: don't care.
U: The contents of the destination
register Y(Q1) remain unchanged.

In rungs 10 and 11, an oscillator timer "TOS_16" is implemented as follows: the input signal IN is taken from I0.4. num = 49 and therefore the timer status bit (or output Q) of this oscillator timer is TQ49. The 16-bit integer variable used to count the rising edge signals connected to the CLK input is TV_H+num and TV_L+num = TV_H+49 and TV_L+49. The preset time 0 PT0 = tcnst0×CLK = 50×10 ms (re_T10ms) = 0.5 s. The preset time 1 PT1 = tcnst1×CLK = 100×10 ms (re_T10ms) = 1 s. As can be seen from rung 11, the timer status bit TQ49 is sent to output Q0.4.

In rungs 12 and 13, an oscillator timer "TOS_16" is implemented as follows: the input signal IN is taken from I0.5. num = 59 and therefore the timer status bit (or output Q) of this oscillator timer is TQ59. The 16-bit integer variable used to count the rising edge signals connected to the CLK input is TV_H+num and TV_L+num = TV_H+59 and TV_L+59. The preset time 0 PT0 = tcnst0×CLK = 100×10 ms (re_T10ms) = 1 s. The preset time 1 PT1 = tcnst1×CLK = 50×10 ms (re_T10ms) = 0.5 s. As can be seen from rung 13, the timer status bit TQ59 is sent to output Q0.5.

In rungs 14 and 15, an oscillator timer "TOS_16" is implemented as follows: the input signal IN is taken from I0.6. num = 69 and therefore the timer status bit (or output Q) of this oscillator timer is TQ69. The 16-bit integer variable used to count the rising edge signals connected to the CLK input is TV_H+num and TV_L+num = TV_H+69 and TV_L+69. The preset time 0 PT0 = tcnst0×CLK = 200×10 ms (re_T10ms) = 2 s. The preset time 1 PT1 = tcnst1×CLK = 100×10 ms (re_T10ms) = 1 s. As can be seen from rung 15, the timer status bit TQ69 is sent to output Q0.6.

In rungs 16 and 17, an oscillator timer "TOS_16" is implemented as follows: the input signal IN is taken from I0.7. num = 79 and therefore the timer status bit (or output Q) of this oscillator timer is TQ79. The 16-bit integer variable used to count the rising edge signals connected to the CLK input is TV_H+num and TV_L+num = TV_H+79 and TV_L+79. The preset time 0 PT0 = tcnst0×CLK = 100×10 ms (re_T10ms) = 1 s. The preset time 1 PT1 = tcnst1×CLK = 200×10 ms (re_T10ms) = 2 s. As can be seen from rung 17, the timer status bit TQ79 is sent to output Q0.7.

In rung 18, a 4x1 byte multiplexer is used to observe the low byte of current count values of one of the four 16-bit variables considered between rungs 10 and 17 from Q2. The following table shows the operation of this 4x1 byte multiplexer.

inputs			output
E	s1	s0	Y
I2.2	I2.1	I2.0	Q2
0	×	×	U
1	0	0	TV_L+49
1	0	1	TV_L+59
1	1	0	TV_L+69
1	1	1	TV_L+79

×: don't care.
U: The contents of the destination
register Y(Q2) remain unchanged.

In rung 19, a 4x1 byte multiplexer is used to observe the high byte of current count values of one of the four 16-bit variables considered between rungs 10 and 17 from Q3. The following table shows the operation of this 4x1 byte multiplexer.

inputs			output
E	s1	s0	Y
I2.2	I2.1	I2.0	Q3
0	×	×	U
1	0	0	TV_H+49
1	0	1	TV_H+59
1	1	0	TV_H+69
1	1	1	TV_H+79

×: don't care.
U: The contents of the destination
register Y(Q3) remain unchanged.

6 Counter Macros

INTRODUCTION

A PLC counter counts up or down until it reaches a limit. When the limit is reached the output is set. In fact, the counting is widely used in PLC programming. Often it is necessary to count different entities, an example of which could be to keep track of how many times a process has been completed or how many products have been produced. In this chapter, the following counter macros are described:

CTU_8 (8 Bit Up Counter),
CTU_16 (16 Bit Up Counter),
CTD_8 (8 Bit Down Counter),
CTD_16 (16 Bit Down Counter),
CTUD_8 (8 Bit Up/Down Counter),
CTUD_16 (16 Bit Up/Down Counter),
GCTUD_8 (8 Bit Generalized Up/Down Counter),
GCTUD_16 (16 Bit Generalized Up/Down Counter).

Counters can be used in a wide range of applications. The definition of 8-bit variables, namely CV_L, CV_L+1, ..., CV_L+79, CV_H, CV_H+1, ..., CV_H+79, C_Q0, C_Q1, ..., C_Q9 and C_QD0, C_QD1, ..., C_QD9, to be used for counter macros, and their allocation in SRAM data memory are both shown in Figures 6.1 and 6.2, respectively. All 8-bit variables defined for counters must be cleared at the beginning of the PLC operation for a proper operation. Therefore, all variables of counter macros are initialized within the macro "initialize" as shown in Figure 6.3. The definition of CQ counter outputs (status bits) to be used for CTU_8, CTU_16, CTD_8, and CTD_16 counter macros are shown in Figure 6.4. The definition of QU (Output Up) counter outputs (up status bits) to be used for CTUD_8 and CTUD_16 counter macros are shown in Figure 6.5. The definition of QL (Output-within-Limits) counter outputs (within-limits status bits) to be used for GCTUD_8 and GCTUD_16 counter macros are shown in Figure 6.6. The definition of QD (Output Down) counter outputs (down status bits) to be used for CTUD_8, CTUD_16, GCTUD_8, and GCTUD_16 counter macros are shown in Figure 6.7.

The individual bits (1-bit variables) of 8-bit SRAM registers C_Q0, C_Q1, ..., C_Q9 are shown in Table 6.1 as CQ counter outputs (status bits) CQx, x = 0, 1, ...,79, in Table 6.2 as QU (Output Up) counter outputs (up status bits) QUx, x = 0, 1, ...,79, and in Table 6.3 as QL (Output-within-Limits) counter outputs (within-limits status bits) QLx, x = 0, 1, ...,79. The individual bits (1-bit variables) of 8-bit SRAM registers C_QD0, C_QD1, ..., C_QD9 are shown in Table 6.4 as QD (Output Down) counter outputs (down status bits) QDx, x = 0, 1, ...,79.

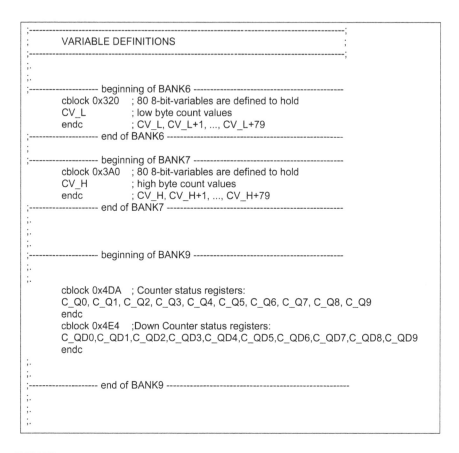

FIGURE 6.1 Definition of 8-bit variables to be used for counter macros.

At any time at most 80 different counters can be used irrespective of the counter type. A unique counter number from 0 to 79 can be assigned to only one of the counter macros CTU_8, CTD_8, CTUD_8, GCTUD_8, CTU_16, CTD_16, CTUD_16, GCTUD_16. All counter macros, share eighty 8-bit variables, namely, CV_L, CV_L+1, ..., CV_L+79. These variables are used to hold current count values (CV) for 8-bit counters and likewise low byte of current count values (CVL) for 16-bit counters. In addition,16-bit counter macros, namely, CTU_16, CTD_16, CTUD_16, GCTUD_16, share eighty 8-bit variables, namely, CV_H, CV_H+1, ..., CV_H+79. These variables are used to hold high byte of current count values (CVH) for 16-bit counters. Since there is one current count value for each counter, do not assign the same number to more than one counter.

The file "PICPLC_PIC16F1847_macros_Bsc.inc", which is downloadable from this book's webpage under the downloads section, contains macros defined for the PIC16F1847-Based PLC explained in this book (Hardware and Basic Concepts). Let us now consider the counter macros. In the following, first of all, a general description is given for the considered counter function and then its 8-bit and 16-bit implementations in the PIC16F1847-Based PLC are provided.

	Bank 6				Bank 7		
320h	CV_L	350h	CV_L+48	3A0h	CV_H	3D0h	CV_H+48
321h	CV_L+1	351h	CV_L+49	3A1h	CV_H+1	3D1h	CV_H+49
322h	CV_L+2	352h	CV_L+50	3A2h	CV_H+2	3D2h	CV_H+50
323h	CV_L+3	353h	CV_L+51	3A3h	CV_H+3	3D3h	CV_H+51
324h	CV_L+4	354h	CV_L+52	3A4h	CV_H+4	3D4h	CV_H+52
325h	CV_L+5	355h	CV_L+53	3A5h	CV_H+5	3D5h	CV_H+53
326h	CV_L+6	356h	CV_L+54	3A6h	CV_H+6	3D6h	CV_H+54
327h	CV_L+7	357h	CV_L+55	3A7h	CV_H+7	3D7h	CV_H+55
328h	CV_L+8	358h	CV_L+56	3A8h	CV_H+8	3D8h	CV_H+56
329h	CV_L+9	359h	CV_L+57	3A9h	CV_H+9	3D9h	CV_H+57
32Ah	CV_L+10	35Ah	CV_L+58	3AAh	CV_H+10	3DAh	CV_H+58
32Bh	CV_L+11	35Bh	CV_L+59	3ABh	CV_H+11	3DBh	CV_H+59
32Ch	CV_L+12	35Ch	CV_L+60	3ACh	CV_H+12	3DCh	CV_H+60
32Dh	CV_L+13	35Dh	CV_L+61	3ADh	CV_H+13	3DDh	CV_H+61
32Eh	CV_L+14	35Eh	CV_L+62	3AEh	CV_H+14	3DEh	CV_H+62
32Fh	CV_L+15	35Fh	CV_L+63	3AFh	CV_H+15	3DFh	CV_H+63
330h	CV_L+16	360h	CV_L+64	3B0h	CV_H+16	3E0h	CV_H+64
331h	CV_L+17	361h	CV_L+65	3B1h	CV_H+17	3E1h	CV_H+65
332h	CV_L+18	362h	CV_L+66	3B2h	CV_H+18	3E2h	CV_H+66
333h	CV_L+19	363h	CV_L+67	3B3h	CV_H+19	3E3h	CV_H+67
334h	CV_L+20	364h	CV_L+68	3B4h	CV_H+20	3E4h	CV_H+68
335h	CV_L+21	365h	CV_L+69	3B5h	CV_H+21	3E5h	CV_H+69
336h	CV_L+22	366h	CV_L+70	3B6h	CV_H+22	3E6h	CV_H+70
337h	CV_L+23	367h	CV_L+71	3B7h	CV_H+23	3E7h	CV_H+71
338h	CV_L+24	368h	CV_L+72	3B8h	CV_H+24	3E8h	CV_H+72
339h	CV_L+25	369h	CV_L+73	3B9h	CV_H+25	3E9h	CV_H+73
33Ah	CV_L+26	36Ah	CV_L+74	3BAh	CV_H+26	3EAh	CV_H+74
33Bh	CV_L+27	36Bh	CV_L+75	3BBh	CV_H+27	3EBh	CV_H+75
33Ch	CV_L+28	36Ch	CV_L+76	3BCh	CV_H+28	3ECh	CV_H+76
33Dh	CV_L+29	36Dh	CV_L+77	3BDh	CV_H+29	3EDh	CV_H+77
33Eh	CV_L+30	36Eh	CV_L+78	3BEh	CV_H+30	3EEh	CV_H+78
33Fh	CV_L+31	36Fh	CV_L+79	3BFh	CV_H+31	3EFh	CV_H+79
340h	CV_L+32			3C0h	CV_H+32		
341h	CV_L+33			3C1h	CV_H+33		
342h	CV_L+34			3C2h	CV_H+34		
343h	CV_L+35			3C3h	CV_H+35		
344h	CV_L+36			3C4h	CV_H+36		
345h	CV_L+37			3C5h	CV_H+37		
346h	CV_L+38			3C6h	CV_H+38		
347h	CV_L+39			3C7h	CV_H+39		
348h	CV_L+40			3C8h	CV_H+40		
349h	CV_L+41			3C9h	CV_H+41		
34Ah	CV_L+42			3CAh	CV_H+42		
34Bh	CV_L+43			3CBh	CV_H+43		
34Ch	CV_L+44			3CCh	CV_H+44		
34Dh	CV_L+45			3CDh	CV_H+45		
34Eh	CV_L+46			3CEh	CV_H+46		
34Fh	CV_L+47			3CFh	CV_H+47		

Bank 6   Bank 7

FIGURE 6.2 Allocation of 8-bit variables in SRAM data memory to be used for the counter macros.

4DAh	C_Q0		4E4h	C_QD0
4DBh	C_Q1		4E5h	C_QD1
4DCh	C_Q2		4E6h	C_QD2
4DDh	C_Q3		4E7h	C_QD3
4DEh	C_Q4		4E8h	C_QD4
4DFh	C_Q5		4E9h	C_QD5
4E0h	C_Q6		4EAh	C_QD6
4E1h	C_Q7		4EBh	C_QD7
4E2h	C_Q8		4ECh	C_QD8
4E3h	C_Q9		4EDh	C_QD9

Bank 9

FIGURE 6.2 Continued

6.1 UP COUNTER (CTU)

The up counter (CTU) can be used to signal when a count has reached a maximum value. The symbol and the algorithm of the up counter (CTU) are shown in Figure 6.8, while its' truth table is given in Table 6.5. The up counter counts the number of "rising edges" (↑) detected at the input CU. PV defines the maximum value for the

```
initialize macro
        local   L1,L2
;·
;·
;·
;·
;·

;-------- clear BANKs from 0 to 12 -------------
        clrf    FSR0L          ; initialize the pointer to the first location
        movlw   b'00100000'    ; of linear memory, i.e., 0x2000
        movwf   FSR0H          ;
        clrw                   ; W = 0
L2      movwi   FSR0++         ; clear INDF0 register
        btfss   FSR0H,2        ; all done?       if not goto L2
        bra     L2             ; if 0x2400 is reached then skip
;-------------------------------

;·
;·
;·
;·
;·
;·
        endm
```

FIGURE 6.3 Initialization of all variables of counter macros within the macro "initialize".

#define	CQ0	C_Q0,0	#define	CQ40	C_Q5,0
#define	CQ1	C_Q0,1	#define	CQ41	C_Q5,1
#define	CQ2	C_Q0,2	#define	CQ42	C_Q5,2
#define	CQ3	C_Q0,3	#define	CQ43	C_Q5,3
#define	CQ4	C_Q0,4	#define	CQ44	C_Q5,4
#define	CQ5	C_Q0,5	#define	CQ45	C_Q5,5
#define	CQ6	C_Q0,6	#define	CQ46	C_Q5,6
#define	CQ7	C_Q0,7	#define	CQ47	C_Q5,7
#define	CQ8	C_Q1,0	#define	CQ48	C_Q6,0
#define	CQ9	C_Q1,1	#define	CQ49	C_Q6,1
#define	CQ10	C_Q1,2	#define	CQ50	C_Q6,2
#define	CQ11	C_Q1,3	#define	CQ51	C_Q6,3
#define	CQ12	C_Q1,4	#define	CQ52	C_Q6,4
#define	CQ13	C_Q1,5	#define	CQ53	C_Q6,5
#define	CQ14	C_Q1,6	#define	CQ54	C_Q6,6
#define	CQ15	C_Q1,7	#define	CQ55	C_Q6,7
#define	CQ16	C_Q2,0	#define	CQ56	C_Q7,0
#define	CQ17	C_Q2,1	#define	CQ57	C_Q7,1
#define	CQ18	C_Q2,2	#define	CQ58	C_Q7,2
#define	CQ19	C_Q2,3	#define	CQ59	C_Q7,3
#define	CQ20	C_Q2,4	#define	CQ60	C_Q7,4
#define	CQ21	C_Q2,5	#define	CQ61	C_Q7,5
#define	CQ22	C_Q2,6	#define	CQ62	C_Q7,6
#define	CQ23	C_Q2,7	#define	CQ63	C_Q7,7
#define	CQ24	C_Q3,0	#define	CQ64	C_Q8,0
#define	CQ25	C_Q3,1	#define	CQ65	C_Q8,1
#define	CQ26	C_Q3,2	#define	CQ66	C_Q8,2
#define	CQ27	C_Q3,3	#define	CQ67	C_Q8,3
#define	CQ28	C_Q3,4	#define	CQ68	C_Q8,4
#define	CQ29	C_Q3,5	#define	CQ69	C_Q8,5
#define	CQ30	C_Q3,6	#define	CQ70	C_Q8,6
#define	CQ31	C_Q3,7	#define	CQ71	C_Q8,7
#define	CQ32	C_Q4,0	#define	CQ72	C_Q9,0
#define	CQ33	C_Q4,1	#define	CQ73	C_Q9,1
#define	CQ34	C_Q4,2	#define	CQ74	C_Q9,2
#define	CQ35	C_Q4,3	#define	CQ75	C_Q9,3
#define	CQ36	C_Q4,4	#define	CQ76	C_Q9,4
#define	CQ37	C_Q4,5	#define	CQ77	C_Q9,5
#define	CQ38	C_Q4,6	#define	CQ78	C_Q9,6
#define	CQ39	C_Q4,7	#define	CQ79	C_Q9,7

FIGURE 6.4 Definition of CQ counter outputs (status bits) to be used for CTU_8, CTU_16, CTD_8, and CTD_16 counter macros.

#define	QU0	C_Q0,0	#define	QU40	C_Q5,0
#define	QU1	C_Q0,1	#define	QU41	C_Q5,1
#define	QU2	C_Q0,2	#define	QU42	C_Q5,2
#define	QU3	C_Q0,3	#define	QU43	C_Q5,3
#define	QU4	C_Q0,4	#define	QU44	C_Q5,4
#define	QU5	C_Q0,5	#define	QU45	C_Q5,5
#define	QU6	C_Q0,6	#define	QU46	C_Q5,6
#define	QU7	C_Q0,7	#define	QU47	C_Q5,7
#define	QU8	C_Q1,0	#define	QU48	C_Q6,0
#define	QU9	C_Q1,1	#define	QU49	C_Q6,1
#define	QU10	C_Q1,2	#define	QU50	C_Q6,2
#define	QU11	C_Q1,3	#define	QU51	C_Q6,3
#define	QU12	C_Q1,4	#define	QU52	C_Q6,4
#define	QU13	C_Q1,5	#define	QU53	C_Q6,5
#define	QU14	C_Q1,6	#define	QU54	C_Q6,6
#define	QU15	C_Q1,7	#define	QU55	C_Q6,7
#define	QU16	C_Q2,0	#define	QU56	C_Q7,0
#define	QU17	C_Q2,1	#define	QU57	C_Q7,1
#define	QU18	C_Q2,2	#define	QU58	C_Q7,2
#define	QU19	C_Q2,3	#define	QU59	C_Q7,3
#define	QU20	C_Q2,4	#define	QU60	C_Q7,4
#define	QU21	C_Q2,5	#define	QU61	C_Q7,5
#define	QU22	C_Q2,6	#define	QU62	C_Q7,6
#define	QU23	C_Q2,7	#define	QU63	C_Q7,7
#define	QU24	C_Q3,0	#define	QU64	C_Q8,0
#define	QU25	C_Q3,1	#define	QU65	C_Q8,1
#define	QU26	C_Q3,2	#define	QU66	C_Q8,2
#define	QU27	C_Q3,3	#define	QU67	C_Q8,3
#define	QU28	C_Q3,4	#define	QU68	C_Q8,4
#define	QU29	C_Q3,5	#define	QU69	C_Q8,5
#define	QU30	C_Q3,6	#define	QU70	C_Q8,6
#define	QU31	C_Q3,7	#define	QU71	C_Q8,7
#define	QU32	C_Q4,0	#define	QU72	C_Q9,0
#define	QU33	C_Q4,1	#define	QU73	C_Q9,1
#define	QU34	C_Q4,2	#define	QU74	C_Q9,2
#define	QU35	C_Q4,3	#define	QU75	C_Q9,3
#define	QU36	C_Q4,4	#define	QU76	C_Q9,4
#define	QU37	C_Q4,5	#define	QU77	C_Q9,5
#define	QU38	C_Q4,6	#define	QU78	C_Q9,6
#define	QU39	C_Q4,7	#define	QU79	C_Q9,7

FIGURE 6.5 Definition of QU (Output-Up) counter outputs (up status bits) to be used for CTUD_8 and CTUD_16 counter macros.

#define	QL0	C_Q0,0	#define	QL40	C_Q5,0
#define	QL1	C_Q0,1	#define	QL41	C_Q5,1
#define	QL2	C_Q0,2	#define	QL42	C_Q5,2
#define	QL3	C_Q0,3	#define	QL43	C_Q5,3
#define	QL4	C_Q0,4	#define	QL44	C_Q5,4
#define	QL5	C_Q0,5	#define	QL45	C_Q5,5
#define	QL6	C_Q0,6	#define	QL46	C_Q5,6
#define	QL7	C_Q0,7	#define	QL47	C_Q5,7
#define	QL8	C_Q1,0	#define	QL48	C_Q6,0
#define	QL9	C_Q1,1	#define	QL49	C_Q6,1
#define	QL10	C_Q1,2	#define	QL50	C_Q6,2
#define	QL11	C_Q1,3	#define	QL51	C_Q6,3
#define	QL12	C_Q1,4	#define	QL52	C_Q6,4
#define	QL13	C_Q1,5	#define	QL53	C_Q6,5
#define	QL14	C_Q1,6	#define	QL54	C_Q6,6
#define	QL15	C_Q1,7	#define	QL55	C_Q6,7
#define	QL16	C_Q2,0	#define	QL56	C_Q7,0
#define	QL17	C_Q2,1	#define	QL57	C_Q7,1
#define	QL18	C_Q2,2	#define	QL58	C_Q7,2
#define	QL19	C_Q2,3	#define	QL59	C_Q7,3
#define	QL20	C_Q2,4	#define	QL60	C_Q7,4
#define	QL21	C_Q2,5	#define	QL61	C_Q7,5
#define	QL22	C_Q2,6	#define	QL62	C_Q7,6
#define	QL23	C_Q2,7	#define	QL63	C_Q7,7
#define	QL24	C_Q3,0	#define	QL64	C_Q8,0
#define	QL25	C_Q3,1	#define	QL65	C_Q8,1
#define	QL26	C_Q3,2	#define	QL66	C_Q8,2
#define	QL27	C_Q3,3	#define	QL67	C_Q8,3
#define	QL28	C_Q3,4	#define	QL68	C_Q8,4
#define	QL29	C_Q3,5	#define	QL69	C_Q8,5
#define	QL30	C_Q3,6	#define	QL70	C_Q8,6
#define	QL31	C_Q3,7	#define	QL71	C_Q8,7
#define	QL32	C_Q4,0	#define	QL72	C_Q9,0
#define	QL33	C_Q4,1	#define	QL73	C_Q9,1
#define	QL34	C_Q4,2	#define	QL74	C_Q9,2
#define	QL35	C_Q4,3	#define	QL75	C_Q9,3
#define	QL36	C_Q4,4	#define	QL76	C_Q9,4
#define	QL37	C_Q4,5	#define	QL77	C_Q9,5
#define	QL38	C_Q4,6	#define	QL78	C_Q9,6
#define	QL39	C_Q4,7	#define	QL79	C_Q9,7

FIGURE 6.6 Definition of QL (Output-within-Limits) counter outputs (within-limits status bits) to be used for GCTUD_8 and GCTUD_16 counter macros.

#define	QD0	C_QD0,0	#define	QD40	C_QD5,0
#define	QD1	C_QD0,1	#define	QD41	C_QD5,1
#define	QD2	C_QD0,2	#define	QD42	C_QD5,2
#define	QD3	C_QD0,3	#define	QD43	C_QD5,3
#define	QD4	C_QD0,4	#define	QD44	C_QD5,4
#define	QD5	C_QD0,5	#define	QD45	C_QD5,5
#define	QD6	C_QD0,6	#define	QD46	C_QD5,6
#define	QD7	C_QD0,7	#define	QD47	C_QD5,7
#define	QD8	C_QD1,0	#define	QD48	C_QD6,0
#define	QD9	C_QD1,1	#define	QD49	C_QD6,1
#define	QD10	C_QD1,2	#define	QD50	C_QD6,2
#define	QD11	C_QD1,3	#define	QD51	C_QD6,3
#define	QD12	C_QD1,4	#define	QD52	C_QD6,4
#define	QD13	C_QD1,5	#define	QD53	C_QD6,5
#define	QD14	C_QD1,6	#define	QD54	C_QD6,6
#define	QD15	C_QD1,7	#define	QD55	C_QD6,7
#define	QD16	C_QD2,0	#define	QD56	C_QD7,0
#define	QD17	C_QD2,1	#define	QD57	C_QD7,1
#define	QD18	C_QD2,2	#define	QD58	C_QD7,2
#define	QD19	C_QD2,3	#define	QD59	C_QD7,3
#define	QD20	C_QD2,4	#define	QD60	C_QD7,4
#define	QD21	C_QD2,5	#define	QD61	C_QD7,5
#define	QD22	C_QD2,6	#define	QD62	C_QD7,6
#define	QD23	C_QD2,7	#define	QD63	C_QD7,7
#define	QD24	C_QD3,0	#define	QD64	C_QD8,0
#define	QD25	C_QD3,1	#define	QD65	C_QD8,1
#define	QD26	C_QD3,2	#define	QD66	C_QD8,2
#define	QD27	C_QD3,3	#define	QD67	C_QD8,3
#define	QD28	C_QD3,4	#define	QD68	C_QD8,4
#define	QD29	C_QD3,5	#define	QD69	C_QD8,5
#define	QD30	C_QD3,6	#define	QD70	C_QD8,6
#define	QD31	C_QD3,7	#define	QD71	C_QD8,7
#define	QD32	C_QD4,0	#define	QD72	C_QD9,0
#define	QD33	C_QD4,1	#define	QD73	C_QD9,1
#define	QD34	C_QD4,2	#define	QD74	C_QD9,2
#define	QD35	C_QD4,3	#define	QD75	C_QD9,3
#define	QD36	C_QD4,4	#define	QD76	C_QD9,4
#define	QD37	C_QD4,5	#define	QD77	C_QD9,5
#define	QD38	C_QD4,6	#define	QD78	C_QD9,6
#define	QD39	C_QD4,7	#define	QD79	C_QD9,7

FIGURE 6.7 Definition of QD (Output-Down) counter outputs (down status bits) to be used for CTUD_8, CTUD_16, GCTUD_8, and GCTUD_16 counter macros.

TABLE 6.1
Individual bits of 8-bit SRAM registers C_Q0, C_Q1, ..., C_Q9 defined as CQ counter outputs (status bits) and used in CTU_8, CTU_16, CTD_8, and CTD_16 counter macros

Address	Name	Bit 7	Bit 6	Bit 5	Bit 4	Bit 3	Bit 2	Bit 1	Bit 0
4DAh	C_Q0	CQ7	CQ6	CQ5	CQ4	CQ3	CQ2	CQ1	CQ0
4DBh	C_Q1	CQ15	CQ14	CQ13	CQ12	CQ11	CQ10	CQ9	CQ8
4DCh	C_Q2	CQ23	CQ22	CQ21	CQ20	CQ19	CQ18	CQ17	CQ16
4DDh	C_Q3	CQ31	CQ30	CQ29	CQ28	CQ27	CQ26	CQ25	CQ24
4DEh	C_Q4	CQ39	CQ38	CQ37	CQ36	CQ35	CQ34	CQ33	CQ32
4DFh	C_Q5	CQ47	CQ46	CQ45	CQ44	CQ43	CQ42	CQ41	CQ40
4E0h	C_Q6	CQ55	CQ54	CQ53	CQ52	CQ51	CQ50	CQ49	CQ48
4E1h	C_Q7	CQ63	CQ62	CQ61	CQ60	CQ59	CQ58	CQ57	CQ56
4E2h	C_Q8	CQ71	CQ70	CQ69	CQ68	CQ67	CQ66	CQ65	CQ64
4E3h	C_Q9	CQ79	CQ78	CQ77	CQ76	CQ75	CQ74	CQ73	CQ72

TABLE 6.2
Individual bits of 8-bit SRAM registers C_Q0, C_Q1, ..., C_Q9 defined as QU counter outputs (up status bits) and used in CTUD_8 and CTUD_16 counter macros

Address	Name	Bit 7	Bit 6	Bit 5	Bit 4	Bit 3	Bit 2	Bit 1	Bit 0
4DAh	C_Q0	QU7	QU6	QU5	QU4	QU3	QU2	QU1	QU0
4DBh	C_Q1	QU15	QU14	QU13	QU12	QU11	QU10	QU9	QU8
4DCh	C_Q2	QU23	QU22	QU21	QU20	QU19	QU18	QU17	QU16
4DDh	C_Q3	QU31	QU30	QU29	QU28	QU27	QU26	QU25	QU24
4DEh	C_Q4	QU39	QU38	QU37	QU36	QU35	QU34	QU33	QU32
4DFh	C_Q5	QU47	QU46	QU45	QU44	QU43	QU42	QU41	QU40
4E0h	C_Q6	QU55	QU54	QU53	QU52	QU51	QU50	QU49	QU48
4E1h	C_Q7	QU63	QU62	QU61	QU60	QU59	QU58	QU57	QU56
4E2h	C_Q8	QU71	QU70	QU69	QU68	QU67	QU66	QU65	QU64
4E3h	C_Q9	QU79	QU78	QU77	QU76	QU75	QU74	QU73	QU72

counter. When $CV < PVmax$, each time the counter is called with a new rising edge (↑) on CU, the count value CV is incremented by one. When the current count value CV is greater than or equal to the PV, the counter output Q is set true (ON – 1). The reset input R can be used to set the output Q false (OFF – 0), and clear the current count value CV to zero. The following two sections explain the implementations of 8-bit and 16-bit up counters, respectively for the PIC16F1847-Based PLC.

6.2 MACRO "CTU_8" (8 BIT UP COUNTER)

The macro "CTU_8" defines eighty 8-bit up counters selected with the num = 0, 1, ..., 79. Table 6.6 shows the symbol of the macro "CTU_8". The macro "CTU_8"

TABLE 6.3

Individual bits of 8-bit SRAM registers C_Q0, C_Q1, ..., C_Q9 defined as QL counter outputs (within-limits status bits) and used in GCTUD_8 and GCTUD_16 counter macros

Address	Name	Bit 7	Bit 6	Bit 5	Bit 4	Bit 3	Bit 2	Bit 1	Bit 0
4DAh	C_Q0	QL7	QL6	QL5	QL4	QL3	QL2	QL1	QL0
4DBh	C_Q1	QL15	QL14	QL13	QL12	QL11	QL10	QL9	QL8
4DCh	C_Q2	QL23	QL22	QL21	QL20	QL19	QL18	QL17	QL16
4DDh	C_Q3	QL31	QL30	QL29	QL28	QL27	QL26	QL25	QL24
4DEh	C_Q4	QL39	QL38	QL37	QL36	QL35	QL34	QL33	QL32
4DFh	C_Q5	QL47	QL46	QL45	QL44	QL43	QL42	QL41	QL40
4E0h	C_Q6	QL55	QL54	QL53	QL52	QL51	QL50	QL49	QL48
4E1h	C_Q7	QL63	QL62	QL61	QL60	QL59	QL58	QL57	QL56
4E2h	C_Q8	QL71	QL70	QL69	QL68	QL67	QL66	QL65	QL64
4E3h	C_Q9	QL79	QL78	QL77	QL76	QL75	QL74	QL73	QL72

TABLE 6.4

Individual bits of 8-bit SRAM registers C_QD0, C_QD1, ..., C_QD9 defined as QD counter outputs (down status bits) and used in CTUD_8, CTUD_16, GCTUD_8, and GCTUD_16 counter macros

Address	Name	Bit 7	Bit 6	Bit 5	Bit 4	Bit 3	Bit 2	Bit 1	Bit 0
4E4h	C_QD0	QD7	QD6	QD5	QD4	QD3	QD2	QD1	QD0
4E5h	C_QD1	QD15	QD14	QD13	QD12	QD11	QD10	QD9	QD8
4E6h	C_QD2	QD23	QD22	QD21	QD20	QD19	QD18	QD17	QD16
4E7h	C_QD3	QD31	QD30	QD29	QD28	QD27	QD26	QD25	QD24
4E8h	C_QD4	QD39	QD38	QD37	QD36	QD35	QD34	QD33	QD32
4E9h	C_QD5	QD47	QD46	QD45	QD44	QD43	QD42	QD41	QD40
4EAh	C_QD6	QD55	QD54	QD53	QD52	QD51	QD50	QD49	QD48
4EBh	C_QD7	QD63	QD62	QD61	QD60	QD59	QD58	QD57	QD56
4ECh	C_QD8	QD71	QD70	QD69	QD68	QD67	QD66	QD65	QD64
4EDh	C_QD9	QD79	QD78	QD77	QD76	QD75	QD74	QD73	QD72

and its flowchart are depicted in Figures 6.9 and 6.10, respectively. CU (count up input), Q (output signal = counter status bit), and R (reset input) are all defined as Boolean variables. The PV (preset count value) is an integer constant and for 8-bit resolution it is chosen any number in the range 1–255. The maximum preset count value for the 8-bit up counter (PVmax) is 255. The up counter outputs C_Q0+num/8,num-(8*(num/8)) are defined by status bits CQ0, CQ1, ..., CQ79 as shown in Figure 6.4. A unique memory bit (Mreg,Mbit) = M0.0, M0.1, ..., M127.7 is used to detect the rising edge signals at the input CU. An 8-bit integer variable "CV_L+num" is used to count the rising edge signals detected at the input CU. Let

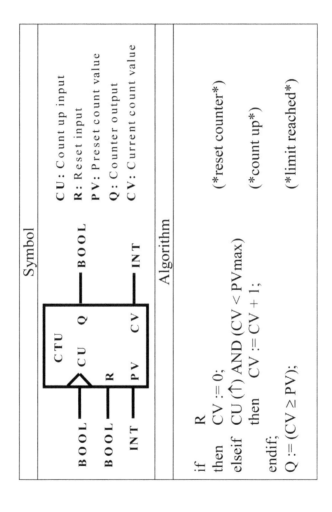

FIGURE 6.8 The symbol and the algorithm of the Up Counter (CTU).

TABLE 6.5
The truth table of the up counter (CTU)

R	CU	Operation
1	×	Clear the current count value CV (i.e., CV:= 0).
0	0	NOP (No Operation)
0	1	NOP
0	↓	NOP
0	↑	If CV < PVmax, then increment the CV (i.e., CV:= CV + 1).
×	×	If CV ≥ PV, then set the output Q true (ON – HIGH), else set the output Q false (OFF – LOW).

×: don't care.

TABLE 6.6
The symbol of the macro "CTU_8"

Symbol	
CTU_8 CU Q R PV CV num Mreg,Mbit	**CU** (count up input, CUreg,CUbit) = 0, 1 **R** (reset input, Rreg,Rbit) = 0, 1 **PV** (preset count value, 8-bit constant) = 1, 2, …, 255 **num** (the unique number of the counter) = 0, 1, …, 79 **Mreg,Mbit** (A unique memory bit, rising edge detector for the CU input) = M0.0, M0.1, …, M127.7 **Q** (counter output – counter status bit) = C_Q0+num/8,num-(8*(num/8)) (num = 0, 1, …, 79) **CV** (8-bit current count value, hold in an 8-bit register) = CV_L+num (num = 0, 1, …, 79) **PVmax** (maximum preset count value for 8-bit counter resolution) = 255

us now briefly consider how the macro "CTU_8" works. If the input signal R is true (ON – 1), then the counter "CV_L+num" is loaded with "00h" and the output signal C_Q0+num/8,num-(8*(num/8)) is forced to be false (OFF – 0). If the input signal R is false (OFF – 0) and the current count value of the counter < PVmax, then with each "rising edge" of the CU, the related counter "CV_L+num" is incremented by one. When the current count value of "CV_L+num" is greater than or equal to the PV, then state-change from OFF to ON is issued for the output signal (counter output – counter status bit) C_Q0+num/8,num-(8*(num/8)). Note that the count up process stops when the current count value of the counter is equal to PVmax. Assumption: The operands "Mreg,Mbit", "CUreg,CUbit", and "Rreg,Rbit" can be in any Bank.

In order to explain how an up counter is set-up by using the macro "CTU_8", let us consider Figure 6.11, where an up counter is obtained with the macro "CTU_8".

```
CTU_8 macro   num, Mreg,Mbit, CUreg,CUbit, Rreg,Rbit, PV
        local   L0,L1,L2,L3,L4,L5
        local   CQR_n,b_n,CV_Ln
;------------------------------------------------------------------------
 if num < 80                ;if num < 80 then carry on, else do not compile.
;------------------------------------------------------------------------
 if (PV > 0)&&(PV < 256);if 0<PV<256 then carry on, else do not compile.
;------------------------------------------------------------------------
CQR_n set    C_Q0+num/8            ;C_Q0 Register number
b_n    set    num-(8*(num/8))       ;C_Q0 bit number
CV_Ln set    CV_L+num              ;CV_L register number
;------------------------------------------------------------------------
;------------------------------------------------;--------------------
; Temp_1,7 := rising edge signal of CUreg,CUbit (re_CU)
;------------------------------------------------;--------------------
        clrf      Temp_1       ;
        banksel CUreg          ;
        btfsc     CUreg,CUbit  ;
        goto      L5           ;
        banksel Mreg           ;
        bsf       Mreg,Mbit    ;
        goto      L4           ;
L5                             ;
        banksel Mreg           ;
        btfss     Mreg,Mbit    ;
        goto      L4           ;
        bcf       Mreg,Mbit    ;
        bsf       Temp_1,7     ; set re_CU
L4                             ;
;------------------------------------------------;--------------------
;IF R THEN CV:= 0;              ;
;------------------------------------------------;--------------------
        banksel Rreg           ;
        btfss     Rreg,Rbit    ;
        goto      L3           ;
        banksel CV_Ln          ;
        clrf      CV_Ln        ;
        goto      L1           ;
L3                             ;
;------------------------------------------------;--------------------
;ELSIF re_CU AND (CV < PVmax) THEN CV:= CV+1;
;------------------------------------------------;--------------------
        banksel CV_Ln          ;
        btfss     Temp_1,7     ;
        goto      L2           ;
        movlw     0xFF         ;
        subwf     CV_Ln,W      ;
        btfsc     STATUS,C     ;
        goto      L2           ;
        incf      CV_Ln,F      ;
```

FIGURE 6.9 The macro "CTU_8".

```
L2                                  ;
;-------------------------------------;----------------------
;Q:= (CV >= PV);                     ;
;-------------------------------------;----------------------
        movlw   PV                  ;
        subwf   CV_Ln,W             ;
        btfss   STATUS,C            ;
        goto    L1                  ;
        banksel CQR_n               ;
        bsf     CQR_n,b_n           ;
        goto    L0                  ;
L1                                  ;
        banksel CQR_n               ;
        bcf     CQR_n,b_n           ;
LO                                  ;
;----------------------------------------------------------------
        else
        error "Make sure that 0 < PV < 256 !"
        endif
;----------------------------------------------------------------
        else
        error "The counter number must be one of 0, 1, 2, ..., 79."
        endif
;----------------------------------------------------------------
        endm
```

FIGURE 6.9　Continued

In this example program we have

num = 57 (decimal)
Mreg,Mbit = M0.0
CU = I0.0
R = I0.1
PV = 5 (decimal)

Therefore, we obtain the following from the macro "CTU_8":

CV_Ln = CV_L register number
CV_Ln = CV_L+num = CV_L+57 = 320h + 57d = 320h + 39h = 359h (This is the
　8-bit register in BANK6 of SRAM memory where the current count value
　of this up counter will be held.)

Note that in the following, the division $\dfrac{num}{8}$ is an integer division.

CQR_n = C_Q0 **R**egister **n**umber

$$CQR\_n = C\_Q0 + num/8 = C\_Q0 + \frac{num}{8} = 4DAh + \frac{57}{8} = 4DAh + 7d = 4E1h$$

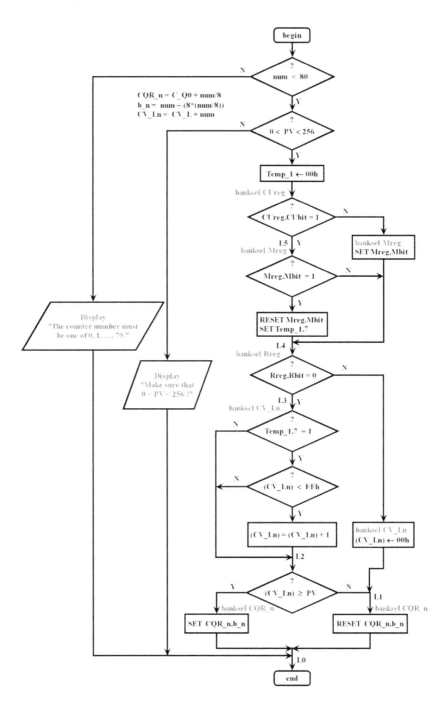

FIGURE 6.10   The flowchart of the macro "CTU_8".

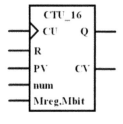

$$;num,Mreg,Mbit,CU,R,PV \qquad ; rung\ 1$$
$$CTU\_8 \quad 57,M0.0,I0.0,I0.1,.5$$

FIGURE 6.11 Example up counter obtained with the macro "CTU_8": a user program and its ladder diagram.

$$b\_n = C\_Q0\ \mathbf{bit\ number}$$

$$b\_n = num - \left(8*\left(num/8\right)\right) = num - \left(8*\left(\frac{num}{8}\right)\right)$$

$$= 57 - \left(8*\left(\frac{57}{8}\right)\right) = 57 - (8*7) = 57 - 56 = 1$$

up counter output (status bit)=CQR_n,b_n=4E1h,1=CQ57 (This is the 1-bit variable in BANK9 of SRAM memory where the state of the up counter output will be held.)

6.3 MACRO "CTU_16" (16 BIT UP COUNTER)

The macro "CTU_16" defines eighty 16-bit up counters selected with the num=0, 1, ..., 79. Table 6.7 shows the symbol of the macro "CTU_16". The macro "CTU_16"

TABLE 6.7

The symbol of the macro "CTU_16"

CTU_16 CU Q R PV CV num Mreg,Mbit	**CU** (count up input, CUreg,CUbit)=0, 1 **R** (reset input, Rreg,Rbit)=0, 1 **PV** (preset count value, 16-bit constant)=1, 2, ..., 65535 **num** (the unique number of the counter)=0, 1, ..., 79 **Mreg,Mbit** (A unique memory bit, rising edge detector for the CU input)=M0.0, M0.1, ..., M127.7 **Q** (counter output – counter status bit) = C_Q0+num/8,num-(8*(num/8)) (num=0, 1, ..., 79) **CV** (16-bit current count value, hold in two 8-bit registers)=CV_H+num & CV_L+num (num=0, 1, ..., 79) **PVmax** (maximum preset count value for 16-bit counter resolution)=65535

---

and its flowchart are depicted in Figures 6.12 and 6.13, respectively. CU (count up input), Q (output signal = counter status bit), and R (reset input) are all defined as Boolean variables. The PV (preset count value) is an integer constant and for 16-bit resolution it is chosen any number in the range 1–65535. The maximum preset count value for the 16-bit up counter (PVmax) is 65535. The up counter outputs C_Q0+num/8,num-(8*(num/8)) are defined by status bits CQ0, CQ1, ..., CQ79 as shown in Figure 6.4. A unique memory bit (Mreg,Mbit) = M0.0, M0.1, ..., M127.7 is used to detect the rising edge signals at the input CU. A 16-bit integer variable CV consisting of two 8-bit variables "CV_H+num&CV_L+num" is used to count the rising edge signals detected at the input CU. "CV_H+num" holds the high byte of the CV, while "CV_L+num" holds the low byte of the CV. Let us now briefly

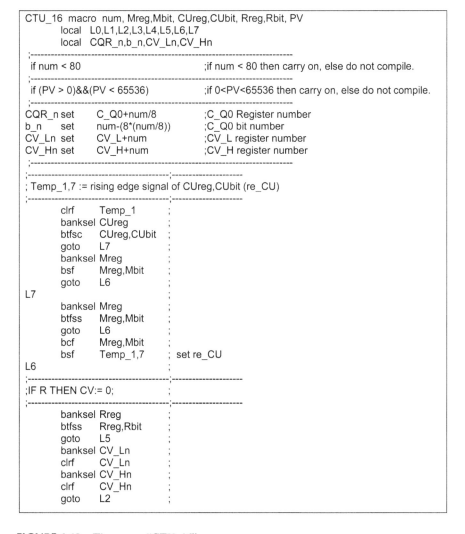

```
CTU_16  macro  num, Mreg,Mbit, CUreg,CUbit, Rreg,Rbit, PV
        local   L0,L1,L2,L3,L4,L5,L6,L7
        local   CQR_n,b_n,CV_Ln,CV_Hn
;-------------------------------------------------------------------
 if num < 80                          ;if num < 80 then carry on, else do not compile.
;-------------------------------------------------------------------
 if (PV > 0)&&(PV < 65536)            ;if 0<PV<65536 then carry on, else do not compile.
;-------------------------------------------------------------------
CQR_n set     C_Q0+num/8              ;C_Q0 Register number
b_n    set    num-(8*(num/8))         ;C_Q0 bit number
CV_Ln set     CV_L+num                ;CV_L register number
CV_Hn set     CV_H+num                ;CV_H register number
;-------------------------------------------------------------------
;--------------------------------------;--------------------
; Temp_1,7 := rising edge signal of CUreg,CUbit (re_CU)
;--------------------------------------;--------------------
        clrf      Temp_1          ;
        banksel CUreg             ;
        btfsc     CUreg,CUbit     ;
        goto      L7              ;
        banksel Mreg              ;
        bsf       Mreg,Mbit       ;
        goto      L6              ;
L7                                ;
        banksel Mreg              ;
        btfss     Mreg,Mbit       ;
        goto      L6              ;
        bcf       Mreg,Mbit       ;
        bsf       Temp_1,7        ; set re_CU
L6                                ;
;--------------------------------------;--------------------
;IF R THEN CV:= 0;                     ;
;--------------------------------------;--------------------
        banksel Rreg              ;
        btfss     Rreg,Rbit       ;
        goto      L5              ;
        banksel CV_Ln             ;
        clrf      CV_Ln           ;
        banksel CV_Hn             ;
        clrf      CV_Hn           ;
        goto      L2              ;
```

FIGURE 6.12   The macro "CTU_16".

```
L5                              ;
;-------------------------------;--------------------------------
;ELSIF re_CU AND (CV < PVmax) THEN CV:= CV+1;
;-------------------------------;--------------------------------
        btfss    Temp_1,7       ;
        goto     L3             ;
        ;-----------------------;------------------------
        movlw    0xFF           ;
        banksel  CV_Hn          ;If CV_Hn < High byte of PVmax
        subwf    CV_Hn,W        ;Then CV < PVmax.
        btfss    STATUS,C       ;Therefore goto L4
        goto     L4             ;Else
        ;-----------------------;------------------------
                                ;If CV_Hn = High byte of PVmax
                                ;Then carry on
        btfss    STATUS,Z       ;skip to see if CV_Ln < Low byte of PVmax.
        goto     L3             ;If CV_Hn =/= High byte of PVmax, Then goto L3.
        ;-----------------------;------------------------
        movlw    0xFF           ;While CV_Hn = High byte of PVmax
        banksel  CV_Ln          ;If CV_Ln < Low byte of PVmax
        subwf    CV_Ln,W        ;Then CV < PVmax.
        btfsc    STATUS,C       ;Therefore skip to L4
        goto     L3             ;Else goto L3
        ;-----------------------;------------------------
L4
        banksel  CV_Ln          ;
        incfsz   CV_Ln,F        ;
        goto     L3             ;
        banksel  CV_Hn          ;
        incf     CV_Hn,F        ;
L3
;-------------------------------;--------------------------------
;Q:= (CV >= PV);                ;
;-------------------------------;--------------------------------
        movlw    HIGH PV        ;
        banksel  CV_Hn          ;
        subwf    CV_Hn,W        ;
        btfss    STATUS,C       ;
        goto     L2             ;
        ;-----------------------;------------------------
        btfss    STATUS,Z       ;
        goto     L1             ;
        ;-----------------------;------------------------
        movlw    LOW PV         ;
        banksel  CV_Ln          ;
        subwf    CV_Ln,W        ;
        btfsc    STATUS,C       ;
        goto     L1             ;
        ;-----------------------;------------------------
```

FIGURE 6.12 Continued

consider how the macro "CTU_16" works. If the input signal R is true (ON – 1), then the counter "CV_H+num&CV_L+num" is loaded with "0000h" and the output signal C_Q0+num/8,num-(8*(num/8)) is forced to be false (OFF – 0). If the input signal R is false (OFF – 0) and the current count value of the counter < PVmax, then with each "rising edge" of the CU, the related counter "CV_H+num&CV_L+num" is incremented by one. When the current count value of "CV_H+num&CV_L+num" is

```
L2
       banksel CQR_n        ;
       bcf       CQR_n,b_n   ;
       goto      L0          ;
L1
       banksel CQR_n        ;
       bsf       CQR_n,b_n   ;
L0                           ;
;----------------------------------------------------------------
else
error "Make sure that 0 < PV < 65536 !"
endif
;----------------------------------------------------------------
else
error "The counter number must be one of 0, 1, 2, ..., 79."
endif
;----------------------------------------------------------------
       endm
```

FIGURE 6.12 Continued

greater than or equal to the PV, then state-change from OFF to ON is issued for the output signal (counter output – counter status bit) C_Q0+num/8,num-(8*(num/8)). Note that the count up process stops when the current count value of the counter is equal to PVmax. Assumption: The operands "Mreg,Mbit", "CUreg,CUbit", and "Rreg,Rbit" can be in any Bank.

In order to explain how an up counter is set-up by using the macro "CTU_16", let us consider Figure 6.14, where an up counter is obtained with the macro "CTU_16".

In this example program we have

num = 79 (decimal)
Mreg,Mbit = M0.0
CU = I0.0
R = I0.1
PV = 5000 (decimal)

Therefore, we obtain the following from the macro "CTU_16":

CV_Ln = CV_L register number
CV_Ln = CV_L+num = CV_L+79 = 320h + 79d = 320h + 4Fh = 36Fh (This is the 8-bit register in BANK6 of SRAM memory where the low byte of current count value of this up counter will be held.)

CV_Hn = CV_H register number
CV_Hn = CV_H+num = CV_H+79 = 3A0h + 79d = 3A0h + 4Fh = 3EFh (This is the 8-bit register in BANK7 of SRAM memory where the high byte of current count value of this up counter will be held.)

Note that in the following, the division $\dfrac{num}{8}$ is an integer division.

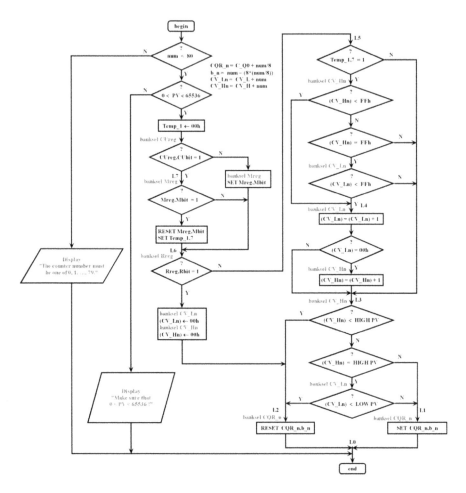

FIGURE 6.13 The flowchart of the macro "CTU_16".

```
      I0.0                    CTU_16
1  ┤├────────────────────▷ CU      Q ──── CQ79
      I0.1
   ┤├─────────────┬──────── R
                  │                        CV_H+79
          5000 ── PV      CV ──── CV_L+79
            79 ── num
           M0.0 ── Mreg,Mbit
```

;num,Mreg,Mbit,CU,R,PV   ; rung 1
CTU_16 79,M0.0,I0.0,I0.1,.5000

FIGURE 6.14 Example up counter obtained with the macro "CTU_16": a user program and its ladder diagram.

CQR_n = C_Q0 **R**egister **n**umber

$$CQR\_n = C\_Q0 + num/8 = C\_Q0 + \frac{num}{8} = 4DAh + \frac{79}{8} = 4DAh + 9d = 4E3h$$

b_n = C_Q0 **b**it **n**umber

$$b\_n = num - \left(8*\left(num/8\right)\right) = num - \left(8*\left(\frac{num}{8}\right)\right)$$

$$= 79 - \left(8*\left(\frac{79}{8}\right)\right) = 79\left(8*9\right) = 79 - 72 = 7$$

up counter output (status bit)=CQR_n,b_n=4E3h,7=CQ79 (This is the 1-bit variable in BANK9 of SRAM memory where the state of the up counter output will be held.)

6.4 DOWN COUNTER (CTD)

The down counter (CTD) can be used to signal when a count has reached zero, on counting down from a preset value. The symbol and the algorithm of the down counter (CTD) are shown in Figure 6.15, while its' truth table is given in Table 6.8. PV defines starting value for the counter. When CV > 0, each time the counter is called with a new rising edge (↑) on CD, the current count value CV is decremented by one, i.e., the counter counts down. When the counter reaches zero, the counter output Q is set true (ON – 1) and the counting-down process stops. The load input LD can be used to clear the output Q to false (OFF – 0), and load the current count value CV with the preset value PV. The following two sections explain the implementations of 8-bit and 16-bit down counters, respectively for the PIC16F1847-Based PLC.

6.5 MACRO "CTD_8" (8 BIT DOWN COUNTER)

The macro "CTD_8" defines eighty 8-bit down counters selected with the num=0, 1, …, 79. Table 6.9 shows the symbol of the macro "CTD_8". The macro "CTD_8" and its flowchart are depicted in Figures 6.16 and 6.17, respectively. CD (count down input), Q (output signal=counter status bit), and LD (load input) are all defined as Boolean variables. The PV (preset count value) is an integer constant (here for 8-bit resolution it is chosen any number in the range 1–255) and is used to define a starting value for the counter. The maximum preset count value for the 8-bit down counter (PVmax) is 255. The counter outputs C_Q0+num/8,num-(8*(num/8)) are defined by status bits CQ0, CQ1, …, CQ79 as shown in Figure 6.4. A unique memory bit (Mreg,Mbit)=M0.0, M0.1, …, M127.7 is used to detect the rising edge signals at the input CD. An 8-bit integer variable "CV_L+num" is used to count down with the rising edges detected at the input CD. Let us now briefly consider how the macro "CTD_8" works. If the input signal LD is true (ON – 1), then the output signal C_Q0+num/8,num-(8*(num/8)) is forced to be false (OFF – 0), and the counter "CV_L+num" is loaded with PV. If the input signal LD is false (OFF – 0) and the

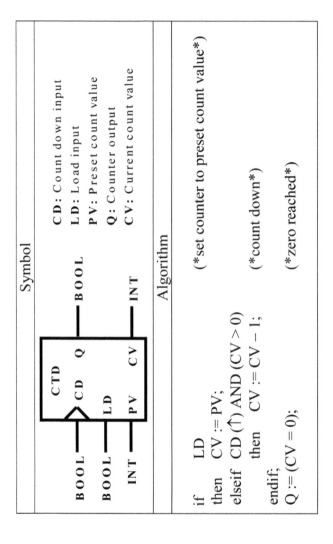

FIGURE 6.15 The symbol and the algorithm of the Down Counter (CTD).

TABLE 6.8
The truth table of the down counter (CTD)

LD	CD	Operation
1	×	Load the current count value CV with the preset count value PV (i.e., CV:= PV).
0	0	NOP (No Operation)
0	1	NOP
0	↓	NOP
0	↑	If CV > 0, then decrement the CV (i.e., CV:= CV − 1).
×	×	If CV = 0, then set the output Q true (ON – HIGH), else set the output Q to false (OFF – LOW).

×: don't care.

TABLE 6.9
The symbol of the macro "CTD_8"

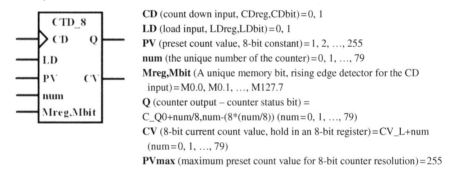

CD (count down input, CDreg,CDbit) = 0, 1

LD (load input, LDreg,LDbit) = 0, 1

PV (preset count value, 8-bit constant) = 1, 2, ..., 255

num (the unique number of the counter) = 0, 1, ..., 79

Mreg,Mbit (A unique memory bit, rising edge detector for the CD input) = M0.0, M0.1, ..., M127.7

Q (counter output – counter status bit) =
C_Q0+num/8,num-(8*(num/8)) (num = 0, 1, ..., 79)

CV (8-bit current count value, hold in an 8-bit register) = CV_L+num (num = 0, 1, ..., 79)

PVmax (maximum preset count value for 8-bit counter resolution) = 255

---

current count value CV > 0, then with each "rising edge" signal at the input CD, the related counter "CV_L+num" is decremented by one. When the count value of "CV_L+num" is equal to zero, then state-change from OFF to ON is issued for the output signal (counter output – counter status bit) C_Q0+num/8,num-(8*(num/8)) and the counting stops. Assumption: The operands "Mreg,Mbit", "CDreg,CDbit", and "LDreg,LDbit" can be in any Bank.

In order to explain how a down counter is set-up by using the macro "CTD_8", let us consider Figure 6.18, where a down counter is obtained with the macro "CTD_8".

In this example program we have

num = 7 (decimal)
Mreg,Mbit = M0.0
CD = I0.0
LD = I0.1
PV = 255 (decimal)

```
CTD_8 macro   num, Mreg,Mbit, CDreg,CDbit, LDreg,LDbit, PV
        local   L0,L1,L2,L3,L4,L5
        local   CQR_n,b_n,CV_Ln
 ;-------------------------------------------------------------------
 if num < 80                      ;if num < 80 then carry on, else do not compile.
 ;-------------------------------------------------------------------
 if (PV > 0)&&(PV < 256)          ;if 0<PV<256 then carry on, else do not compile.
 ;-------------------------------------------------------------------
CQR_n set      C_Q0+num/8          ;C_Q0 Register number
b_n    set     num-(8*(num/8))     ;C_Q0 bit number
CV_Ln set      CV_L+num            ;CV_L register number
 ;-------------------------------------------------------------------
 ;--------------------------------------------;--------------------
 ; Temp_1,7 := rising edge signal of CDreg,CDbit (re_CD)
 ;--------------------------------------------;--------------------
        clrf    Temp_1           ;
        banksel CDreg            ;
        btfsc   CDreg,CDbit      ;
        goto    L5               ;
        banksel Mreg             ;
        bsf     Mreg,Mbit        ;
        goto    L4               ;
L5                               ;
        banksel Mreg             ;
        btfss   Mreg,Mbit        ;
        goto    L4               ;
        bcf     Mreg,Mbit        ;
        bsf     Temp_1,7         ; set re_CD
L4                               ;
 ;--------------------------------------------;--------------------
 ;IF LD THEN CV:= PV;             ;
 ;--------------------------------------------;--------------------
        banksel LDreg            ;
        btfss   LDreg,LDbit      ;
        goto    L3               ;
        movlw   PV               ;
        banksel CV_Ln            ;
        movwf   CV_Ln            ;
        goto    L2               ;
L3                               ;
 ;--------------------------------------------;--------------------
 ;ELSIF re_CD AND (CV > PVmin) THEN CV:= CV-1;
 ;--------------------------------------------;--------------------
        banksel CV_Ln            ;
        btfss   Temp_1,7         ;
        goto    L2               ;
        movf    CV_Ln,F          ;
        btfsc   STATUS,Z         ;
        goto    L2               ;
        decf    CV_Ln,F          ;
```

FIGURE 6.16   The macro "CTD_8".

```
L2                                    ;
;-----------------------------------;--------------------
;Q:= (CV = 0);                         ;
;-----------------------------------;--------------------
        movf    CV_Ln,F       ;
        btfss   STATUS,Z      ;
        goto    L1            ;
        banksel CQR_n         ;
        bsf     CQR_n,b_n     ;
        goto    L0            ;
L1                            ;
        banksel CQR_n         ;
        bcf     CQR_n,b_n     ;
L0                            ;
;--------------------------------------------------------
   else
   error "Make sure that 0 < PV < 256 !"
   endif
;--------------------------------------------------------
   else
   error "The counter number must be one of 0, 1, 2, ..., 79."
   endif
;--------------------------------------------------------
        endm
```

FIGURE 6.16   Continued

Therefore, we obtain the following from the macro "CTD_8":

CV_Ln=CV_L register number

CV_Ln=CV_L+num=CV_L+7=320h+7d=327h (This is the 8-bit register in BANK6 of SRAM memory where the current count value of this down counter will be held.)

Note that in the following, the division $\dfrac{num}{8}$ is an integer divison.

CQR_n = C_Q0 **R**egister **n**umber

$$CQR\_n = C\_Q0 + num/8 = C\_Q0 + \frac{num}{8} = 4DAh + \frac{7}{8} = 4DAh + 0 = 4DAh$$

$$b\_n = C\_Q0 \ \textbf{b}it \ \textbf{n}umber$$

$$b\_n = num - \left(8*(num/8)\right) = num - \left(8*\left(\frac{num}{8}\right)\right)$$

$$= 7 - \left(8*\left(\frac{7}{8}\right)\right) = 7 - (8*0) = 7 - 0 = 7$$

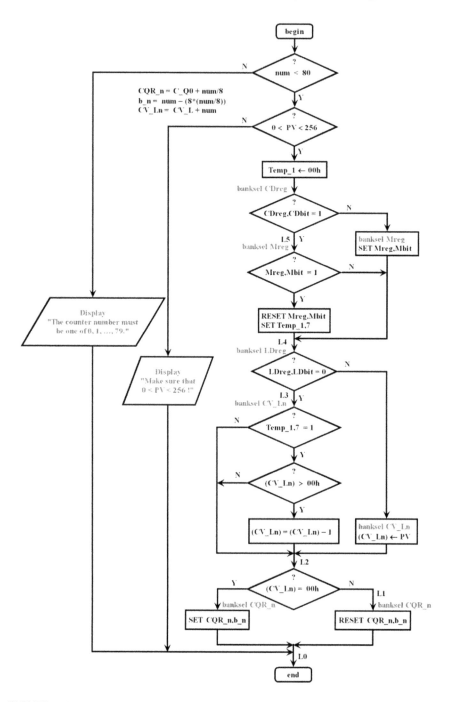

FIGURE 6.17   The flowchart of the macro "CTD_8".

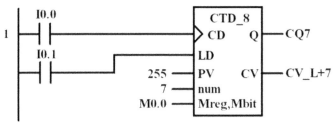

;num,Mreg,Mbit,CD,LD,PV   ; rung 1
CTD_8   7,M0.0,I0.0,I0.1,.255

FIGURE 6.18 Example down counter obtained with the macro "CTD_8": a user program and its ladder diagram.

down counter output (status bit) = CQR_n,b_n = 4DAh,7 = CQ7 (This is the 1-bit variable in BANK9 of SRAM memory where the state of the down counter output will be held.)

6.6   MACRO "CTD_16" (16 BIT DOWN COUNTER)

The macro "CTD_16" defines eighty 16-bit down counters selected with the num = 0, 1, …, 79. Table 6.10 shows the symbol of the macro "CTD_16". The macro "CTD8" and its flowchart are depicted in Figures 6.19 and 6.20, respectively. CD (count down input), Q (output signal = counter status bit), and LD (load input) are all defined as Boolean variables. The PV (preset count value) is an integer constant and for 16-bit resolution it is chosen any number in the range 1–65535. The maximum preset count value for the 8-bit up counter (PVmax) is 65535. The counter outputs C_ Q0+num/8,num-(8*(num/8)) are defined by status bits CQ0, CQ1, …, CQ79 as shown in Figure 6.4. A unique memory bit (Mreg,Mbit) = M0.0, M0.1, …, M127.7 is used to

TABLE 6.10
The symbol of the macro "CTD_16"

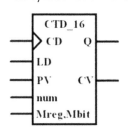

CD (count down input, CDreg,CDbit) = 0, 1
LD (load input, LDreg,LDbit) = 0, 1
PV (preset count value, 16-bit constant) = 1, 2, …, 65535
num (the unique number of the counter) = 0, 1, …, 79
Mreg,Mbit (A unique memory bit, rising edge detector for the CD
    input) = M0.0, M0.1, …, M127.7
Q (counter output – counter status bit) =
C_Q0+num/8,num-(8*(num/8)) (num = 0, 1, …, 79)
CV (16-bit current count value hold in two 8-bit registers) = CV_H+num
    & CV_L+num (num = 0, 1, …, 79)
PVmax (maximum preset count value for 16-bit counter
    resolution) = 65535

```
CTD_16  macro  num, Mreg,Mbit, CDreg,CDbit, LDreg,LDbit, PV
        local  L0,L1,L2,L3,L4,L5
        local  CQR_n,b_n,CV_Ln,CV_Hn
;-----------------------------------------------------------------
if num < 80                    ;if num < 80 then carry on, else do not compile.
;-----------------------------------------------------------------
if (PV > 0)&&(PV < 65536)      ;if 0<PV<65536 then carry on, else do not compile.
;-----------------------------------------------------------------
CQR_n set    C_Q0+num/8          ;C_Q0 Register number
b_n   set    num-(8*(num/8))     ;C_Q0 bit number
CV_Ln set    CV_L+num            ;CV_L register number
CV_Hn set    CV_H+num            ;CV_H register number
;-----------------------------------------------------------------
;------------------------------------------;--------------------
; Temp_1,7 := rising edge signal of CDreg,CDbit (re_CD)
;------------------------------------------;--------------------
        clrf     Temp_1          ;
        banksel CDreg            ;
        btfsc   CDreg,CDbit      ;
        goto    L5               ;
        banksel Mreg             ;
        bsf     Mreg,Mbit        ;
        goto    L4               ;
L5                               ;
        banksel Mreg             ;
        btfss   Mreg,Mbit        ;
        goto    L4               ;
        bcf     Mreg,Mbit        ;
        bsf     Temp_1,7         ; set re_CD
L4                               ;
;------------------------------------------;--------------------
;IF LD THEN CV:= PV;             ;
;------------------------------------------;--------------------
        banksel LDreg            ;
        btfss   LDreg,LDbit      ;
        goto    L3               ;
        movlw   LOW PV           ;
        banksel CV_Ln            ;
        movwf   CV_Ln            ;
        movlw   HIGH PV          ;
        banksel CV_Hn            ;
        movwf   CV_Hn            ;
        goto    L2               ;
```

FIGURE 6.19   The macro "CTD_16".

detect the rising edge signals at the input CD. A 16-bit integer variable CV consisting of two 8-bit variables "CV_H+num&CV_L+num" is used to count down with
the rising edge signals detected at the input CD. "CV_H+num" holds the high byte
of the CV, while "CV_L+num" holds the low byte of the CV. Let us now briefly consider how the macro "CTD_16" works. If the input signal LD is true (ON – 1), then
the output signal C_Q0+num/8,num-(8*(num/8)) is forced to be false (OFF – 0), and
the counter "CV_H+num&CV_L+num" is loaded with PV. If the input signal LD is
false (OFF – 0) and the current count value CV > 0, then with each "rising edge" signal at the input CD, the related counter "CV_H+num&CV_L+num" is decremented
by one. When the count value of "CV_H+num&CV_L+num" is equal to zero, then
state-change from OFF to ON is issued for the output signal (counter output – counter

```
L3                          ;
;------------------------------;-------------------
;ELSIF re_CD AND (CV > PVmin) THEN CV:= CV-1;
;------------------------------;-------------------
        btfss   Temp_1,7    ;
        goto    L2          ;
        banksel CV_Ln       ;
        movf    CV_Ln,W     ;
        banksel CV_Hn       ;
        iorwf   CV_Hn,W     ;
        btfsc   STATUS,Z    ;
        goto    L2          ;
        banksel CV_Ln       ;
        decf    CV_Ln,F     ;
        movf    CV_Ln,W     ;
        xorlw   0xFF        ;
        btfss   STATUS,Z    ;
        goto    L2          ;
        banksel CV_Hn       ;
        decf    CV_Hn,F     ;
L2                          ;
;------------------------------;-------------------
;Q:= (CV = 0);               ;
;------------------------------;-------------------
        banksel CV_Ln       ;
        movf    CV_Ln,W     ;
        banksel CV_Hn       ;
        iorwf   CV_Hn,W     ;
        btfss   STATUS,Z    ;
        goto    L1          ;
        banksel CQR_n       ;
        bsf     CQR_n,b_n   ;
        goto    L0          ;
L1                          ;
        banksel CQR_n       ;
        bcf     CQR_n,b_n   ;
L0                          ;
  ;----------------------------------------------------------------
  else
  error "Make sure that 0 < PV < 65536 !"
  endif
  ;----------------------------------------------------------------
  else
  error "The counter number must be one of 0, 1, 2, ..., 79."
  endif
  ;----------------------------------------------------------------
        endm
```

FIGURE 6.19   Continued

status bit) C_Q0+num/8,num-(8*(num/8)) and the counting stops. Assumption: The operands "Mreg,Mbit", "CDreg,CDbit", and "LDreg,LDbit" can be in any Bank.

In order to explain how a down counter is set-up by using the macro "CTD_16", let us consider Figure 6.21, where a down counter is obtained with the macro "CTD_16".

In this example program we have

num = 66 (decimal)
Mreg,Mbit = M0.0

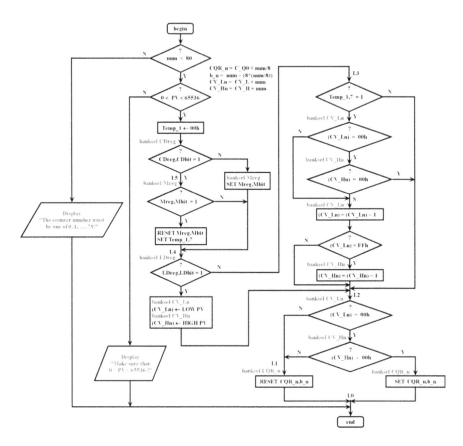

FIGURE 6.20   The flowchart of the macro "CTD_16".

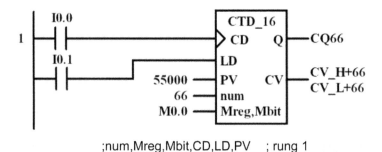

;num,Mreg,Mbit,CD,LD,PV   ; rung 1
CTD_16   66,M0.0,I0.0,I0.1,.55000

FIGURE 6.21   Example down counter obtained with the macro "CTD_16": a user program and its ladder diagram.

CD = I0.0
LD = I0.1
PV = 55000 (decimal)

Therefore, we obtain the following from the macro "CTD_16":

CV_Ln = CV_L register number
CV_Ln = CV_L+num = CV_L+66 = 320h+66d = 320h+42h = 362h (This is the 8-bit register in BANK6 of SRAM memory where the low byte of current count value of this down counter will be held.)

CV_Hn = CV_H register number
CV_Hn = CV_H+num = CV_H+66 = 3A0h+66d = 3A0h+42h = 3E2h (This is the 8-bit register in BANK7 of SRAM memory where the high byte of current count value of this down counter will be held.)

Note that in the following, the division $\dfrac{num}{8}$ is an integer divison.

CQR_n = C_Q0 **R**egister **n**umber

$$CQR\_n = C\_Q0 + num/8 = C\_Q0 + \frac{num}{8} = 4DAh + \frac{66}{8} = 4DAh + 8d = 4E2h$$

b_n = C_Q0 **b**it **n**umber

$$b\_n = num - \left(8 * \left(num/8\right)\right) = num - \left(8 * \left(\frac{num}{8}\right)\right)$$

$$= 66 - \left(8 * \left(\frac{66}{8}\right)\right) = 66 - \left(8 * 8\right) = 66 - 64 = 2$$

down counter output (status bit) = CQR_n,b_n = 4E2h,2 = CQ66 (This is the 1-bit variable in BANK9 of SRAM memory where the state of the down counter output will be held.)

6.7   UP/DOWN COUNTER (CTUD)

The up/down counter (CTUD) has two inputs CU and CD. It can be used to both count up on one input and count down on the other. The symbol and the algorithm of the up/down counter (CTUD) are shown in Figure 6.22, while its' truth table is given in Table 6.11. When CV < PVmax and a "rising edge" (↑) signal is detected at the input CU, the current count value CV is incremented by one, i.e., the up/down counter counts up. When CV > 0 and a "rising edge" (↑) signal is detected at the input CD, the current count value CV is decremented by one, i.e., the up/down counter counts down. The PV defines the preset count value for the counter. When the current count

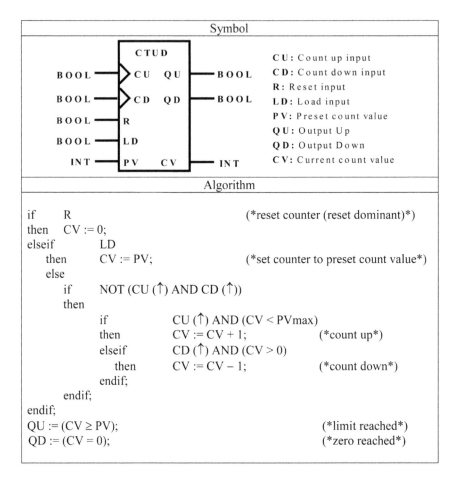

FIGURE 6.22 The symbol and the algorithm of the Up/Down Counter (CTUD).

value CV is greater than or equal to the PV, the QU (Output Up) counter output is set true (ON – 1). When the current count value CV is equal to 0, the QD (Output Down) counter output is set true (ON – 1). The reset input R can be used to clear the current count value CV. The load input LD can be used to load the count value CV with the preset count value PV. When the counter reaches the maximum count value PVmax, the counting up stops. Likewise, when the counter reaches zero, the counting down stops. If a rising edge signal is detected at the same time at inputs CU and CD, then no operation is done. The following two sections explain the implementations of 8-bit and 16-bit up/down counters, respectively for the PIC16F1847-Based PLC.

6.8 MACRO "CTUD_8" (8 BIT UP/DOWN COUNTER)

The macro "CTUD_8" defines eighty 8-bit up/down counters selected with the num=0, 1, …, 79. Table 6.12 shows the symbol of the macro "CTUD_8". The macro "CTUD_8" and its flowchart are depicted in Figures 6.23 and 6.24, respectively.

TABLE 6.11
The truth table of the up/down counter (CTUD)

R	LD	CU	CD	Operation
1	×	×	×	Clear the current count value CV to zero (i.e., CV:= 0).
0	1	×	×	Load the current count value CV with the preset count value PV (i.e., CV:= PV).
0	0	0	0	NOP (No Operation)
0	0	0	1	NOP
0	0	1	0	NOP
0	0	1	1	NOP
0	0	0	↓	NOP
0	0	1	↓	NOP
0	0	↓	0	NOP
0	0	↓	1	NOP
0	0	↓	↓	NOP
0	0	↑	↑	NOP
0	0	↑	0	If CV < PVmax, then increment the CV (i.e., CV:= CV + 1).
0	0	0	↑	If CV > 0, then decrement the CV (i.e., CV:= CV − 1).
×	×	×	×	If CV ≥ PV, then set the output QU true (ON – HIGH), else set the output QU false (OFF – LOW).
×	×	×	×	If CV = 0, then set the output QD true (ON – HIGH), else set the output QD false (OFF – LOW).

×: don't care.

CU (count up input), CD (count down input), QU (Output Up counter output), QD (Output Down counter output), R (reset input), and LD (load input) are all defined as Boolean variables. The PV (preset count value) is an integer constant and for 8-bit resolution it is chosen any number in the range 1–255. The maximum preset count value for the 8-bit up counter (PVmax) is 255. The QU (Output Up) counter outputs C_Q0+num/8,num-(8*(num/8)) are defined by up status bits QU0, QU1, ..., QU79 as shown in Figure 6.5. The QD (Output Down) counter outputs C_QD0+num/8,num-(8*(num/8)) are defined by down status bits QD0, QD1, ..., QD79 as shown in Figure 6.7. A unique memory bit (MregU,MbitU) = M0.0, M0.1, ..., M127.7 is used to detect the rising edge signals at the input CU. Likewise, another unique memory bit (MregD,MbitD) = M0.0, M0.1, ..., M127.7 is used to detect the rising edge signals at the input CD. An 8-bit integer variable "CV_L+num" is used to count up with the rising edge signals detected at the input CU and count down with the rising edge signals detected at the input CD. Let us now briefly consider how the macro "CTUD_8" works. If the input signal R is true (ON – 1), then the counter "CV_L+num" is loaded with "00h". If the input signal R is false (OFF – 0) and the input signals LD is true (ON – 1), then the counter "CV_L+num" is loaded with the PV. If both input signal R and LD are false (OFF – 0) and the rising edge signals at inputs CU and CD do not occur at the same time, then either the count up operation or the count down operation can take place. In this case, if the current count value CV < FFh, then with each

TABLE 6.12
The symbol of the macro "CTUD_8"

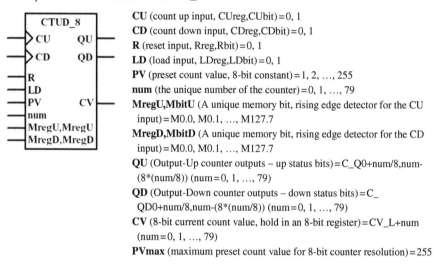

CU (count up input, CUreg,CUbit)=0, 1
CD (count down input, CDreg,CDbit)=0, 1
R (reset input, Rreg,Rbit)=0, 1
LD (load input, LDreg,LDbit)=0, 1
PV (preset count value, 8-bit constant)=1, 2, ..., 255
num (the unique number of the counter)=0, 1, ..., 79
MregU,MbitU (A unique memory bit, rising edge detector for the CU input)=M0.0, M0.1, ..., M127.7
MregD,MbitD (A unique memory bit, rising edge detector for the CD input)=M0.0, M0.1, ..., M127.7
QU (Output-Up counter outputs – up status bits)=C_Q0+num/8,num-(8*(num/8)) (num=0, 1, ..., 79)
QD (Output-Down counter outputs – down status bits)=C_QD0+num/8,num-(8*(num/8)) (num=0, 1, ..., 79)
CV (8-bit current count value, hold in an 8-bit register)=CV_L+num (num=0, 1, ..., 79)
PVmax (maximum preset count value for 8-bit counter resolution)=255

"rising edge" signal detected at the input CU, the related counter "CV_L+num" is incremented by one. Similarly, in this case, if the current count value CV > 0, then with each "rising edge" signal detected at the input CD, the related counter "CV_L+num" is decremented by one. If the count value of "CV_L+num" is greater than or equal to the PV, then state-change from OFF to ON is issued for the QU (Output Up) counter output C_Q0+num/8,num-(8*(num/8)). If the count value of "CV_L+num" is equal to 0, then state-change from OFF to ON is issued for the QD (Output Down) counter output C_QD0+num/8,num-(8*(num/8)). The count up process stops when the CV=FFh. Likewise, the count down process stops when the CV=0. Assumption: The operands "MregU,MbitU", "MregD,MbitD", "CUreg,CUbit", "CDreg,CDbit", "Rreg,Rbit", and "LDreg,LDbit" can be in any Bank.

In order to explain how an up/down counter is set-up by using the macro "CTUD_8", let us consider Figure 6.25, where an up/down counter is obtained with the macro "CTUD_8".

In this example program we have

num = 34 (decimal)
MregU,MbitU=M0.0, MregD,MbitD=M0.1
CU=I0.0, CD=I0.1, R=I0.2, LD=I0.3
PV = 100 (decimal)

Therefore, we obtain the following from the macro "CTUD_8":

CV_Ln=CV_L register number

```
CTUD_8  macro  num, MregU,MbitU, MregD,MbitD, CUreg,CUbit, CDreg,CDbit,
Rreg,Rbit, LDreg,LDbit, PV
        local   L1,L2,L3,L4,L5,L6,L7,L8,L9,L10,L11,L12,L13
        local   QUR_n,QDR_n,b_n,CV_Ln
  ;----------------------------------------------------------------------------
  if num < 80                     ;if num < 80 then carry on, else do not compile.
  ;----------------------------------------------------------------------------
  if (PV > 0)&&(PV < 256)         ;if 0<PV<256 then carry on, else do not compile.
  ;----------------------------------------------------------------------------
QUR_n set     C_Q0+num/8              ;QU Register number
QDR_n set     C_QD0+num/8             ;QD Register number
b_n    set    num-(8*(num/8))         ;C_Q0 bit number
CV_Ln set     CV_L+num                ;CV_L register number
  ;----------------------------------------------------------------------------
        clrf      Temp_1       ;
  ;------------------------------------------;--------------------
  ; Temp_1,7 := rising edge signal of CUreg,CUbit (re_CU)
  ;------------------------------------------;--------------------
        banksel CUreg         ;
        btfsc     CUreg,CUbit  ;
        goto      L13          ;
        banksel MregU         ;
        bsf       MregU,MbitU  ;
        goto      L12          ;
L13                           ;
        banksel MregU         ;
        btfss     MregU,MbitU  ;
        goto      L12          ;
        bcf       MregU,MbitU  ;
        bsf       Temp_1,7     ; set re_CU
L12                           ;
  ;------------------------------------------;--------------------
  ; Temp_1,6 := rising edge signal of CDreg,CDbit (re_CD)
  ;------------------------------------------;--------------------
        banksel CDreg         ;
        btfsc     CDreg,CDbit  ;
        goto      L11          ;
        banksel MregD         ;
        bsf       MregD,MbitD  ;
        goto      L10          ;
L11                           ;
        banksel MregD         ;
        btfss     MregD,MbitD  ;
        goto      L10          ;
        bcf       MregD,MbitD  ;
        bsf       Temp_1,6     ; set re_CD
```

FIGURE 6.23 The macro "CTUD_8".

```
L10                              ;
;-------------------------------------;------------------------
;IF R THEN CV:= 0;                ;
;-------------------------------------;------------------------
        banksel Rreg             ;
        btfss   Rreg,Rbit        ;
        goto    L9               ;
        banksel CV_Ln            ;
        clrf    CV_Ln            ;
        goto    L5               ;
;-------------------------------------;------------------------
;ELSIF LD THEN CV:= PV;           ;
;-------------------------------------;------------------------
L9                               ;
        banksel LDreg            ;
        btfss   LDreg,LDbit      ;
        goto    L8               ;
        movlw   PV               ;
        banksel CV_Ln            ;
        movwf   CV_Ln            ;
        goto    L5               ;
L8                               ;
;-------------------------------------;------------------------
;IF NOT (re_CU AND re_CD) THEN    ;
;-------------------------------------;------------------------
        btfss   Temp_1,7         ; If (re_CU AND re_CD),
        goto    L7               ; then goto  L5,
        btfss   Temp_1,6         ; else goto  L7.
        goto    L7               ;
        goto    L5               ;
L7                               ;
;-------------------------------------;------------------------
;ELSIF re_CU AND (CV < PVmax) THEN CV:= CV+1;
;-------------------------------------;------------------------
        banksel CV_Ln            ;
        btfss   Temp_1,7         ;
        goto    L6               ;
        movlw   0xFF             ;
        subwf   CV_Ln,W          ;
        btfsc   STATUS,C         ;
        goto    L6               ;
        incf    CV_Ln,F          ;
```

FIGURE 6.23 Continued

CV_Ln = CV_L+num = CV_L+34 = 320h + 34d = 320h + 22h = 342h (This is the 8-bit register in BANK6 of SRAM memory where the current count value of this up/down counter will be held.)

Note that in the following, the division $\dfrac{num}{8}$ is an integer divison.

QUR_n = QU **R**egister **n**umber

$$QUR\_n = C\_Q0 + num/8 = C\_Q0 + \frac{num}{8} = 4DAh + \frac{34}{8} = 4DAh + 4d = 4DEh$$

```
;-----------------------------:---------------------
;ELSIF re_CD AND (CV > 0) THEN CV:= CV-1;
;-----------------------------:---------------------
L6      btfss    Temp_1,6      ;
        goto     L5            ;
        movf     CV_Ln,F       ;
        btfsc    STATUS,Z      ;
        goto     L5            ;
        decf     CV_Ln,F       ;
L5                             ;
;-----------------------------:---------------------
;QU:= (CV >= PV);             ;
;-----------------------------:---------------------
        movlw    PV            ;
        banksel  CV_Ln         ;
        subwf    CV_Ln,W       ;
        btfss    STATUS,C      ;
        goto     L4            ;
        banksel  QUR_n         ;
        bsf      QUR_n,b_n     ;
        goto     L3            ;
L4                             ;
        banksel  QUR_n         ;
        bcf      QUR_n,b_n     ;
L3                             ;
;-----------------------------:---------------------
;QD:= (CV = 0);              ;
;-----------------------------:---------------------
        banksel  CV_Ln         ;
        movf     CV_Ln,F       ;
        btfss    STATUS,Z      ;
        goto     L2            ;
        banksel  QDR_n         ;
        bsf      QDR_n,b_n     ;
        goto     L1            ;
L2                             ;
        banksel  QDR_n         ;
        bcf      QDR_n,b_n     ;
L1                             ;
;--------------------------------------------------------------------
 else
 error "Make sure that 0 < PV < 256 !"
 endif
;--------------------------------------------------------------------
 else
 error "The counter number must be one of 0, 1, 2, ..., 79."
 endif
;--------------------------------------------------------------------
        endm
```

FIGURE 6.23 Continued

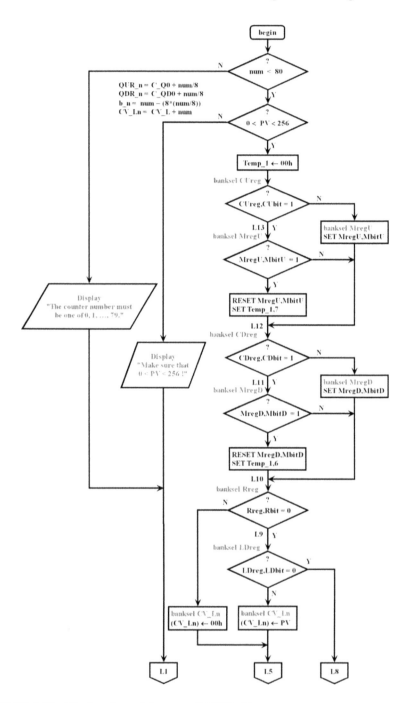

FIGURE 6.24   The flowchart of the macro "CTUD_8".

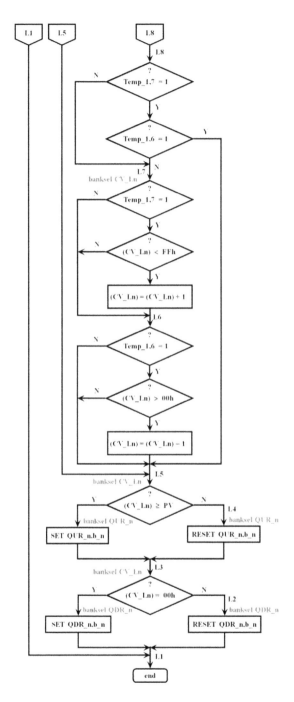

FIGURE 6.24 Continued

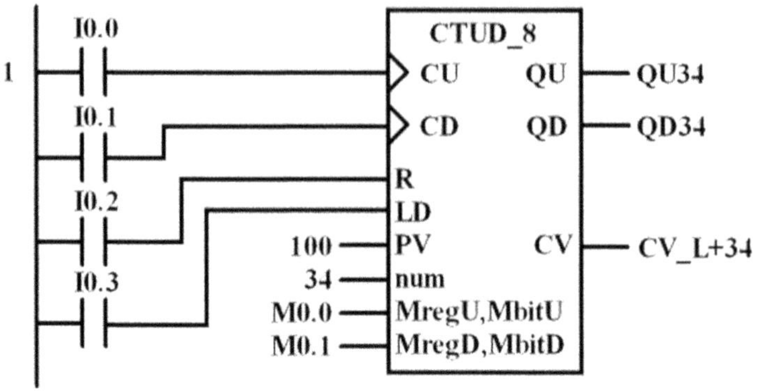

; rung 1
;num,MregU,MbitU,MregD,MbitD,CU,CD,R,LD,PV
CTUD_8   34,M0.0,M0.1,I0.0,I0.1,I0.2,I0.3,.100

FIGURE 6.25 Example up/down counter obtained with the macro "CTUD_8": a user program and its ladder diagram.

$$QDR\_n = QD \textbf{ R}egister \textbf{ n}umber$$

$$QDR\_n = C\_QD0 + num/8 = C\_QD0 + \frac{num}{8}$$

$$= 4E4h + \frac{34}{8} = 4E4h + 4d = 4E8h$$

$$b\_n = C\_Q0 \text{ or } C\_QD0 \textbf{ b}it \textbf{ n}umber$$

$$b\_n = num - \left(8 * \left(num/8\right)\right) = num - \left(8 * \left(\frac{num}{8}\right)\right)$$

$$= 34 - \left(8 * \left(\frac{34}{8}\right)\right) = 34 - \left(8 * 4\right) = 34 - 32 = 2$$

up/down counter QU (Output Up) counter output (up status bit)=QUR_n,b_n=4
   DEh,2=QU34 (This is the 1-bit variable in BANK9 of SRAM memory where
   the state of the QU (Output Up) counter output of this counter will be held.)
up/down counter QD (Output Down) counter output (down status bit)=QDR
   _n,b_n=4E8h,2=QD34 (This is the 1-bit variable in BANK9 of SRAM
   memory where the state of the QD (Output Down) counter output of this
   counter will be held.)

6.9   MACRO "CTUD_16" (16 BIT UP/DOWN COUNTER)

The macro "CTUD_16" defines eighty 16-bit up/down counters selected with the
num=0, 1, …, 79. Table 6.13 shows the symbol of the macro "CTUD_16". The macro
"CTUD_16" and its flowchart are depicted in Figures 6.26 and 6.27, respectively.
CU (count up input), CD (count down input), QU (Output Up counter output), QD

TABLE 6.13
The symbol of the macro "CTUD_16"

CTUD_16 — CU, QU; CD, QD; R; LD; PV, CV; num; MregU,MbitU; MregD,MbitD	**CU** (count up input, CUreg,CUbit)=0, 1
	CD (count down input, CDreg,CDbit)=0, 1
	R (reset input, Rreg,Rbit)=0, 1
	LD (load input, LDreg,LDbit)=0, 1
	PV (preset count value, 16-bit constant)=1, 2, ..., 65535
	num (the unique number of the counter)=0, 1, ..., 79
	MregU,MbitU (A unique memory bit, rising edge detector for the CU input)=M0.0, M0.1, ..., M127.7
	MregD,MbitD (A unique memory bit, rising edge detector for the CD input)=M0.0, M0.1, ..., M127.7
	QU (Output-Up counter outputs – up status bits) = C_Q0+num/8,num-(8*(num/8)) (num=0, 1, ..., 79)
	QD (Output-Down counter outputs – down status bits) = C_QD0+num/8,num-(8*(num/8)) (num=0, 1, ..., 79)
	CV (16-bit count value, hold in two 8-bit registers) = CV_H+num & CV_L+num (num=0, 1, ..., 79)
	PVmax (maximum preset count value for 16-bit counter resolution)=65535

(Output Down counter output), R (reset input), and LD (load input) are all defined as Boolean variables. The PV (preset count value) is an integer constant and for 16-bit resolution it is chosen any number in the range 1–65535. The maximum preset count value for the 16-bit up/down counter (PVmax) is 65535. The QU (Output Up) counter outputs C_Q0+num/8,num-(8*(num/8)) are defined by up status bits QU0, QU1, ..., QU79 as shown in Figure 6.5. The QD (Output Down) counter outputs C_QD0+num/8,num-(8*(num/8)) are defined by down status bits QD0, QD1, ..., QD79 as shown in Figure 6.7. A unique memory bit (MregU,MbitU)=M0.0, M0.1, ..., M127.7 is used to detect the rising edge signals at the input CU. Likewise, another unique memory bit (MregD,MbitD)=M0.0, M0.1, ..., M127.7 is used to detect the rising edge signals at the input CD. A 16-bit integer variable "CV_H+num&CV_L+num" is used to count up with the rising edge signals detected at the input CU and count down with the rising edge signals detected at the input CD. "CV_H+num" holds the high byte of the CV, while "CV_L+num" holds the low byte of the CV.

Let us now briefly consider how the macro "CTUD_16" works. If the input signal R is true (ON – 1), then the counter "CV_H+num&CV_L+num" is loaded with "0000h". If the input signal R is false (OFF – 0) and the input signal LD is true (ON – 1), then the counter "CV_H+num&CV_L+num" is loaded with the PV. If both input signals R and LD are false (OFF – 0) and the rising edge signals at inputs CU and CD do not occur at the same time, then either the count up operation or the count down operation can take place. In this case, if the current count value CV < FFFFh, then with each "rising edge" signal detected at the input CU, the related counter "CV_H+num&CV_L+num" is incremented by one. Similarly, in this case, if the current count value CV > 0, then with each "rising edge" signal detected at the input CD, the related counter "CV_H+num&CV_L+num" is

```
CTUD_16  macro  num, MregU,MbitU, MregD,MbitD, CUreg,CUbit, CDreg,CDbit,
Rreg,Rbit, LDreg,LDbit, PV
         local   L0,L1,L2,L3,L4,L5,L6,L7,L8,L9,L10,L11,L12,L13,L14
         local   QUR_n,QDR_n,b_n,CV_Ln,CV_Hn
;-------------------------------------------------------------------
 if num < 80                        ;if num < 80 then carry on, else do not compile.
;-------------------------------------------------------------------
 if (PV > 0)&&(PV < 65536)    ;if 0<PV<65536 then carry on, else do not compile.
;-------------------------------------------------------------------
QUR_n set     C_Q0+num/8           ;QU Register number
QDR_n set     C_QD0+num/8          ;QD Register number
b_n     set     num-(8*(num/8))       ;C_Q0 bit number
CV_Ln set     CV_L+num             ;CV_L register number
CV_Hn set     CV_H+num             ;CV_H register number
         ;-------------------------------------------------------------------
         clrf     Temp_1       ;
;----------------------------------------;--------------------
; Temp_1,7 := rising edge signal of CUreg,CUbit (re_CU)
;----------------------------------------;--------------------
         banksel CUreg        ;
         btfsc    CUreg,CUbit  ;
         goto     L14          ;
         banksel MregU        ;
         bsf      MregU,MbitU  ;
         goto     L13          ;
L14                           ;
         banksel MregU        ;
         btfss    MregU,MbitU  ;
         goto     L13          ;
         bcf      MregU,MbitU  ;
         bsf      Temp_1,7     ; set re_CU
L13                           ;
;----------------------------------------;--------------------
; Temp_1,6 := rising edge signal of CDreg,CDbit (re_CD)
;----------------------------------------;--------------------
         banksel CDreg        ;
         btfsc    CDreg,CDbit  ;
         goto     L12          ;
         banksel MregD        ;
         bsf      MregD,MbitD  ;
         goto     L11          ;
L12                           ;
         banksel MregD        ;
         btfss    MregD,MbitD  ;
         goto     L11          ;
         bcf      MregD,MbitD  ;
         bsf      Temp_1,6     ; set re_CD
L11                           ;
```

FIGURE 6.26 The macro "CTUD_16".

decremented by one. If the count value of "CV_H+num&CV_L+num" is greater than or equal to the PV, then state-change from OFF to ON is issued for the QU (Output Up) counter output C_Q0+num/8,num-(8*(num/8)). If the count value of "CV_H+num&CV_L+num" is equal to 0, then state-change from OFF to ON is issued for the QD (Output Down) counter output C_QD0+num/8,num-(8*(num/8)). The count up process stops when the CV=FFFFh. Likewise, the count down process stops when the CV=0. Assumption: The operands "MregU,MbitU", "MregD,MbitD", "CUreg,CUbit", "CDreg,CDbit", "Rreg,Rbit", and "LDreg,LDbit" can be in any Bank.

```
;----------------------------------------;--------------------
;IF R THEN CV:= 0;                       ;
;----------------------------------------;--------------------
        banksel Rreg            ;
        btfss    Rreg,Rbit      ;
        goto     L10            ;
        banksel CV_Ln           ;
        clrf     CV_Ln          ;
        banksel CV_Hn           ;
        clrf     CV_Hn          ;
        goto     L5             ;
;----------------------------------------;--------------------
;ELSIF LD THEN CV:= PV;                  ;
;----------------------------------------;--------------------
L10                             ;
        banksel LDreg           ;
        btfss    LDreg,LDbit    ;
        goto     L9             ;
        movlw   LOW PV          ;
        banksel CV_Ln           ;
        movwf   CV_Ln           ;
        movlw   HIGH PV         ;
        banksel CV_Hn           ;
        movwf   CV_Hn           ;
        goto     L5             ;
L9                              ;
;----------------------------------------;--------------------
;IF NOT (re_CU AND re_CD) THEN           ;
;----------------------------------------;--------------------
        btfss    Temp_1,7       ; If (re_CU AND re_CD),
        goto     L8             ; then goto  L5,
        btfss    Temp_1,6       ; else goto  L8.
        goto     L8             ;
        goto     L5             ;
L8                              ;
;----------------------------------------;--------------------
;ELSIF re_CU AND (CV < PVmax) THEN CV:= CV+1;
;----------------------------------------;--------------------
        btfss    Temp_1,7       ;
        goto     L6             ;
        ;-------------------------------;--------------------
        movlw   0xFF            ;
        banksel CV_Hn           ;If CV_Hn < High byte of PVmax
        subwf   CV_Hn,W         ;Then CV < PVmax.
        btfss    STATUS,C       ;Therefore goto L7
        goto     L7             ;Else
        ;-------------------------------;--------------------
                                ;If CV_Hn = High byte of PVmax
                                ;Then carry on
        btfss    STATUS,Z       ;skip to see if CV_Ln < Low byte of PVmax.
        goto     L6             ;If CV_Hn =/= High byte of PVmax, Then goto L6.
```

FIGURE 6.26   Continued

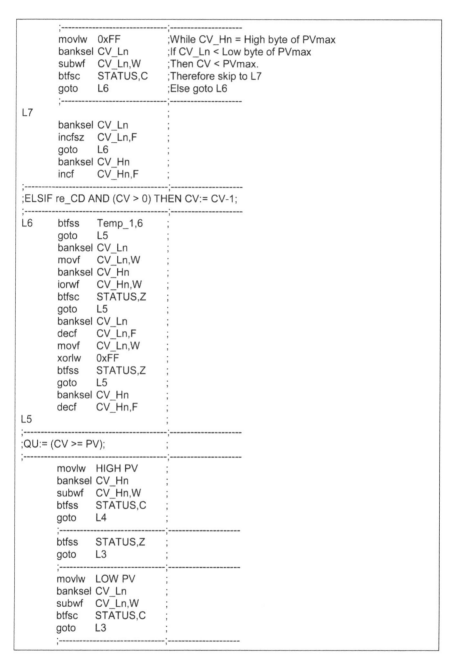

```
        movlw   0xFF        ;While CV_Hn = High byte of PVmax
        banksel CV_Ln       ;If CV_Ln < Low byte of PVmax
        subwf   CV_Ln,W     ;Then CV < PVmax.
        btfsc   STATUS,C    ;Therefore skip to L7
        goto    L6          ;Else goto L6
        ;---------------------------------
L7
        banksel CV_Ln
        incfsz  CV_Ln,F
        goto    L6
        banksel CV_Hn
        incf    CV_Hn,F
;------------------------------------------------
;ELSIF re_CD AND (CV > 0) THEN CV:= CV-1;
;------------------------------------------------
L6      btfss   Temp_1,6
        goto    L5
        banksel CV_Ln
        movf    CV_Ln,W
        banksel CV_Hn
        iorwf   CV_Hn,W
        btfsc   STATUS,Z
        goto    L5
        banksel CV_Ln
        decf    CV_Ln,F
        movf    CV_Ln,W
        xorlw   0xFF
        btfss   STATUS,Z
        goto    L5
        banksel CV_Hn
        decf    CV_Hn,F
L5
;------------------------------------------------
;QU:= (CV >= PV);
;------------------------------------------------
        movlw   HIGH PV
        banksel CV_Hn
        subwf   CV_Hn,W
        btfss   STATUS,C
        goto    L4
        ;---------------------------------
        btfss   STATUS,Z
        goto    L3
        ;---------------------------------
        movlw   LOW PV
        banksel CV_Ln
        subwf   CV_Ln,W
        btfsc   STATUS,C
        goto    L3
        ;---------------------------------
```

FIGURE 6.26 Continued

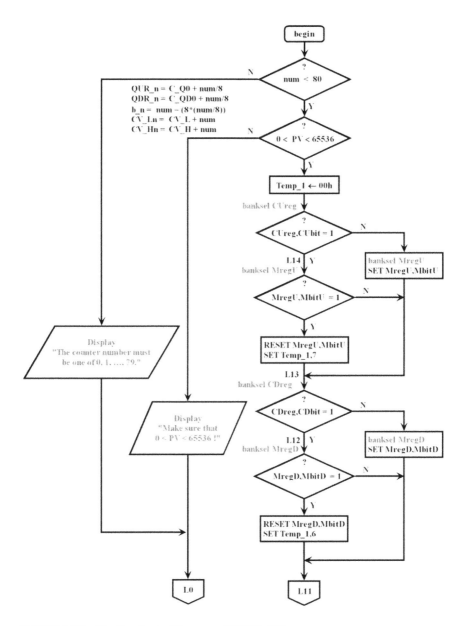

FIGURE 6.27 The flowchart of the macro "CTUD_16".

In order to explain how an up/down counter is set-up by using the macro "CTUD_16", let us consider Figure 6.28, where an up/down counter is obtained with the macro "CTUD_16".

In this example program we have

num = 18 (decimal), MregU,MbitU = M0.0, MregD,MbitD = M0.1
CU = I0.0, CD = I0.1, R = I0.2, LD = I0.3, PV = 10000 (decimal)

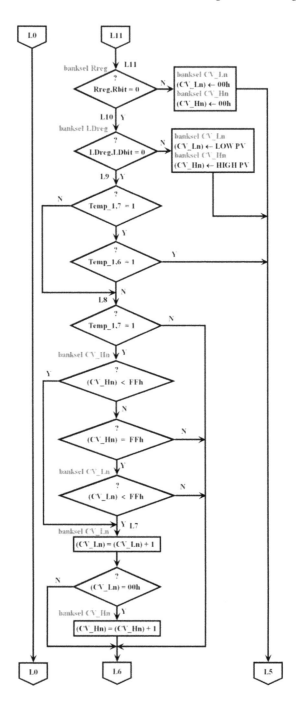

FIGURE 6.27   Continued

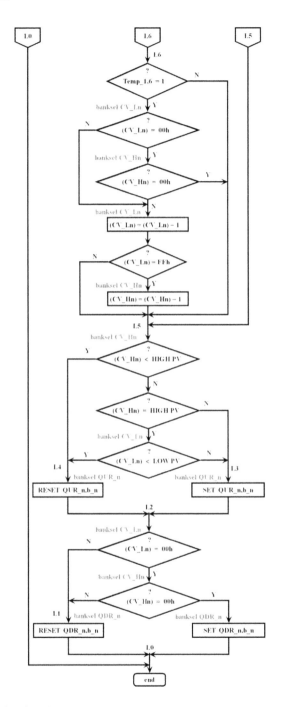

FIGURE 6.27 Continued

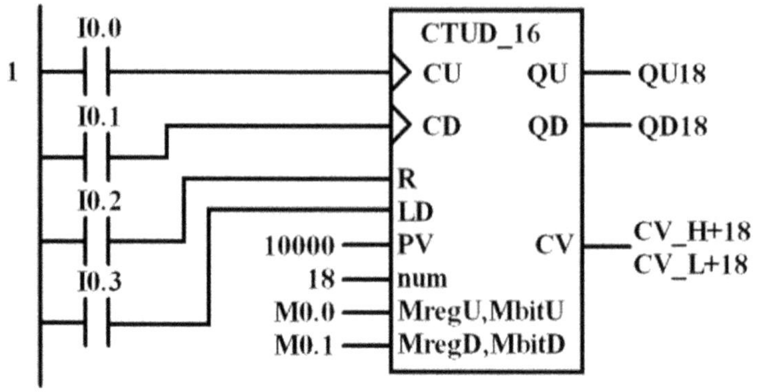

; rung 1
;num,MregU,MbitU,MregD,MbitD,CU,CD,R,LD,PV
CTUD_16   18,M0.0,M0.1,I0.0,I0.1,I0.2,I0.3,.10000

FIGURE 6.28 Example up/down counter obtained with the macro "CTUD_16": a user program and its ladder diagram.

Therefore, we obtain the following from the macro "CTUD_16":

CV_Ln=CV_L register number
CV_Ln=CV_L+num=CV_L+18=320h+18d=320h+12h=332h (This is the
   8-bit register in BANK6 of SRAM memory where the low byte of current
   count value of this up/down counter will be held.)

CV_Hn=CV_H register number
CV_Hn=CV_H+num=CV_H+18=3A0h+18d=3A0h+12h=3B2h (This is
   the 8-bit register in BANK7 of SRAM memory where the high byte of current
   count value of this up/down counter will be held.)

Note that in the following, the division $\dfrac{num}{8}$ is an integer divison.

QDR_n = QD **R**egister **n**umber

$$QDR\_n = C\_QD0 + num/8 = C\_QD0 + \frac{num}{8} = 4E4h + \frac{18}{8} = 4E4h + 2d = 4E6h$$

b_n = C_Q0 or C_QD0 bit **n**umber

$$b\_n = num - \left(8*\left(num/8\right)\right) = num - \left(8*\left(\frac{num}{8}\right)\right)$$

$$= 18 - \left(8*\left(\frac{18}{8}\right)\right) = 18 - \left(8*2\right) = 18 - 16 = 2$$

up/down counter QU (Output Up) counter output (up status bit)=QUR_n,b_n
=4DCh,2=QU18 (This is the 1-bit variable in BANK9 of SRAM memory
where the state of the QU (Output Up) counter output of this counter will
be held.)

up/down counter QD (Output Down) counter output (down status bit)=QDR
_n,b_n=4E6h,2=QD18 (This is the 1-bit variable in BANK9 of SRAM
memory where the state of the QD (Output Down) counter output of this
counter will be held.)

6.10   GENERALIZED UP/DOWN COUNTER (GCTUD)

The generalized up/down counter (GCTUD) has two inputs for count up and count
down, namely, CU and CD. The symbol and the algorithm of the generalized up/
down counter (GCTUD) are shown in Figure 6.29, while its' truth table is given in

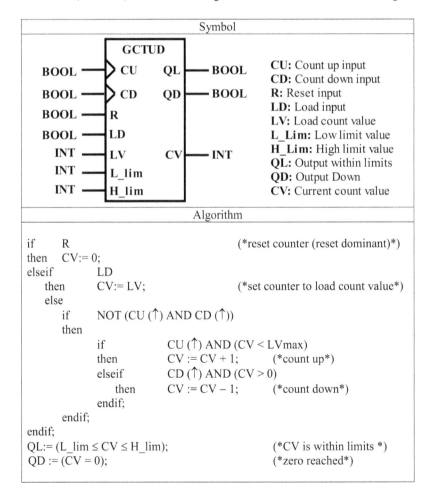

FIGURE 6.29 The symbol and the algorithm of the Generalized Up/Down Counter
(GCTUD).

TABLE 6.14

The truth table of the generalized up/down counter (GCTUD)

R	LD	CU	CD	Operation
1	×	×	×	Clear the current count value CV to zero (i.e., CV:= 0).
0	1	×	×	Load the current count value CV with the load count value LV (i.e., CV:= LV).
0	0	0	0	NOP (No Operation)
0	0	0	1	NOP
0	0	1	0	NOP
0	0	1	1	NOP
0	0	0	↓	NOP
0	0	1	↓	NOP
0	0	↓	0	NOP
0	0	↓	1	NOP
0	0	↓	↓	NOP
0	0	↑	↑	NOP
0	0	↑	0	If CV < LVmax, then increment the CV (i.e., CV:= CV + 1).
0	0	0	↑	If CV > 0, then decrement the CV (i.e., CV:= CV − 1).
×	×	×	×	If (L_lim ≤ CV ≤ H_lim), then set the output QL true (ON – HIGH), else set the output QL false (OFF – LOW).
×	×	×	×	If CV = 0, then set the output QD true (ON – HIGH), else set the output QD false (OFF – LOW).

×: don't care.

Table 6.14. The LV defines the load count value for the generalized up/down counter. The LVmax defines the maximum load count value for the counter. The LVmax is defined with the counter resolution such as 8-bit, 16-bit, etc. For example, when the counter resolution is 8-bit (respectively 16-bit), the LVmax is 255 (respectively 65535). The reset input R can be used to clear the current count value CV. The load input LD can be used to load the count value CV with the LV. When the inputs R and LD are not active and rising edge signals at the inputs CU and CD do not occur at the same time, the generalized up/down counter counts up or down. When the current count value CV < LVmax and a "rising edge" (↑) signal is detected at the input CU, the current count value CV is incremented by one, i.e., the generalized up/down counter counts up. When CV > 0 and a "rising edge" (↑) signal is detected at the input CD, the current count value CV is decremented by one, i.e., the generalized up/down counter counts down. The state of the QL (Output-within-Limits) counter output is defined based on two limit values L_Lim (low limit value), and H_Lim (High limit value). The QL is set true (ON – 1) when L_lim ≤ CV ≤ H_lim. When the current count value CV is equal to 0, the QD (Output Down) counter output is set true (ON – 1). When the counter reaches the LVmax, the counting-up process stops. Likewise, when the counter reaches zero, the counting-down process stops. The following two sections explain the implementations of 8-bit and 16-bit generalized up/down counters, respectively for the PIC16F1847-Based PLC.

TABLE 6.15
The symbol of the macro "GCTUD_8"

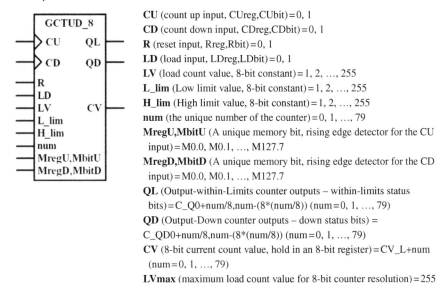

CU (count up input, CUreg,CUbit) = 0, 1
CD (count down input, CDreg,CDbit) = 0, 1
R (reset input, Rreg,Rbit) = 0, 1
LD (load input, LDreg,LDbit) = 0, 1
LV (load count value, 8-bit constant) = 1, 2, …, 255
L_lim (Low limit value, 8-bit constant) = 1, 2, …, 255
H_lim (High limit value, 8-bit constant) = 1, 2, …, 255
num (the unique number of the counter) = 0, 1, …, 79
MregU,MbitU (A unique memory bit, rising edge detector for the CU
    input) = M0.0, M0.1, …, M127.7
MregD,MbitD (A unique memory bit, rising edge detector for the CD
    input) = M0.0, M0.1, …, M127.7
QL (Output-within-Limits counter outputs – within-limits status
    bits) = C_Q0+num/8,num-(8*(num/8)) (num = 0, 1, …, 79)
QD (Output-Down counter outputs – down status bits) =
    C_QD0+num/8,num-(8*(num/8)) (num = 0, 1, …, 79)
CV (8-bit current count value, hold in an 8-bit register) = CV_L+num
    (num = 0, 1, …, 79)
LVmax (maximum load count value for 8-bit counter resolution) = 255

6.11   MACRO "GCTUD_8" (GENERALIZED 8 BIT UP/DOWN COUNTER)

The macro "GCTUD_8" defines eighty 8-bit generalized up/down counters selected with the num = 0, 1, …, 79. Table 6.15 shows the symbol of the macro "GCTUD_8". The macro "GCTUD_8" and its flowchart are depicted in Figures 6.30 and 6.31, respectively. CU (count up input), CD (count down input), QL (Output-within-Limits counter output), QD (Output Down counter output), R (reset input), and LD (load input) are all defined as Boolean variables. The LV (load count value) is an integer constant and for 8-bit resolution it is chosen any number in the range 1–255. The maximum load count value for the 8-bit up counter (LVmax) is 255. The QL (Output-within-Limits) counter outputs C_Q0+num/8,num-(8*(num/8)) are defined by within-limits status bits QL0, QL1, …, QL79 as shown in Figure 6.6. The QD (Output Down) counter outputs C_QD0+num/8,num-(8*(num/8)) are defined by down status bits QD0, QD1, …, QD79 as shown in Figure 6.7. A unique memory bit (MregU,MbitU) = M0.0, M0.1, …, M127.7 is used to detect the rising edge signals at the input CU. Likewise, another unique memory bit (MregD,MbitD) = M0.0, M0.1, …, M127.7 is used to detect the rising edge signals at the input CD. An 8-bit integer variable "CV_L+num" is used to count up with the rising edge signals detected at the input CU and count down with the rising edge signals detected at the input CD.

Let us now briefly consider how the macro "GCTUD_8" works. If the input signal R is true (ON – 1), then the counter "CV_L+num" is loaded with "00h". If the input signal R is false (OFF – 0) and the input signal LD is true (ON – 1), then the counter "CV_L+num" is loaded with the LV. If both input signals R and LD are false

```
GCTUD_8  macro  num, MregU,MbitU, MregD,MbitD, CUreg,CUbit, CDreg,CDbit, Rreg,Rbit,
LDreg,LDbit, LV, L_lim, H_lim
          local   L1,L2,L3,L4,L5,L6,L7,L8,L9,L10,L11,L12,L13
          local   CQR_n,QDR_n,b_n,CV_Ln
;-------------------------------------------------------------------------
  if (num<80)&&(L_lim <= H_lim)&&(LV > 0)&&(LV < 256)
    ;if (num<80)AND(L_lim <= H_lim)AND(0<LV<256) then carry on, else do not compile.
;-------------------------------------------------------------------------
CQR_n set     C_Q0+num/8              ;C_Q0 Register number
QDR_n set     C_QD0+num/8             ;QD Register number
b_n    set    num-(8*(num/8))         ;C_Q0 bit number
CV_Ln set     CV_L+num                ;CV_L register number
;-------------------------------------------------------------------------
          clrf    Temp_1       ;
;------------------------------------------;--------------------------
; Temp_1,7 := rising edge signal of CUreg,CUbit (re_CU)
;------------------------------------------;--------------------------
          banksel CUreg        ;
          btfsc   CUreg,CUbit  ;
          goto    L13          ;
          banksel MregU        ;
          bsf     MregU,MbitU  ;
          goto    L12          ;
L13                            ;
          banksel MregU        ;
          btfss   MregU,MbitU  ;
          goto    L12          ;
          bcf     MregU,MbitU  ;
          bsf     Temp_1,7     ; set re_CU
L12                            ;
;------------------------------------------;--------------------------
; Temp_1,6 := rising edge signal of CDreg,CDbit (re_CD)
;------------------------------------------;--------------------------
          banksel CDreg        ;
          btfsc   CDreg,CDbit  ;
          goto    L11          ;
          banksel MregD        ;
          bsf     MregD,MbitD  ;
          goto    L10          ;
L11                            ;
          banksel MregD        ;
          btfss   MregD,MbitD  ;
          goto    L10          ;
          bcf     MregD,MbitD  ;
          bsf     Temp_1,6     ; set re_CD
```

FIGURE 6.30   The macro "GCTUD_8".

(OFF – 0) and rising edge signals at inputs CU and CD do not occur at the same time, then either the count up operation or the count down operation can take place. In this case, if the current count value CV < FFh, then with each "rising edge" signal detected at the input CU, the related counter "CV_L+num" is incremented by one. Similarly, in this case, if the current count value CV > 0, then with each "rising edge" signal detected at the input CD, the related counter "CV_L+num" is decremented by one. If L_lim ≤ (CV_L+num) ≤ H_lim, then state-change from OFF to ON is issued for the QL (Output-within-Limits) counter output C_Q0+num/8,num-(8*(num/8)). If the count value of "CV_L+num" is equal to 0, then state-change from OFF to ON is issued for the QD (Output Down) counter output C_QD0+num/8,num-(8*(num/8)).

```
L10                              ;
;----------------------------------;------------------------
;IF R THEN CV:= 0;                ;
;----------------------------------;------------------------
        banksel  Rreg            ;
        btfss    Rreg,Rbit       ;
        goto     L9              ;
        banksel  CV_Ln           ;
        clrf     CV_Ln           ;
        goto     L5              ;
;----------------------------------;------------------------
;ELSIF LD THEN CV:= PV;           ;
;----------------------------------;------------------------
L9                               ;
        banksel  LDreg           ;
        btfss    LDreg,LDbit     ;
        goto     L8              ;
        movlw    LV              ;
        banksel  CV_Ln           ;
        movwf    CV_Ln           ;
        goto     L5              ;
L8                               ;
;----------------------------------;------------------------
;IF NOT (re_CU AND re_CD) THEN
;----------------------------------;------------------------
        btfss    Temp_1,7        ; If (re_CU AND re_CD),
        goto     L7              ; then goto  L5,
        btfss    Temp_1,6        ; else goto  L7.
        goto     L7              ;
        goto     L5              ;
L7                               ;
;----------------------------------;------------------------
;ELSIF re_CU AND (CV < PVmax) THEN CV:= CV+1;
;----------------------------------;------------------------
        banksel  CV_Ln           ;
        btfss    Temp_1,7        ;
        goto     L6              ;
        movlw    0xFF            ;
        subwf    CV_Ln,W         ;
        btfsc    STATUS,C        ;
        goto     L6              ;
        incf     CV_Ln,F         ;
;----------------------------------;------------------------
;ELSIF re_CD AND (CV > 0) THEN CV:= CV-1;
;----------------------------------;------------------------
L6      btfss    Temp_1,6        ;
        goto     L5              ;
        movf     CV_Ln,F         ;
        btfsc    STATUS,Z        ;
        goto     L5              ;
        decf     CV_Ln,F         ;
```

FIGURE 6.30 Continued

The count-up process stops when the CV = FFh. Likewise, the count-down process stops when the CV = 0. Assumption 1: The operands "MregU,MbitU", "MregD,MbitD", "CUreg,CUbit", "CDreg,CDbit", "Rreg,Rbit", and "LDreg,LDbit" can be in any Bank. Assumption 2: L_lim ≤ H_lim.

In order to explain how a generalized up/down counter is set-up by using the macro "GCTUD_8", let us consider Figure 6.32, where a generalized up/down counter is obtained with the macro "GCTUD_8".

```
L5                              ;
;------------------------------------;----------------------
;QL:= (L_lim <= CV <= H_lim);         ;
;------------------------------------;----------------------
        banksel CV_Ln           ;
        movlw   L_lim           ;
        subwf   CV_Ln,W         ;
        btfss   STATUS,C        ;
        goto    L4              ;
        movfw   CV_Ln           ;
        sublw   H_lim           ;
        btfss   STATUS,C        ;
        goto    L4              ;
        banksel CQR_n           ;
        bsf     CQR_n,b_n       ;
        goto    L3              ;
L4                              ;
        banksel CQR_n           ;
        bcf     CQR_n,b_n       ;
L3                              ;
;------------------------------------;----------------------
;QD:= (CV = 0);                       ;
;------------------------------------;----------------------
        banksel CV_Ln           ;
        movf    CV_Ln,F         ;
        btfss   STATUS,Z        ;
        goto    L2              ;
        banksel QDR_n           ;
        bsf     QDR_n,b_n       ;
        goto    L1              ;
L2                              ;
        banksel QDR_n           ;
        bcf     QDR_n,b_n       ;
L1                              ;
    endif                       ;
;----------------------------------------------------------
    if    (num >= 80)
    error "The counter number must be one of 0, 1, ...., 79."
    endif
;----------------------------------------------------------
    if    !((LV > 0)&&(LV < 256))
    error "Make sure that 0 < LV < 256 !"
    endif
;----------------------------------------------------------
    if    (L_lim > H_lim)
    error "Low limit (L_lim) value cannot be greater than High limit (H_lim) value!"
    endif
;----------------------------------------------------------
        endm
```

FIGURE 6.30 Continued

In this example program we have

num = 29 (decimal)
MregU,MbitU = M0.0
MregD,MbitD = M0.1
CU = I0.0
CD = I0.1

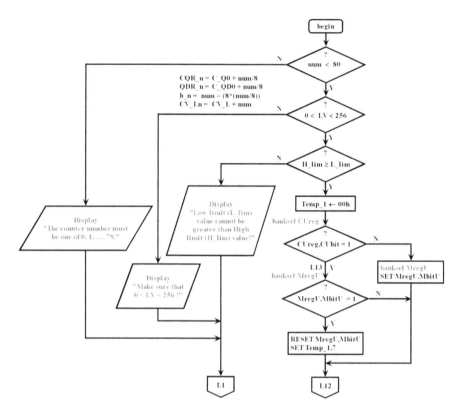

FIGURE 6.31   The flowchart of the macro "GCTUD_8".

R = I0.2
LD = I0.3
LV = 5 (decimal)
L_lim = 10 (decimal)
H_lim = 20 (decimal)

Therefore, we obtain the following from the macro "GCTUD_8":

CV_Ln = CV_L register number
CV_Ln = CV_L+num = CV_L+29 = 320h + 29d = 320h + 1Dh = 33Dh   (This is the 8-bit register in BANK6 of SRAM memory where the current count value of this generalized up/down counter will be held.)

Note that in the following, the division $\dfrac{num}{8}$ is an integer division.

CQR_n = CQ **R**egister **n**umber

$$CQR\_n = C\_Q0 + num/8 = C\_Q0 + \dfrac{num}{8} = 4DAh + \dfrac{29}{8} = 4DAh + 3d = 4DDh$$

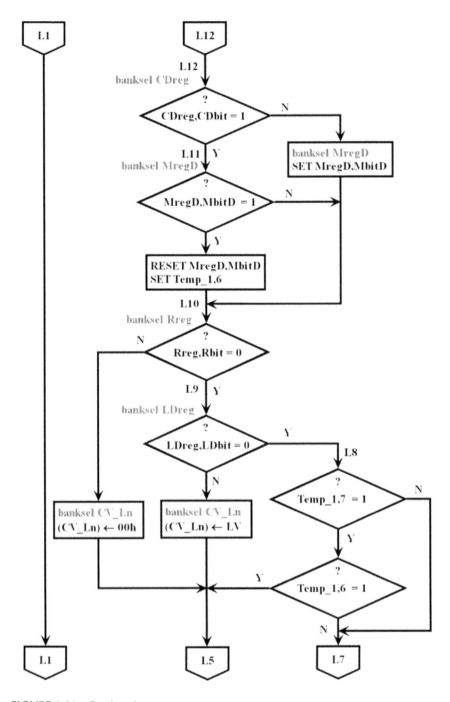

FIGURE 6.31 Continued

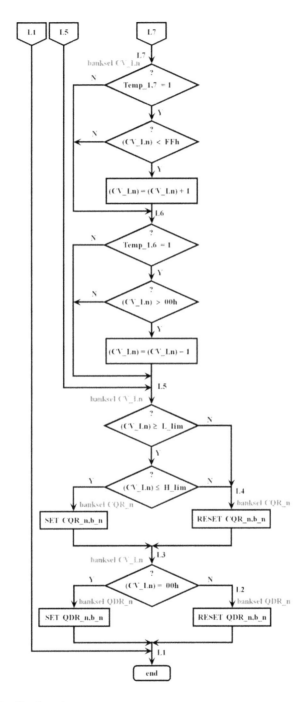

FIGURE 6.31   Continued

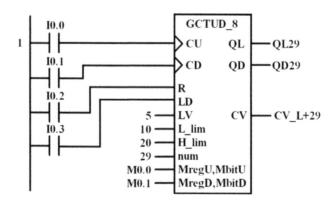

```
                        ;num,MregU,MbitU,MregD,MbitD,CU,CD,R,LD,LV,L_lim,H_lim       ; rung 1
GCTUD_8                 29,M0.0,M0.1,I0.0,I0.1,I0.2,I0.3,.5,.10,.20
```

FIGURE 6.32 Example generalized up/down counter obtained with the macro "GCTUD_8": a user program and its ladder diagram.

$$QDR\_n = QD\ \mathbf{R}egister\ \mathbf{n}umber$$

$$QDR\_n = C\_QD0 + num/8 = C\_QD0 + \frac{num}{8}$$

$$= 4E4h + \frac{29}{8} = 4E4h + 3d = 4E7h$$

$$b\_n = C\_Q0\ or\ C\_QD0\ \mathbf{b}it\ \mathbf{n}umber$$

$$b\_n = num - \left(8*\left(num/8\right)\right) = num - \left(8*\left(\frac{num}{8}\right)\right)$$

$$= 29 - \left(8*\left(\frac{29}{8}\right)\right) = 29 - \left(8*3\right) = 29 - 24 = 5$$

generalized up/down counter QL (Output-within-Limits) counter output (within-limits status bit) = CQR_n,b_n = 4DDh,5 = QL29 (This is the 1-bit variable in BANK9 of SRAM memory where the state of the QL (Output-within-Limits) counter output of this counter will be held.)

generalized up/down counter QD (Output Down) counter output (down status bit) = QDR_n,b_n = 4E7h,5 = QD29 (This is the 1-bit variable in BANK9 of SRAM memory where the state of the QD (Output Down) counter output of this counter will be held.)

6.12 MACRO "GCTUD_16" (GENERALIZED 16 BIT UP/DOWN COUNTER)

The macro "GCTUD_16" defines eighty 16-bit generalized up/down counters selected with the num = 0, 1, ..., 79. Table 6.16 shows the symbol of the macro

TABLE 6.16
The symbol of the macro "GCTUD_16"

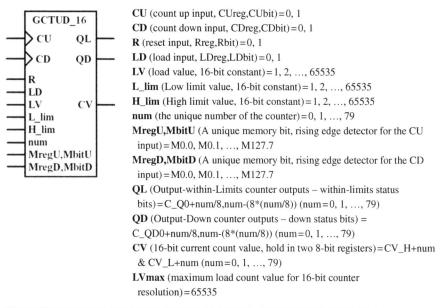

CU (count up input, CUreg,CUbit)=0, 1

CD (count down input, CDreg,CDbit)=0, 1

R (reset input, Rreg,Rbit)=0, 1

LD (load input, LDreg,LDbit)=0, 1

LV (load value, 16-bit constant)=1, 2, ..., 65535

L_lim (Low limit value, 16-bit constant)=1, 2, ..., 65535

H_lim (High limit value, 16-bit constant)=1, 2, ..., 65535

num (the unique number of the counter)=0, 1, ..., 79

MregU,MbitU (A unique memory bit, rising edge detector for the CU
input)=M0.0, M0.1, ..., M127.7

MregD,MbitD (A unique memory bit, rising edge detector for the CD
input)=M0.0, M0.1, ..., M127.7

QL (Output-within-Limits counter outputs – within-limits status
bits)=C_Q0+num/8,num-(8*(num/8)) (num=0, 1, ..., 79)

QD (Output-Down counter outputs – down status bits) =
C_QD0+num/8,num-(8*(num/8)) (num=0, 1, ..., 79)

CV (16-bit current count value, hold in two 8-bit registers)=CV_H+num
& CV_L+num (num=0, 1, ..., 79)

LVmax (maximum load count value for 16-bit counter
resolution)=65535

"GCTUD_16". The macro "GCTUD_16" and its flowchart are depicted in Figures 6.33 and 6.34, respectively. CU (count up input), CD (count down input), QL (Output-within-Limits counter output), QD (Output Down counter output), R (reset input), and LD (load input) are all defined as Boolean variables. The LV (load count value) is an integer constant and for 16-bit resolution it is chosen any number in the range 1–65535. The maximum load count value for the 16-bit up counter (LVmax) is 65535. The QL (Output-within-Limits) counter outputs C_Q0+num/8,num-(8*(num/8)) are defined by within-limits status bits QL0, QL1, ..., QL79 as shown in Figure 6.6. The QD (Output Down) counter outputs C_QD0+num/8,num-(8*(num/8)) are defined by down status bits QD0, QD1, ..., QD79 as shown in Figure 6.7. A unique memory bit (MregU,MbitU)=M0.0, M0.1, ..., M127.7 is used to detect the rising edge signals at the input CU. Likewise, another unique memory bit (MregD,MbitD)=M0.0, M0.1, ..., M127.7 is used to detect the rising edge signals at the input CD. A 16-bit integer variable "CV_H+num&CV_L+num" is used to count up with the rising edge signals detected at the input CU and count down with the rising edge signals detected at the input CD. "CV_H+num" holds the high byte of the CV, while "CV_L+num" holds the low byte of the CV.

Let us now briefly consider how the macro "GCTUD_16" works. If the input signal R is true (ON – 1), then the counter "CV_H+num&CV_L+num" is loaded with "0000h". If the input signal R is false (OFF – 0) and the input signal LD is true (ON – 1), then the counter "CV_H+num&CV_L+num" is loaded with the LV. If both input signals R and LD are false (OFF – 0) and rising edge signals for

```
GCTUD_16 macro num, MregU,MbitU, MregD,MbitD, CUreg,CUbit, CDreg,CDbit, Rreg,Rbit,
LDreg,LDbit, LV, L_lim, H_lim
         local   L0,L1,L2,L3,L4,L5,L6,L7,L8,L9,L10,L11,L12,L13,L14,L15,L16,L17
         local   CQR_n,QDR_n,b_n,CV_Ln,CV_Hn
;------------------------------------------------------------------------
 if (num<80)&&(L_lim <= H_lim)&&(LV > 0)&&(LV < 65536)
 ;if (num<80)AND(L_lim <= H_lim)AND(0<LV<65536) then carry on, else do not compile.
;------------------------------------------------------------------------
CQR_n set      C_Q0+num/8            ;C_Q0 Register number
QDR_n set      C_QD0+num/8           ;QD Register number
b_n   set      num-(8*(num/8))       ;C_Q0 bit number
CV_Ln set      CV_L+num              ;CV_L register number
CV_Hn set      CV_H+num              ;CV_H register number
;------------------------------------------------------------------------
         clrf    Temp_1       ;
;--------------------------------------;---------------------
; Temp_1,7 := rising edge signal of CUreg,CUbit (re_CU)
;--------------------------------------;---------------------
         banksel CUreg         ;
         btfsc   CUreg,CUbit   ;
         goto    L17           ;
         banksel MregU         ;
         bsf     MregU,MbitU   ;
         goto    L16           ;
L17                            ;
         banksel MregU         ;
         btfss   MregU,MbitU   ;
         goto    L16           ;
         bcf     MregU,MbitU   ;
         bsf     Temp_1,7      ; set re_CU
L16                            ;
;--------------------------------------;---------------------
; Temp_1,6 := rising edge signal of CDreg,CDbit (re_CD)
;--------------------------------------;---------------------
         banksel CDreg         ;
         btfsc   CDreg,CDbit   ;
         goto    L15           ;
         banksel MregD         ;
         bsf     MregD,MbitD   ;
         goto    L14           ;
L15                            ;
         banksel MregD         ;
         btfss   MregD,MbitD   ;
         goto    L14           ;
         bcf     MregD,MbitD   ;
         bsf     Temp_1,6      ; set re_CD
L14                            ;
;--------------------------------------;---------------------
;IF R THEN CV:= 0;             ;
;--------------------------------------;---------------------
```

FIGURE 6.33   The macro "GCTUD_16".

inputs CU and CD do not occur at the same time, then either the count up operation or the count down operation can take place. In this case, if the current count
value CV < FFFFh, then with each "rising edge" signal detected at the input CU,
the related counter "CV_H+num&CV_L+num" is incremented by one. Similarly,
in this case, if the current count value CV > 0, then with each "rising edge" signal

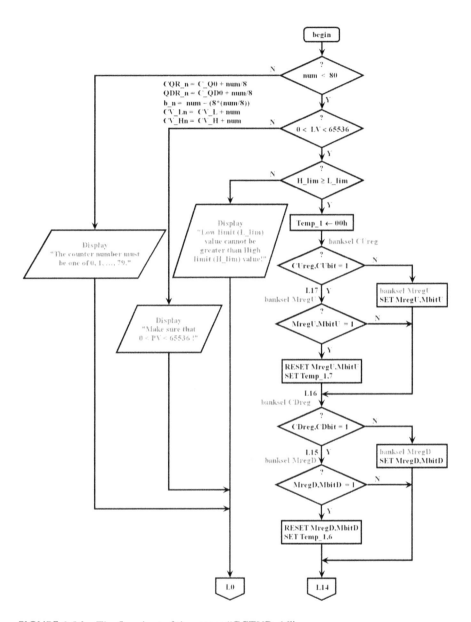

FIGURE 6.34 The flowchart of the macro "GCTUD_16".

In order to explain how a generalized up/down counter is set-up by using the macro "GCTUD_16", let us consider Figure 6.35, where a generalized up/down counter is obtained with the macro "GCTUD_16".

In this example program we have

num = 43 (decimal)
MregU,MbitU = M0.0

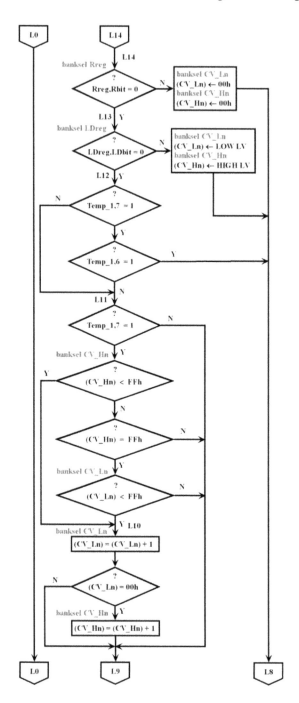

FIGURE 6.34   Continued

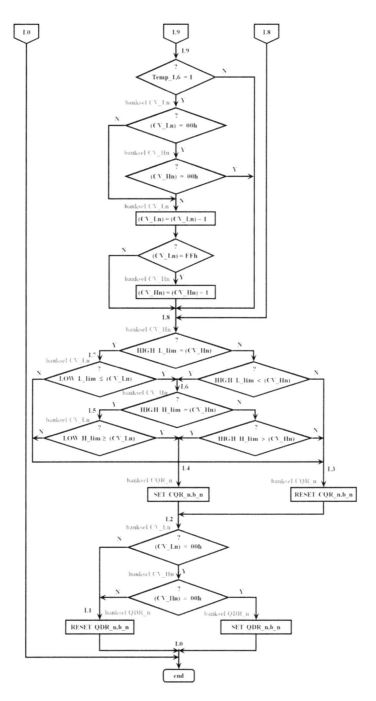

FIGURE 6.34 Continued

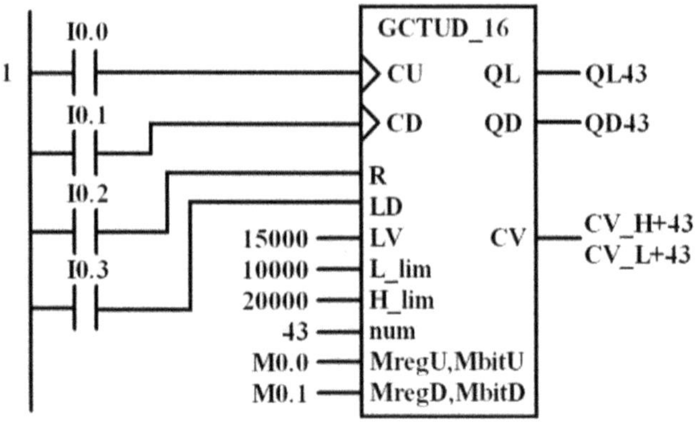

```
                ;num,MregU,MbitU,MregD,MbitD,CU,CD,R,LD,LV,L_lim,H_lim      ; rung 1
GCTUD_16        43,M0.0,M0.1,I0.0,I0.1,I0.2,I0.3,.15000,.10000,.20000
```

FIGURE 6.35 Example generalized up/down counter obtained with the macro "GCTUD_16": a user program and its ladder diagram.

MregD,MbitD = M0.1
CU = I0.0
CD = I0.1
R = I0.2
LD = I0.3
LV = 15000 (decimal)
L_lim = 10000 (decimal)
H_lim = 20000 (decimal)

Therefore, we obtain the following from the macro "GCTUD_16":

CV_Ln = CV_L register number
CV_Ln = CV_L+num = CV_L+43 = 320h + 43d = 320h + 2Bh = 34Bh   (This is the 8-bit register in BANK6 of SRAM memory where the low byte of current count value of this generalized up/down counter will be held.)

CV_Hn = CV_H register number
CV_Hn = CV_H+num = CV_H+43 = 3A0h + 43d = 3A0h + 2Bh = 3CBh   (This is the 8-bit register in BANK7 of SRAM memory where the high byte of current count value of this generalized up/down counter will be held.)

Note that in the following, the division $\dfrac{num}{8}$ is an integer divison.

CQR_n = CQ **R**egister **n**umber

$$CQR\_n = C\_Q0 + num/8 = C\_Q0 + \frac{num}{8} = 4DAh + \frac{43}{8} = 4DAh + 5d = 4DFh$$

QDR_n = QD **R**egister **n**umber

$$QDR\_n = C\_QD0 + num/8 = C\_QD0 + \frac{num}{8}$$

$$= 4E4h + \frac{43}{8} = 4E4h + 5d = 4E9h$$

b_n = C_Q0 or C_QD0 **b**it **n**umber

$$b\_n = num - \left(8*(num/8)\right) = num - \left(8*\left(\frac{num}{8}\right)\right)$$

$$= 43 - \left(8*\left(\frac{43}{8}\right)\right) = 43 - (8*5) = 43 - 40 = 3$$

generalized up/down counter QL (Output-within-Limits) counter output (within-limits status bit)=CQR_n,b_n=4DFh,3=QL43 (This is the 1-bit variable in BANK9 of SRAM memory where the state of the QL (Output-within-Limits) counter output of this counter will be held.)

generalized up/down counter QD (Output Down) counter output (down status bit)=QDR_n,b_n=4E9h,3=QD43 (This is the 1-bit variable in BANK9 of SRAM memory where the state of the QD (Output Down) counter output of this counter will be held.)

6.13   EXAMPLES FOR COUNTER MACROS

Up to now in this chapter, we have seen counter macros developed for the PIC16F1847-Based PLC. It is now time to consider some examples related to these macros. Before you can run the example programs considered here, you are expected to construct your own PIC16F1847-Based PLC hardware by using the necessary PCB files, and by producing your PCBs, with their components. For an effective use of examples, all example programs considered in this book are allocated within the file "PICPLC_PIC16F1847_user_Bsc.inc", which is downloadable from this book's webpage under the downloads section. Initially all example programs are commented out by putting a semicolon ';' in front of each line. When you would like to test one of the example programs you must uncomment each line of the example program by following the steps shown below:

1. Highlight the block of source lines you want to uncomment by dragging the mouse with the left mouse button held down over these lines. With default coloring in MPLAB X IDE you will see now green characters on a blue background.
2. Release the mouse button.
3. Press Ctrl/Shift/C or Press "Alt", "S" and "M" keys in succession or from the toolbar "Source" menu, select "Toggle Comment". Now a semicolon

will be removed from all selected source lines. With default coloring you will see red characters on a white background.

Then, you can run the project by pressing the symbol from the toolbar. Next, the MPLAB X IDE will produce the "PICPLC_PIC16F1847.X.production.hex" file for the project. Then the MPLAB X IDE will be connected to the PICkit3 programmer and finally it will program the PIC16F1847 microcontroller within the CPU board of the PIC16F1847-Based PLC. During these steps make sure that in the CPU board of the PIC16F1847-Based PLC, the 4PDT switch is in "PROG" position and the power switch is in "OFF" position. After loading the program file to the PIC16F1847 microcontroller, switch the 4PDT in "RUN" and the power switch in "ON" position. Finally, you are ready to test the example program. Warning: When you finish your study with an example and try to take a look at another example, do not forget to comment the current example program before uncommenting another one. In other words, make sure that only one example program is uncommented and tested at the same time. Otherwise if you somehow leave more than one example uncommented, the example you are trying to test probably will not function as expected since it may try to access the same resources that are being used and changed by other examples.

Please check the accuracy of each program by cross-referencing it with the related macros.

6.13.1 EXAMPLE 6.1

Example 6.1 shows the usage of following up counter macros: CTU_8 and CTU_16. The user program of Example 6.1 is shown in Figure 6.36 . The schematic diagram

```
user_program_1   macro;
;--- PLC codes to be allocated in the "user_program_1" macro start from here ---
;__Example_Bsc_6.1

                        ;num,Mreg,Mbit,CU,R,PV      ;rung 1
        CTU_8           0,M0.0,I0.0,I0.1,.5
        in_out          CQ0,Q0.0                    ;rung 2

                        ;num,Mreg,Mbit,CU,R,PV      ;rung 3
        CTU_8           10,M0.1,I0.2,I0.3,.10
        in_out          CQ10,Q0.1                   ;rung 4

                        ;num,Mreg,Mbit,CU,R,PV      ;rung 5
        CTU_8           20,M0.2,I0.4,I0.5,.15
        in_out          CQ20,Q0.2                   ;rung 6

                        ;num,Mreg,Mbit,CU,R,PV      ;rung 7
        CTU_8           30,M0.3,I0.6,I0.7,.20
        in_out          CQ30,Q0.3                   ;rung 8

        ld              I1.2                         ;rung 9
                        ;s1,s0,R3,R2,R1,R0,OUT
        B_mux_4_1_E I1.1,I1.0,CV_L+30,CV_L+20,CV_L+10,CV_L,Q1
```

FIGURE 6.36   The user program of "Example 6.1".

```
                    ;num,Mreg,Mbit,CU,R,PV
        CTU_16      49,M0.4,I2.0,I2.1,.30               ;rung 10
        in_out      CQ49,Q0.4                          ;rung 11

        ld          I2.2                               ;rung 12
        and         T_2ms
        out         M1.1

                    ;num,Mreg,Mbit,CU,R,PV
        CTU_16      59,M0.5,M1.1,I2.3,.2000            ;rung 13
        in_out      CQ59,Q0.5                          ;rung 14

        ld          I2.4                               ;rung 15
        and         T_2ms
        out         M1.2

                    ;num,Mreg,Mbit,CU,R,PV
        CTU_16      69,M0.6,M1.2,I2.5,.30000           ;rung 16
        in_out      CQ69,Q0.6                          ;rung 17

        ld          I2.6                               ;rung 18
        and         T_2ms
        out         M1.3

                    ;num,Mreg,Mbit,CU,R,PV
        CTU_16      79,M0.7,M1.3,I2.7,.65535           ;rung 19
        in_out      CQ79,Q0.7                          ;rung 20

        ld          I3.2                               ;rung 21
                    ;s1,s0,R3,R2,R1,R0,OUT
        B_mux_4_1_E I3.1,I3.0,CV_L+79,CV_L+69,CV_L+59,CV_L+49,Q2

        ld          I3.2                               ;rung 22
                    ;s1,s0,R3,R2,R1,R0,OUT
        B_mux_4_1_E I3.1,I3.0,CV_H+79,CV_H+69,CV_H+59,CV_H+49,Q3
;
;--- PLC codes to be allocated in the "user_program_1" macro end here ----------
        endm
```

FIGURE 6.36 Continued

and the ladder diagram of Example 6.1 are depicted in Figure 6.37(a) and in Figure 6.37(b) respectively. When the project file of the PIC16F1847-Based PLC is open in the MPLAB X IDE, from the file "PICPLC_PIC16F1847_user_Bsc.inc" if you uncomment Example 6.1 and run the project by pressing the symbol ▷ from the tool-bar, then the PIC16F1847 microcontroller within the CPU board of the PIC16F1847-Based PLC will be programmed. After loading the program file to the PIC16F1847 microcontroller, switch the 4PDT in "RUN" and the power switch in "ON" position. Next you can test the operation of this example.

In rungs 1 and 2, an up counter "CTU_8" is implemented as follows: the count up input CU is taken from I0.0, while the reset input R is taken from I0.1. The unique memory bit used to detect rising edge signals at the CU input is Mreg,Mbit=M0.0. num=0 and therefore the counter status bit (or

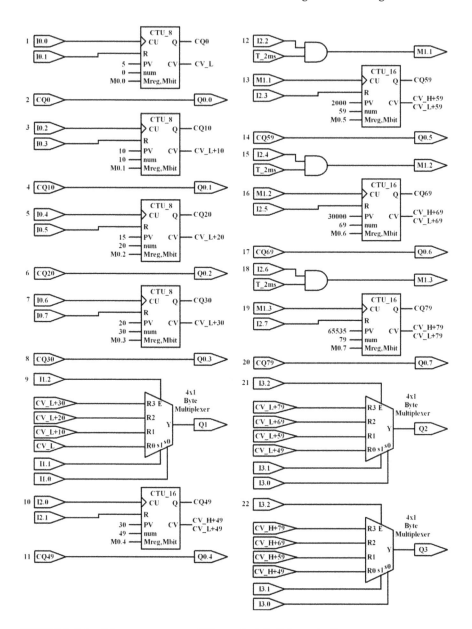

FIGURE 6.37A The user program of "Example 6.1": Schematic diagram.

output Q) of this up counter is CQ0. The 8-bit integer variable used to count the rising edge signals at the CU input is CV_L+num=CV_L. The preset count value PV=5. As can be seen from rung 2, the state of the counter status bit CQ0 is sent to output Q0.0.

In rungs 3 and 4, an up counter "CTU_8" is implemented as follows: the count up input CU is taken from I0.2, while the reset input R is taken from

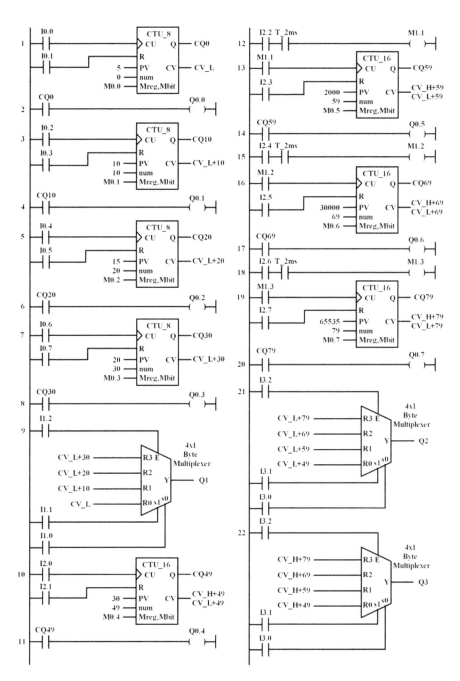

FIGURE 6.37B The user program of "Example 6.1": Ladder diagram.

I0.3. The unique memory bit used to detect rising edge signals at the CU input is Mreg,Mbit=M0.1. num=10 and therefore the counter status bit (or output Q) of this up counter is CQ10. The 8-bit integer variable used to count the rising edge signals at the CU input is CV_L+num=CV_L+10. The preset count value PV=10. As can be seen from **rung 4**, the state of the counter status bit CQ10 is sent to output Q0.1.

In rungs 5 and 6, an up counter "CTU_8" is implemented as follows: the count up input CU is taken from I0.4, while the reset input R is taken from I0.5. The unique memory bit used to detect rising edge signals at the CU input is Mreg,Mbit=M0.2. num=20 and therefore the counter status bit (or output Q) of this up counter is CQ20. The 8-bit integer variable used to count the rising edge signals at the CU input is CV_L+num=CV_L+20. The preset count value PV=15. As can be seen from **rung 6**, the state of the counter status bit CQ20 is sent to output Q0.2.

In rungs 7 and 8, an up counter "CTU_8" is implemented as follows: the count up input CU is taken from I0.6, while the reset input R is taken from I0.7. The unique memory bit used to detect rising edge signals at the CU input is Mreg,Mbit=M0.3. num=30 and therefore the counter status bit (or output Q) of this up counter is CQ30. The 8-bit integer variable used to count the rising edge signals at the CU input is CV_L+num=CV_L+30. The preset count value PV=20. As can be seen from **rung 8**, the state of the counter status bit CQ30 is sent to output Q0.3.

In rung 9, a 4×1 Byte Multiplexer (Chapter 4 of the "Intermediate Concepts" book) is used to observe the current count values of one of the four 8-bit variables considered in the previous rungs from Q1. The following table shows the operation of this 4x1 Byte multiplexer.

inputs			output
E	s1	s0	Y
I1.2	I1.1	I1.0	Q1
0	×	×	U
1	0	0	CV_L
1	0	1	CV_L+10
1	1	0	CV_L+20
1	1	1	CV_L+30

×: don't care.
U: The contents of the destination register Y(Q1) remain unchanged.

In rungs 10 and 11, an up counter "CTU_16" is implemented as follows: the count up input CU is taken from I2.0, while the reset input R is taken from I2.1. The unique memory bit used to detect rising edge signals at the CU input is Mreg,Mbit=M0.4. num=49 and therefore the counter status bit (or output Q) of this up counter is CQ49. The 16-bit integer variable used to count the rising edge signals at the CU input is

CV_H+num&CV_L+num=CV_H+49&CV_L+49. The preset count value PV=30. As can be seen from **rung 11**, the state of the counter status bit CQ49 is sent to output Q0.4.

In rung 12, Memory bit M1.1 is obtained by ANDing the digital input I2.2 with the reference timing signal T_2ms. This means that when I2.2 is ON, M1.1=T_2ms. This organization is carried out to speed up the count up process for the next up counter.

In rungs 13 and 14, an up counter "CTU_16" is implemented as follows: the count up input CU is taken from M1.1, while the reset input R is taken from I2.3. The unique memory bit used to detect rising edge signals at the CU input is Mreg,Mbit=M0.5. num=59 and therefore the counter status bit (or output Q) of this up counter is CQ59. The 16-bit integer variable used to count the rising edge signals at the CU input is CV_H+num&CV_L+num=CV_H+59&CV_L+59. The preset count value PV=2000. As can be seen from **rung 14**, the state of the counter status bit CQ59 is sent to output Q0.5.

In rung 15, Memory bit M1.2 is obtained by ANDing the digital input I2.4 with the reference timing signal T_2ms. This means that when I2.4 is ON, M1.2=T_2ms. This organization is carried out to speed up the count up process for the next up counter.

In rungs 16 and 17, an up counter "CTU_16" is implemented as follows: the count up input CU is taken from M1.2, while the reset input R is taken from I2.5. The unique memory bit used to detect rising edge signals at the CU input is Mreg,Mbit=M0.6. num=69 and therefore the counter status bit (or output Q) of this up counter is CQ69. The 16-bit integer variable used to count the rising edge signals at the CU input is CV_H+num&CV_L+num=CV_H+69&CV_L+69. The preset count value PV=30000. As can be seen from **rung 17**, the state of the counter status bit CQ69 is sent to output Q0.6.

In rung 18, Memory bit M1.3 is obtained by ANDing the digital input I2.6 with the reference timing signal T_2ms. This means that when I2.6 is ON, M1.3=T_2ms. This organization is carried out to speed up the count up process for the next up counter.

In rungs 19 and 20, an up counter "CTU_16" is implemented as follows: the count up input CU is taken from M1.3, while the reset input R is taken from I2.7. The unique memory bit used to detect rising edge signals at the CU input is Mreg,Mbit=M0.7. num=79 and therefore the counter status bit (or output Q) of this up counter is CQ79. The 16-bit integer variable used to count the rising edge signals at the CU input is CV_H+num&CV_L+num=CV_H+79&CV_L+79. The preset count value PV=65535. As can be seen from **rung 20**, the state of the counter status bit CQ79 is sent to output Q0.7.

In rung 21, a 4x1 Byte Multiplexer (Chapter 4 of the "Intermediate Concepts" book) is used to observe the low byte of current count values of one of the four 16-bit variables considered between rungs 10 and 20 from Q2. The following table shows the operation of this 4x1 Byte multiplexer.

inputs			output
E	s1	s0	Y
I3.2	I3.1	I3.0	Q2
0	×	×	U
1	0	0	CV_L+49
1	0	1	CV_L+59
1	1	0	CV_L+69
1	1	1	CV_L+79

×: don't care. U: The contents of the destination register Y(Q2) remain unchanged.

In rung 22, a 4x1 Byte Multiplexer (Chapter 4 of the "Intermediate Concepts" book) is used to observe the high byte of current count values of one of the four 16-bit variables considered between rungs 10 and 20 from Q3. The following table shows the operation of this 4x1 Byte multiplexer.

inputs			output
E	s1	s0	Y
I3.2	I3.1	I3.0	Q3
0	×	×	U
1	0	0	CV_H+49
1	0	1	CV_H+59
1	1	0	CV_H+69
1	1	1	CV_H+79

×: don't care. U: The contents of the destination register Y(Q3) remain unchanged.

6.13.2 EXAMPLE 6.2

Example 6.2 shows the usage of following down counter macros: CTD_8 and CTD_16. The user program of Example 6.2 is shown in Figure 6.38. The schematic diagram and the ladder diagram of Example 6.2 are depicted in Figure 6.39(a) and in Figure 6.39(b) respectively. When the project file of the PIC16F1847-Based PLC is open in the MPLAB X IDE, from the file "PICPLC_PIC16F1847_user_Bsc. inc" if you uncomment Example 6.2 and run the project by pressing the symbol ▷ from the toolbar, then the PIC16F1847 microcontroller within the CPU board of the PIC16F1847-Based PLC will be programmed. After loading the program file to the PIC16F1847 microcontroller, switch the 4PDT in "RUN" and the power switch in "ON" position. Next you can test the operation of this example.

> **In rungs 1 and 2**, a down counter "CTD_8" is implemented as follows: the count down input CD is taken from I0.0, while the load input LD is taken from I0.1. The unique memory bit used to detect rising edge signals at the CD input is Mreg,Mbit=M0.0. num=0 and therefore the counter status bit (or output Q) of this down counter is CQ0. The 8-bit integer variable

```
user_program_1  macro;
;--- PLC codes to be allocated in the "user_program_1" macro start from here ---
;__Example_Bsc_6.2

                     ;num,Mreg,Mbit,CD,LD,PV        ;rung 1
        CTD_8        0,M0.0,I0.0,I0.1,.5
        in_out       CQ0,Q0.0                       ;rung 2

                     ;num,Mreg,Mbit,CD,LD,PV        ;rung 3
        CTD_8        10,M0.1,I0.2,I0.3,.10
        in_out       CQ10,Q0.1                      ;rung 4

                     ;num,Mreg,Mbit,CD,LD,PV        ;rung 5
        CTD_8        20,M0.2,I0.4,I0.5,.15
        in_out       CQ20,Q0.2                      ;rung 6

                     ;num,Mreg,Mbit,CD,LD,PV        ;rung 7
        CTD_8        30,M0.3,I0.6,I0.7,.20
        in_out       CQ30,Q0.3                      ;rung 8

        ld           I1.2                           ;rung 9
                     ;s1,s0,R3,R2,R1,R0,OUT
        B_mux_4_1_E  I1.1,I1.0,CV_L+30,CV_L+20,CV_L+10,CV_L,Q1
```

FIGURE 6.38 The user program of "Example 6.2".

used to count down with the rising edge signals at the CD input is CV_L+num=CV_L. The preset count value PV=5. As can be seen from **rung 2**, the state of the counter status bit CQ0 is sent to output Q0.0.

In rungs 3 and 4, a down counter "CTD_8" is implemented as follows: the count down input CD is taken from I0.2, while the load input LD is taken from I0.3. The unique memory bit used to detect rising edge signals at the CD input is Mreg,Mbit=M0.1. num=10 and therefore the counter status bit (or output Q) of this down counter is CQ10. The 8-bit integer variable used to count down with the rising edge signals at the CD input is CV_L+num=CV_L+10. The preset count value PV=10. As can be seen from **rung 4**, the state of the counter status bit CQ10 is sent to output Q0.1.

In rungs 5 and 6, a down counter "CTD_8" is implemented as follows: the count down input CD is taken from I0.4, while the load input LD is taken from I0.5. The unique memory bit used to detect rising edge signals at the CD input is Mreg,Mbit=M0.2. num=20 and therefore the counter status bit (or output Q) of this down counter is CQ20. The 8-bit integer variable used to count down with the rising edge signals at the CD input is CV_L+num=CV_L+20. The preset count value PV=15. As can be seen from **rung 6**, the state of the counter status bit CQ20 is sent to output Q0.2.

In rungs 7 and 8, a down counter "CTD_8" is implemented as follows: the count down input CD is taken from I0.6, while the load input LD is taken from I0.7. The unique memory bit used to detect rising edge signals at the CD input is Mreg,Mbit=M0.3. num=30 and therefore the counter status bit (or output Q) of this down counter is CQ30. The 8-bit integer variable used to count down with the rising edge signals at the CD input is CV_L+num=CV_L+30. The preset count value PV=20. As can be seen from **rung 8**, the state of the counter status bit CQ30 is sent to output Q0.3.

```
                          ;num,Mreg,Mbit,CD,LD,PV
        CTD_16            49,M0.4,I2.0,I2.1,.30              ;rung 10
        in_out            CQ49,Q0.4                         ;rung 11

        ld                I2.2                              ;rung 12
        and               T_2ms
        out               M1.1

                          ;num,Mreg,Mbit,CD,LD,PV
        CTD_16            59,M0.5,M1.1,I2.3,.2000           ;rung 13
        in_out            CQ59,Q0.5                         ;rung 14

        ld                I2.4                              ;rung 15
        and               T_2ms
        out               M1.2

                          ;num,Mreg,Mbit,CD,LD,PV
        CTD_16            69,M0.6,M1.2,I2.5,.30000          ;rung 16
        in_out            CQ69,Q0.6                         ;rung 17

        ld                I2.6                              ;rung 18
        and               T_2ms
        out               M1.3

                          ;num,Mreg,Mbit,CD,LD,PV
        CTD_16            79,M0.7,M1.3,I2.7,.65535          ;rung 19
        in_out            CQ79,Q0.7                         ;rung 20

        ld                I3.2                              ;rung 21
                          ;s1,s0,R3,R2,R1,R0,OUT
        B_mux_4_1_E  I3.1,I3.0,CV_L+79,CV_L+69,CV_L+59,CV_L+49,Q2

        ld                I3.2                              ;rung 22
                          ;s1,s0,R3,R2,R1,R0,OUT
        B_mux_4_1_E  I3.1,I3.0,CV_H+79,CV_H+69,CV_H+59,CV_H+49,Q3
    ;
    ;--- PLC codes to be allocated in the "user_program_1" macro end here ----------
        endm
```

FIGURE 6.38   Continued

In rung 9, a 4x1 Byte Multiplexer (Chapter 4 of the "Intermediate Concepts" book) is used to observe the current count values of one of the four 8-bit variables considered in the previous rungs from Q1. The following table shows the operation of this 4x1 Byte multiplexer.

inputs			output
E	s1	s0	Y
I1.2	I1.1	I1.0	Q1
0	×	×	U
1	0	0	CV_L
1	0	1	CV_L+10
1	1	0	CV_L+20
1	1	1	CV_L+30

×: don't care.
U: The contents of the destination register Y(Q1) remain unchanged.

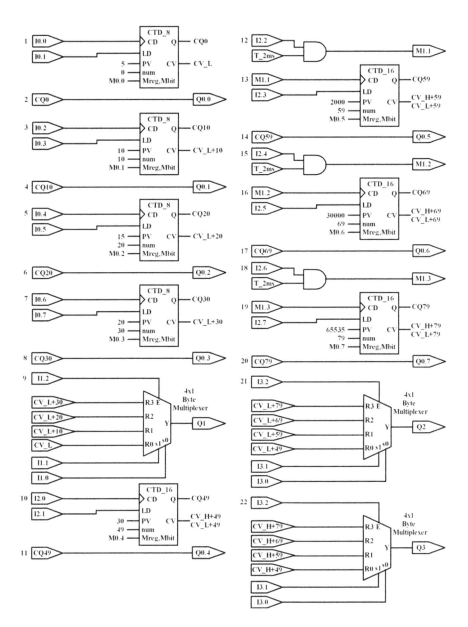

FIGURE 6.39A The user program of "Example 6.2": Schematic diagram.

In rungs 10 and 11, a down counter "CTD_16" is implemented as follows: the count down input CD is taken from I2.0, while the load input LD is taken from I2.1. The unique memory bit used to detect rising edge signals at the CD input is Mreg,Mbit=M0.4. num=49 and therefore the counter status bit (or output Q) of this down counter is CQ49. The 16-bit integer variable used to count down with the rising edge signals at the CD input is

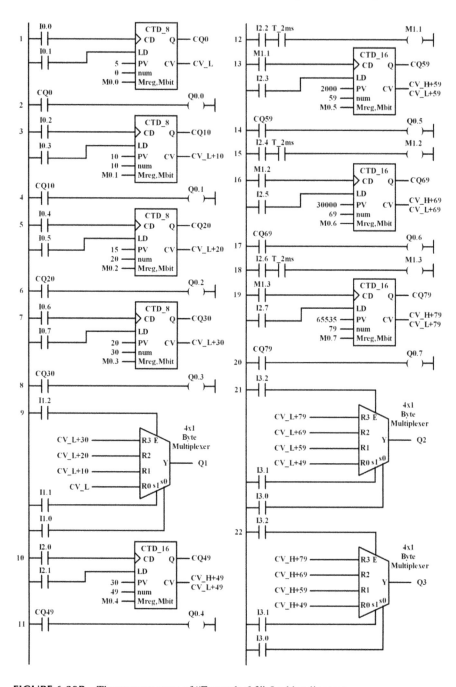

FIGURE 6.39B The user program of "Example 6.2": Ladder diagram.

CV_H+num&CV_L+num=CV_H+49&CV_L+49. The preset count value PV=30. As can be seen from rung 11, the state of the counter status bit CQ49 is sent to output Q0.4.

In rung 12, Memory bit M1.1 is obtained by ANDing the digital input I2.2 with the reference timing signal T_2ms. This means that when I2.2 is ON, M1.1=T_2ms. This organization is carried out to speed up the count down process for the next down counter.

In rungs 13 and 14, a down counter "CTD_16" is implemented as follows: the count down input CD is taken from M1.1, while the load input LD is taken from I2.3. The unique memory bit used to detect rising edge signals at the CD input is Mreg,Mbit=M0.5. num=59 and therefore the counter status bit (or output Q) of this down counter is CQ59. The 16-bit integer variable used to count down with the rising edge signals at the CD input is CV_H+num&CV_L+num=CV_H+59&CV_L+59. The preset count value PV=2000. As can be seen from **rung 14**, the state of the counter status bit CQ59 is sent to output Q0.5.

In rung 15, Memory bit M1.2 is obtained by ANDing the digital input I2.4 with the reference timing signal T_2ms. This means that when I2.4 is ON, M1.2=T_2ms. This organization is carried out to speed up the count down process for the next down counter.

In rungs 16 and 17, a down counter "CTD_16" is implemented as follows: the count down input CD is taken from M1.2, while the load input LD is taken from I2.5. The unique memory bit used to detect rising edge signals at the CD input is Mreg,Mbit=M0.6. num=69 and therefore the counter status bit (or output Q) of this down counter is CQ69. The 16-bit integer variable used to count down with the rising edge signals at the CD input is CV_H+num&CV_L+num=CV_H+69&CV_L+69. The preset count value PV=30000. As can be seen from **rung 17**, the state of the counter status bit CQ69 is sent to output Q0.6.

In rung 18, Memory bit M1.3 is obtained by ANDing the digital input I2.6 with the reference timing signal T_2ms. This means that when I2.6 is ON, M1.3=T_2ms. This organization is carried out to speed up the count down process for the next down counter.

In rungs 19 and 20, a down counter "CTD_16" is implemented as follows: the count down input CD is taken from M1.3, while the load input LD is taken from I2.7. The unique memory bit used to detect rising edge signals at the CD input is Mreg,Mbit=M0.7. num=79 and therefore the counter status bit (or output Q) of this down counter is CQ79. The 16-bit integer variable used to count down with the rising edge signals at the CD input is CV_H+num&CV_L+num=CV_H+79&CV_L+79. The preset count value PV=65535. As can be seen from **rung 20**, the state of the counter status bit CQ79 is sent to output Q0.7.

In rung 21, a 4x1 Byte Multiplexer (Chapter 4 of the "Intermediate Concepts" book) is used to observe the low byte of current count values of one of the four 16-bit variables considered between rungs 10 and 20 from Q2. The following table shows the operation of this 4x1 Byte multiplexer.

inputs			output
E	s1	s0	Y
I3.2	I3.1	I3.0	Q2
0	×	×	U
1	0	0	CV_L+49
1	0	1	CV_L+59
1	1	0	CV_L+69
1	1	1	CV_L+79

×: don't care. U: The contents of the destination register Y(Q2) remain unchanged.

In rung 22, a 4x1 Byte Multiplexer (Chapter 4 of the "Intermediate Concepts" book) is used to observe the high byte of current count values of one of the four 16-bit variables considered between rungs 10 and 20 from Q3. The following table shows the operation of this 4x1 Byte multiplexer.

inputs			output
E	s1	s0	Y
I3.2	I3.1	I3.0	Q3
0	×	×	U
1	0	0	CV_H+49
1	0	1	CV_H+59
1	1	0	CV_H+69
1	1	1	CV_H+79

×: don't care. U: The contents of the destination register Y(Q3) remain unchanged.

6.13.3   EXAMPLE 6.3

Example 6.3 shows the usage of the up/down counter macro CTUD_8. The user program of Example 6.3 is shown in Figure 6.40. The schematic diagram and the ladder diagram of Example 6.3 are depicted in Figure 6.41(a) and in Figure 6.41(b) respectively. When the project file of the PIC16F1847-Based PLC is open in the MPLAB X IDE, from the file "PICPLC_PIC16F1847_user_Bsc.inc" if you uncomment Example 6.3 and run the project by pressing the symbol ▷ from the toolbar, then the PIC16F1847 microcontroller within the CPU board of the PIC16F1847-Based PLC

```
user_program_1   macro
;
;--- PLC codes to be allocated in the "user_program_1" macro start from here ---
;__Example_Bsc_6.3

                    ;num,MregU,MbitU,MregD,MbitD,CU,CD,R,LD,PV
        CTUD_8      0,M0.0,M0.1,I0.0,I0.1,I0.2,I0.3,.5                         ;rung 1
        in_out      QU0,Q0.0                                                  ;rung 2
        in_out      QD0,Q0.1                                                  ;rung 3

                    ;num,MregU,MbitU,MregD,MbitD,CU,CD,R,LD,PV
        CTUD_8      10,M0.2,M0.3,I0.4,I0.5,I0.6,I0.7,.10                      ;rung 4
        in_out      QU10,Q0.2                                                 ;rung 5
        in_out      QD10,Q0.3                                                 ;rung 6

                    ;num,MregU,MbitU,MregD,MbitD,CU,CD,R,LD,PV
        CTUD_8      20,M0.4,M0.5,I1.0,I1.1,I1.2,I1.3,.15                      ;rung 7
        in_out      QU20,Q0.4                                                 ;rung 8
        in_out      QD20,Q0.5                                                 ;rung 9

                    ;num,MregU,MbitU,MregD,MbitD,CU,CD,R,LD,PV
        CTUD_8      79,M0.6,M0.7,I1.4,I1.5,I1.6,I1.7,.20                      ;rung 10
        in_out      QU79,Q0.6                                                 ;rung 11
        in_out      QD79,Q0.7                                                 ;rung 12

        ld          I2.2                                                     ;rung 13
                    ;s1,s0,R3,R2,R1,R0,OUT
        B_mux_4_1_E I2.1,I2.0,CV_L+79,CV_L+20,CV_L+10,CV_L,Q1

;
;--- PLC codes to be allocated in the "user_program_1" macro end here ----------
        endm
```

FIGURE 6.40   The user program of "Example 6.3".

will be programmed. After loading the program file to the PIC16F1847 microcontroller, switch the 4PDT in "RUN" and the power switch in "ON" position. Next you can test the operation of this example.

In rungs 1, 2 and 3, an up/down counter "CTUD_8" is implemented as follows: the count up input CU is taken from I0.0, the count down input CD is taken from I0.1, the reset input R is taken from I0.2 and the load input LD is taken from I0.3. The unique memory bit used to detect rising edge signals at the CU input is MregU,MbitU = M0.0. The unique memory bit used to detect rising edge signals at the CD input is MregD,MbitD = M0.1. num = 0 and therefore the up status bit of this up/down counter is QU0 and the down status bit of this up/down counter is QD0. The 8-bit integer variable used to count up and count down is CV_L+num = CV_L. The preset count value PV = 5. As can be seen from rung 2, the state of the up status bit QU0 is sent to output Q0.0. Likewise, in rung 3, the state of the down status bit QD0 is sent to output Q0.1.

In rungs 4, 5 and 6, an up/down counter "CTUD_8" is implemented as follows: the count up input CU is taken from I0.4, the count down input CD is taken from I0.5, the reset input R is taken from I0.6 and the load input LD is taken from

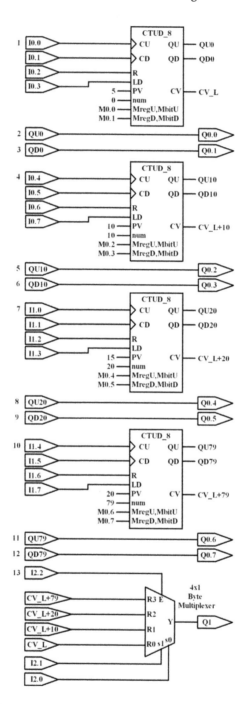

FIGURE 6.41A   The user program of "Example 6.3": Schematic diagram.

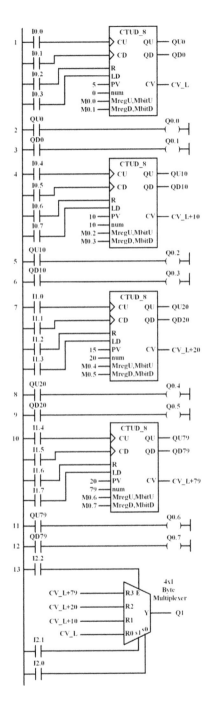

FIGURE 6.41B   The user program of "Example 6.3": Ladder diagram.

I0.7. The unique memory bit used to detect rising edge signals at the CU input is MregU,MbitU=M0.2. The unique memory bit used to detect rising edge signals at the CD input is MregD,MbitD=M0.3. num=10 and therefore the up status bit of this up/down counter is QU10 and the down status bit of this up/down counter is QD10. The 8-bit integer variable used to count up and count down is CV_L+num=CV_L+10. The preset count value PV=10. As can be seen from **rung 5**, the state of the up status bit QU10 is sent to output Q0.2. Likewise, in **rung 6**, the state of the down status bit QD10 is sent to output Q0.3.

In rungs 7, 8 and 9, an up/down counter "CTUD_8" is implemented as follows: the count up input CU is taken from I1.0, the count down input CD is taken from I1.1, the reset input R is taken from I1.2 and the load input LD is taken from I1.3. The unique memory bit used to detect rising edge signals at the CU input is MregU,MbitU=M0.4. The unique memory bit used to detect rising edge signals at the CD input is MregD,MbitD=M0.5. num=20 and therefore the up status bit of this up/down counter is QU20 and the down status bit of this up/down counter is QD20. The 8-bit integer variable used to count up and count down is CV_L+num=CV_L+20. The preset count value PV=15. As can be seen from **rung 8**, the state of the up status bit QU20 is sent to output Q0.4. Likewise, in **rung 9**, the state of the down status bit QD20 is sent to output Q0.5.

In rungs 10, 11 and 12, an up/down counter "CTUD_8" is implemented as follows: the count up input CU is taken from I1.4, the count down input CD is taken from I1.5, the reset input R is taken from I1.6 and the load input LD is taken from I1.7. The unique memory bit used to detect rising edge signals at the CU input is MregU,MbitU=M0.6. The unique memory bit used to detect rising edge signals at the CD input is MregD,MbitD=M0.7. num=79 and therefore the up status bit of this up/down counter is QU79 and the down status bit of this up/down counter is QD79. The 8-bit integer variable used to count up and count down is CV_L+num=CV_L+79. The preset count value PV=20. As can be seen from **rung 11**, the state of the up status bit QU79 is sent to output Q0.6. Likewise, in **rung 12**, the state of the down status bit QD79 is sent to output Q0.7.

In rung 13, a 4x1 Byte Multiplexer (Chapter 4 of the "Intermediate Concepts" book) is used to observe the current count values of one of the four 8-bit variables considered in the previous rungs from Q1. The following table shows the operation of this 4x1 Byte multiplexer.

	inputs		output
E	s1	s0	Y
I2.2	I2.1	I2.0	Q1
0	×	×	U
1	0	0	CV_L
1	0	1	CV_L+10
1	1	0	CV_L+20
1	1	1	CV_L+79

×: don't care. U: The contents of the destination register Y(Q1) remain unchanged.

6.13.4 EXAMPLE 6.4

Example 6.4 shows the usage of the up/down counter macro CTUD_16. The user program of Example 6.4 is shown in Figure 6.42. The schematic diagram and the ladder diagram of Example 6.4 are depicted in Figure 6.43(a) and in Figure 6.43(b) respectively. When the project file of the PIC16F1847-Based PLC is open in the MPLAB X IDE, from the file "PICPLC_PIC16F1847_user_Bsc.inc" if you uncomment Example 6.4 and run the project by pressing the symbol ▷ from the toolbar, then the PIC16F1847 microcontroller within the CPU board of the PIC16F1847-Based PLC will be programmed. After loading the program file to the PIC16F1847 microcontroller, switch the 4PDT in "RUN" and the power switch in "ON" position. Next you can test the operation of this example.

In rungs 1, 2 and 3, an up/down counter "CTUD_16" is implemented as follows: the count up input CU is taken from I0.0, the count down input CD is taken from I0.1, the reset input R is taken from I0.2 and the load input LD is taken from I0.3. The unique memory bit used to detect rising edge signals at the CU input is MregU,MbitU = M0.0. The unique memory bit used to detect rising edge signals at the CD input is MregD,MbitD = M0.1. num = 9 and therefore the up status bit of this up/down counter is QU9 and the down status bit of this up/down counter is QD9. The 16-bit integer variable used to count up and count down is CV_H+num&CV_L+num = CV_H+9&CV_L+9. The preset count value PV = 20. As can be seen from rung 2, the state of the up status bit QU9 is sent to output Q0.0. Likewise, in rung 3, the state of the down status bit QD9 is sent to output Q0.1.

In rungs 4, 5 and 6, an up/down counter "CTUD_16" is implemented as follows: the count up input CU is taken from I0.4, the count down input CD is taken from I0.5, the reset input R is taken from I0.6 and the load input LD is taken from I0.7. The unique memory bit used to detect rising edge signals at the CU input is MregU,MbitU = M0.2. The unique memory bit used to detect rising edge signals at the CD input is MregD,MbitD = M0.3. num = 49 and therefore the up status bit of this up/down counter is QU49 and the down status bit of this up/down counter is QD49. The 16-bit integer variable used to count up and count down is CV_H+num&CV_L+num = CV_H+49&CV_L+49. The preset count value PV = 30. As can be seen from **rung 5**, the state of the up status bit QU49 is sent to output Q0.2. Likewise, in **rung 6**, the state of the down status bit QD49 is sent to output Q0.3.

In rung 7, Memory bit M1.0 is obtained by ANDing the digital input I1.0 with the reference timing signal T_2ms. This means that when I1.0 is ON, M1.0 = T_2ms. This organization is carried out to speed up the count up process for the next up/down counter.

In rung 8, Memory bit M1.1 is obtained by ANDing the digital input I1.1 with the reference timing signal T_2ms. This means that when I1.1 is ON, M1.1 = T_2ms. This organization is carried out to speed up the count down process for the next up/down counter.

```
user_program_1   macro
;--- PLC codes to be allocated in the "user_program_1" macro start from here ---
;__Example_Bsc_6.4
                      ;num,MregU,MbitU,MregD,MbitD,CU,CD,R,LD,PV
         CTUD_16      9,M0.0,M0.1,I0.0,I0.1,I0.2,I0.3,.20                  ;rung 1
         in_out       QU9,Q0.0                                            ;rung 2
         in_out       QD9,Q0.1                                            ;rung 3

                      ;num,MregU,MbitU,MregD,MbitD,CU,CD,R,LD,PV
         CTUD_16      49,M0.2,M0.3,I0.4,I0.5,I0.6,I0.7,.30                 ;rung 4
         in_out       QU49,Q0.2                                           ;rung 5
         in_out       QD49,Q0.3                                           ;rung 6

         ld           I1.0                                                ;rung 7
         and          T_2ms
         out          M1.0

         ld           I1.1                                                ;rung 8
         and          T_2ms
         out          M1.1

                      ;num,MregU,MbitU,MregD,MbitD,CU,CD,R,LD,PV
         CTUD_16      59,M0.4,M0.5,M1.0,M1.1,I1.2,I1.3,.2000              ;rung 9
         in_out       QU59,Q0.4                                           ;rung 10
         in_out       QD59,Q0.5                                           ;rung 11

         ld           I1.4                                                ;rung 12
         and          T_2ms
         out          M1.4

         ld           I1.5                                                ;rung 13
         and          T_2ms
         out          M1.5

                      ;num,MregU,MbitU,MregD,MbitD,CU,CD,R,LD,PV
         CTUD_16      69,M0.6,M0.7,M1.4,M1.5,I1.6,I1.7,.30000             ;rung 14
         in_out       QU69,Q0.6                                           ;rung 15
         in_out       QD69,Q0.7                                           ;rung 16


         ld           I2.2                                                ;rung 17
                      ;s1,s0,R3,R2,R1,R0,OUT
         B_mux_4_1_E  I2.1,I2.0,CV_L+69,CV_L+59,CV_L+49,CV_L+9,Q1

         ld           I2.2                                                ;rung 18
                      ;s1,s0,R3,R2,R1,R0,OUT
         B_mux_4_1_E  I2.1,I2.0,CV_H+69,CV_H+59,CV_H+49,CV_H+9,Q2
;
;--- PLC codes to be allocated in the "user_program_1" macro end here ----------
         endm
```

FIGURE 6.42   The user program of "Example 6.4".

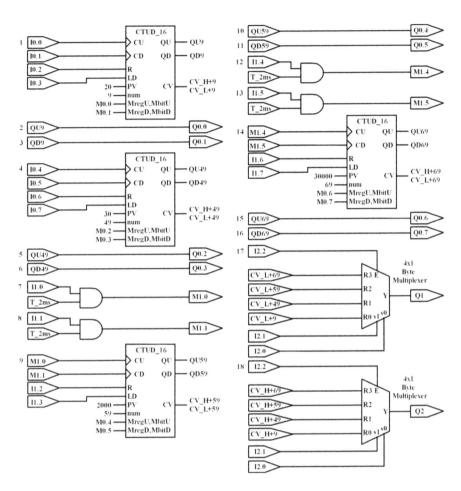

FIGURE 6.43A   The user program of "Example 6.4": Schematic diagram.

In rungs 9, 10 and 11, an up/down counter "CTUD_16" is implemented as follows: the count up input CU is taken from M1.0, the count down input CD is taken from M1.1, the reset input R is taken from I1.2 and the load input LD is taken from I1.3. The unique memory bit used to detect rising edge signals at the CU input is MregU,MbitU = M0.4. The unique memory bit used to detect rising edge signals at the CD input is MregD,MbitD = M0.5. num = 59 and therefore the up status bit of this up/down counter is QU59 and the down status bit of this up/down counter is QD59. The 16-bit integer variable used to count up and count down is CV_H+num&CV_L+num = CV_H+59&CV_L+59. The preset count value PV = 2000. As can be seen from **rung 10**, the state of the up status bit QU59 is sent to output Q0.4. Likewise, in **rung 11**, the state of the down status bit QD59 is sent to output Q0.5.

In rung 12, Memory bit M1.4 is obtained by ANDing the digital input I1.4 with the reference timing signal T_2ms. This means that when I1.4 is ON,

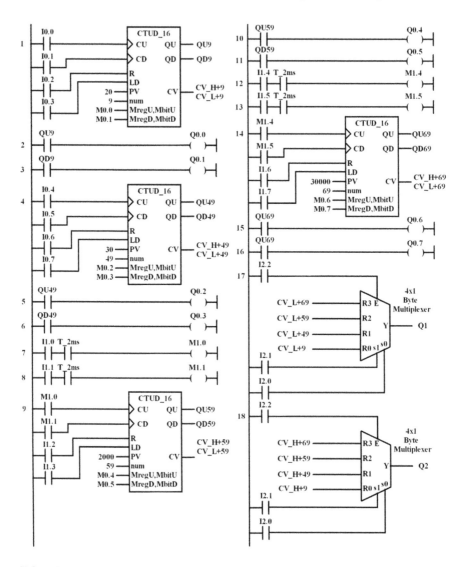

FIGURE 6.43B   The user program of "Example 6.4": Ladder diagram.

M1.4=T_2ms. This organization is carried out to speed up the count up process for the next up/down counter.

In rung 13, Memory bit M1.5 is obtained by ANDing the digital input I1.5 with the reference timing signal T_2ms. This means that when I1.5 is ON, M1.5=T_2ms. This organization is carried out to speed up the count down process for the next up/down counter.

In rungs 14, 15 and 16, an up/down counter "CTUD_16" is implemented as follows: the count up input CU is taken from M1.4, the count down input CD is taken from M1.5, the reset input R is taken from I1.6 and the load input LD is taken from I1.7. The unique memory bit used to detect rising edge signals at the CU input is MregU,MbitU=M0.6. The unique memory bit used to detect rising edge signals at the CD input is MregD,MbitD=M0.7. num=69 and therefore the up status bit of this up/ down counter is QU69 and the down status bit of this up/down counter is QD69. The 16-bit integer variable used to count up and count down is CV_H+num&CV_L+num=CV_H+69&CV_L+69. The preset count value PV=30000. As can be seen from **rung 15**, the state of the up status bit QU69 is sent to output Q0.6. Likewise, in **rung 16**, the state of the down status bit QD69 is sent to output Q0.7.

In rung 17, a 4x1 Byte Multiplexer (Chapter 4 of the "Intermediate Concepts" book) is used to observe the low byte of current count values of one of the four 16-bit variables considered in previous rungs from Q1. The following table shows the operation of this 4x1 Byte multiplexer.

	inputs		output
E	s1	s0	Y
I2.2	I2.1	I2.0	Q1
0	×	×	U
1	0	0	CV_L+9
1	0	1	CV_L+49
1	1	0	CV_L+59
1	1	1	CV_L+69

×: don't care.
U: The contents of the destination register
Y(Q1) remain unchanged.

In rung 18, a 4x1 Byte Multiplexer (Chapter 4 of the "Intermediate Concepts" book) is used to observe the high byte of current count values of one of the four 16-bit variables considered in previous rungs from Q2. The following table shows the operation of this 4x1 Byte multiplexer.

	inputs		output
E	s1	s0	Y
I2.2	I2.1	I2.0	Q2
0	×	×	U
1	0	0	CV_H+9
1	0	1	CV_H+49
1	1	0	CV_H+59
1	1	1	CV_H+69

×: don't care.
U: The contents of the destination register
Y(Q2) remain unchanged.

6.13.5 EXAMPLE 6.5

Example 6.5 shows the usage of the generalized up/down counter macro GCTUD_8. The user program of Example 6.5 is shown in Figure 6.44. The schematic diagram and the ladder diagram of Example 6.5 are depicted in Figure 6.45(a) and in Figure 6.45(b) respectively. When the project file of the PIC16F1847-Based PLC is open in the MPLAB X IDE, from the file "PICPLC_PIC16F1847_user_Bsc.inc" if you uncomment Example 6.5 and run the project by pressing the symbol ▷ from the toolbar, then the PIC16F1847 microcontroller within the CPU board of the PIC16F1847-Based PLC will be programmed. After loading the program file to the PIC16F1847 microcontroller, switch the 4PDT in "RUN" and the power switch in "ON" position. Next you can test the operation of this example.

In rungs 1, 2 and 3, a generalized up/down counter "GCTUD_8" is implemented as follows: the count up input CU is taken from I0.0, the count down input CD is taken from I0.1, the reset input R is taken from I0.2 and the load input LD is taken from I0.3. The load count value is LV = 5. The low limit value is L_lim = 10, while the high limit value is H_lim = 20. The

```
user_program_1  macro
;
;--- PLC codes to be allocated in the "user_program_1" macro start from here ---
;__Example_Bsc_6.5

                     ;num,MregU,MbitU,MregD,MbitD,CU,CD,R,LD,LV,L_lim,H_lim
        GCTUD_8      5,M0.0,M0.1,I0.0,I0.1,I0.2,I0.3,.5,.10,.20         ;rung 1
        in_out       QL5,Q0.0                                          ;rung 2
        in_out       QD5,Q0.1                                          ;rung 3

                     ;num,MregU,MbitU,MregD,MbitD,CU,CD,R,LD,LV,L_lim,H_lim
        GCTUD_8      25,M0.2,M0.3,I0.4,I0.5,I0.6,I0.7,.20,.15,.30      ;rung 4
        in_out       QL25,Q0.2                                         ;rung 5
        in_out       QD25,Q0.3                                         ;rung 6

                     ;num,MregU,MbitU,MregD,MbitD,CU,CD,R,LD,LV,L_lim,H_lim
        GCTUD_8      50,M0.4,M0.5,I1.0,I1.1,I1.2,I1.3,.15,.20,.40      ;rung 7
        in_out       QL50,Q0.4                                         ;rung 8
        in_out       QD50,Q0.5                                         ;rung 9

                     ;num,MregU,MbitU,MregD,MbitD,CU,CD,R,LD,LV,L_lim,H_lim
        GCTUD_8      75,M0.6,M0.7,I1.4,I1.5,I1.6,I1.7,.30,.25,.50      ;rung 10
        in_out       QL75,Q0.6                                         ;rung 11
        in_out       QD75,Q0.7                                         ;rung 12

        ld           I2.2                                              ;rung 13
                     ;s1,s0,R3,R2,R1,R0,OUT
        B_mux_4_1_E  I2.1,I2.0,CV_L+75,CV_L+50,CV_L+25,CV_L+5,Q1
;
;--- PLC codes to be allocated in the "user_program_1" macro end here ----------
        endm
```

FIGURE 6.44 The user program of "Example 6.5".

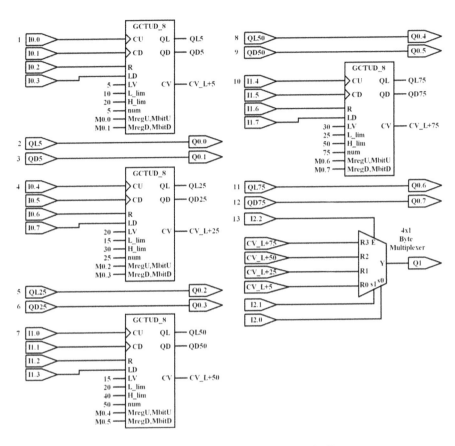

FIGURE 6.45A   The user program of "Example 6.5": Schematic diagram.

unique memory bit used to detect rising edge signals at the CU input is
MregU,MbitU = M0.0. The unique memory bit used to detect rising edge
signals at the CD input is MregD,MbitD = M0.1. num = 5 and therefore the
within-limits status bit of this generalized up/down counter is QL5 and the
down status bit of this generalized up/down counter is QD5. The 8-bit integer variable used to count up and count down is CV_L+num = CV_L+5. As
can be seen from **rung 2**, the state of the within-limits status bit QL5 is sent
to output Q0.0. Likewise, in **rung 3**, the state of the down status bit QD5 is
sent to output Q0.1.

In rungs 4, 5 and 6, a generalized up/down counter "GCTUD_8" is implemented as follows: the count up input CU is taken from I0.4, the count
down input CD is taken from I0.5, the reset input R is taken from I0.6 and
the load input LD is taken from I0.7. The load count value is LV = 20. The
low limit value is L_lim = 15, while the high limit value is H_lim = 30. The
unique memory bit used to detect rising edge signals at the CU input is
MregU,MbitU = M0.2. The unique memory bit used to detect rising edge
signals at the CD input is MregD,MbitD = M0.3. num = 25 and therefore the

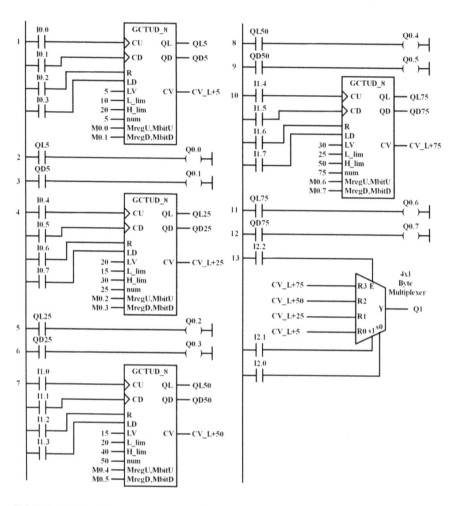

FIGURE 6.45B The user program of "Example 6.5": Ladder diagram.

within-limits status bit of this generalized up/down counter is QL25 and the down status bit of this generalized up/down counter is QD25. The 8-bit integer variable used to count up and count down is CV_L+num=CV_L+25. As can be seen from **rung 5**, the state of the within-limits status bit QL25 is sent to output Q0.2. Likewise, in **rung 6**, the state of the down status bit QD25 is sent to output Q0.3.

In rungs 7, 8 and 9, a generalized up/down counter "GCTUD_8" is implemented as follows: the count up input CU is taken from I1.0, the count down input CD is taken from I1.1, the reset input R is taken from I1.2 and the load input LD is taken from I1.3. The load count value is LV = 15. The low limit value is L_lim = 20, while the high limit value is H_lim = 40. The unique memory bit used to detect rising edge signals at the CU input is MregU,MbitU = M0.4. The unique memory bit used to detect rising edge signals at the CD input is MregD,MbitD = M0.5. num = 50 and therefore the within-limits status bit of this generalized up/down counter is QL50 and the down status bit of this generalized up/down counter is QD50. The 8-bit integer variable used to count up and count down is CV_L+num=CV_L+50. As can be seen from **rung 8**, the state of the within-limits status bit QL50 is sent to output Q0.4. Likewise, in **rung 9**, the state of the down status bit QD50 is sent to output Q0.5.

In rungs 10, 11 and 12, a generalized up/down counter "GCTUD_8" is implemented as follows: the count up input CU is taken from I1.4, the count down input CD is taken from I1.5, the reset input R is taken from I1.6 and the load input LD is taken from I1.7. The load count value is LV=30. The low limit value is L_lim=25, while the high limit value is H_lim=50. The unique memory bit used to detect rising edge signals at the CU input is MregU,MbitU=M0.6. The unique memory bit used to detect rising edge signals at the CD input is MregD,MbitD=M0.7. num=75 and therefore the within-limits status bit of this generalized up/down counter is QL75 and the down status bit of this generalized up/down counter is QD75. The 8-bit integer variable used to count up and count down is CV_L+num=CV_L+75. As can be seen from **rung 11**, the state of the within-limits status bit QL75 is sent to output Q0.6. Likewise, in **rung 12**, the state of the down status bit QD75 is sent to output Q0.7.

In rung 13, a 4x1 Byte Multiplexer (Chapter 4 of the "Intermediate Concepts" book) is used to observe the current count values of one of the four 8-bit variables considered in the previous rungs from Q1. The following table shows the operation of this 4x1 Byte multiplexer.

	inputs		output
E	s1	s0	Y
I2.2	I2.1	I2.0	Q1
0	×	×	U
1	0	0	CV_L+5
1	0	1	CV_L+25
1	1	0	CV_L+50
1	1	1	CV_L+75

×: don't care.
U: The contents of the destination register
Y(Q1) remain unchanged.

6.13.6   EXAMPLE 6.6

Example 6.6 shows the usage of the generalized up/down counter macro GCTUD_16. The user program of Example 6.6 is shown in Figure 6.46. The schematic diagram and the ladder diagram of Example 6.6 are depicted in Figure 6.47(a) and in Figure 6.47(b) respectively. When the project file of the PIC16F1847-Based PLC is open in the MPLAB X IDE, from the file "PICPLC_PIC16F1847_user_Bsc.inc" if you uncomment Example 6.6 and run the project by pressing the symbol ▷ from

```
user_program_1   macro
;
;--- PLC codes to be allocated in the "user_program_1" macro start from here ---
;__Example_Bsc_6.6
                        ;num,MregU,MbitU,MregD,MbitD,CU,CD,R,LD,LV,L_lim,H_lim
        GCTUD_16        19,M0.0,M0.1,I0.0,I0.1,I0.2,I0.3,.15,.20,.30              ;rung 1
        in_out          QL19,Q0.0                                                ;rung 2
        in_out          QD19,Q0.1                                                ;rung 3

                        ;num,MregU,MbitU,MregD,MbitD,CU,CD,R,LD,LV,L_lim,H_lim
        GCTUD_16        49,M0.2,M0.3,I0.4,I0.5,I0.6,I0.7,.30,.20,.50             ;rung 4
        in_out          QL49,Q0.2                                                ;rung 5
        in_out          QD49,Q0.3                                                ;rung 6

        ld              I1.0                                                     ;rung 7
        and             T_2ms
        out             M1.0

        ld              I1.1                                                     ;rung 8
        and             T_2ms
        out             M1.1

                        ;num,MregU,MbitU,MregD,MbitD,CU,CD,R,LD,LV,L_lim,H_lim
        GCTUD_16        59,M0.4,M0.5,M1.0,M1.1,I1.2,I1.3,.100,.200,.300         ;rung 9
        in_out          QL59,Q0.4                                                ;rung 10
        in_out          QD59,Q0.5                                                ;rung 11

        ld              I1.4                                                     ;rung 12
        and             T_2ms
        out             M1.4

        ld              I1.5                                                     ;rung 13
        and             T_2ms
        out             M1.5

                        ;num,MregU,MbitU,MregD,MbitD,CU,CD,R,LD,LV,L_lim,H_lim
        GCTUD_16        79,M0.6,M0.7,M1.4,M1.5,I1.6,I1.7,.65510,.65520,.65530;rung 14
        in_out          QL79,Q0.6                                                ;rung 15
        in_out          QD79,Q0.7                                                ;rung 16


        ld              I2.2                                                     ;rung 17
                        ;s1,s0,R3,R2,R1,R0,OUT
        B_mux_4_1_E  I2.1,I2.0,CV_L+79,CV_L+59,CV_L+49,CV_L+19,Q1

        ld              I2.2                                                     ;rung 18
                        ;s1,s0,R3,R2,R1,R0,OUT
        B_mux_4_1_E  I2.1,I2.0,CV_H+79,CV_H+59,CV_H+49,CV_H+19,Q2
;
;--- PLC codes to be allocated in the "user_program_1" macro end here ----------
        endm
```

FIGURE 6.46   The user program of "Example 6.6".

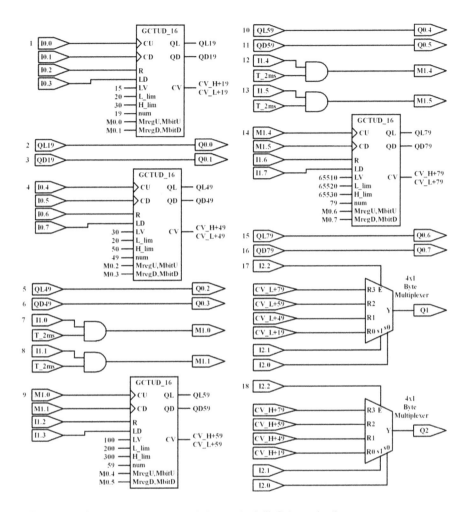

FIGURE 6.47A The user program of "Example 6.6": Schematic diagram.

the toolbar, then the PIC16F1847 microcontroller within the CPU board of the PIC16F1847-Based PLC will be programmed. After loading the program file to the PIC16F1847 microcontroller, switch the 4PDT in "RUN" and the power switch in "ON" position. Next you can test the operation of this example.

In rungs 1, 2 and 3, a generalized up/down counter "GCTUD_16" is implemented as follows: the count up input CU is taken from I0.0, the count down input CD is taken from I0.1, the reset input R is taken from I0.2 and the load input LD is taken from I0.3. The load count value is LV = 15. The low limit value is L_lim = 20, while the high limit value is H_lim = 30. The unique memory bit used to detect rising edge signals at the CU input is MregU,MbitU = M0.0. The unique memory bit used to detect rising edge signals at the CD input is MregD,MbitD = M0.1. num = 19 and therefore the

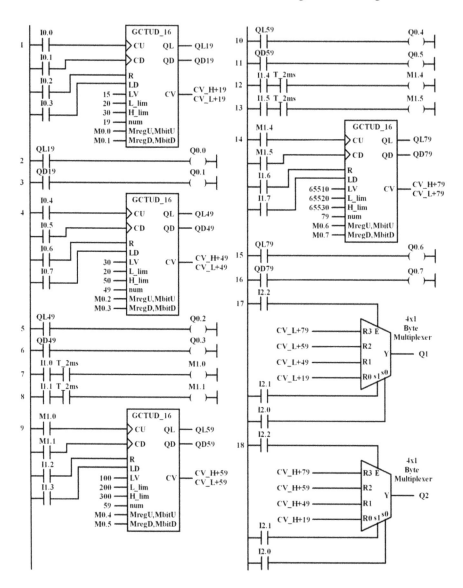

FIGURE 6.47B   The user program of "Example 6.6": Ladder diagram.

within-limits status bit of this generalized up/down counter is CQ19 and the down status bit of this generalized up/down counter is QL19. The 16-bit integer variable used to count up and count down is CV_H+num&CV_ L+num=CV_H+19&CV_L+19. As can be seen from **rung 2**, the state of the within-limits status bit CQ19 is sent to output Q0.0. Likewise, in **rung 3**, the state of the down status bit QL19 is sent to output Q0.1.

In rungs 4, 5 and 6, a generalized up/down counter "GCTUD_16" is implemented as follows: the count up input CU is taken from I0.4, the count down input CD is taken from I0.5, the reset input R is taken from I0.6 and the load input LD is taken from I0.7. The load count value is LV = 30. The low limit value is L_lim = 20, while the high limit value is H_lim = 50. The unique memory bit used to detect rising edge signals at the CU input is MregU,MbitU = M0.2. The unique memory bit used to detect rising edge signals at the CD input is MregD,MbitD = M0.3. num = 49 and therefore the within-limits status bit of this generalized up/down counter is QL49 and the down status bit of this generalized up/down counter is QD49. The 16-bit integer variable used to count up and count down is CV_H+num&CV_L+num = CV_H+49&CV_L+49. As can be seen from **rung 5**, the state of the within-limits status bit QL49 is sent to output Q0.2. Likewise, in **rung 6**, the state of the down status bit QD49 is sent to output Q0.3.

In rung 7, Memory bit M1.0 is obtained by ANDing the digital input I1.0 with the reference timing signal T_2ms. This means that when I1.0 is ON, M1.0 = T_2ms. This organization is carried out to speed up the count up process for the next generalized up/down counter.

In rung 8, Memory bit M1.1 is obtained by ANDing the digital input I1.1 with the reference timing signal T_2ms. This means that when I1.1 is ON, M1.1 = T_2ms. This organization is carried out to speed up the count down process for the next generalized up/down counter.

In rungs 9, 10 and 11, a generalized up/down counter "GCTUD_16" is implemented as follows: the count up input CU is taken from M1.0, the count down input CD is taken from M1.1, the reset input R is taken from I1.2 and the load input LD is taken from I1.3. The load count value is LV = 100. The low limit value is L_lim = 200, while the high limit value is H_lim = 300. The unique memory bit used to detect rising edge signals at the CU input is MregU,MbitU = M0.4. The unique memory bit used to detect rising edge signals at the CD input is MregD,MbitD = M0.5. num = 59 and therefore the within-limits status bit of this generalized up/down counter is QL59 and the down status bit of this generalized up/down counter is QD59. The 16-bit integer variable used to count up and count down is CV_H+num&CV_L+num = CV_H+59&CV_L+59. As can be seen from **rung 10**, the state of the within-limits status bit QL59 is sent to output Q0.4. Likewise, in **rung 11**, the state of the down status bit QD59 is sent to output Q0.5.

In rung 12, Memory bit M1.4 is obtained by ANDing the digital input I1.4 with the reference timing signal T_2ms. This means that when I1.4 is ON, M1.4 = T_2ms. This organization is carried out to speed up the count up process for the next generalized up/down counter.

In rung 13, Memory bit M1.5 is obtained by ANDing the digital input I1.5 with the reference timing signal T_2ms. This means that when I1.5 is ON, M1.5 = T_2ms. This organization is carried out to speed up the count down process for the next generalized up/down counter.

In rungs 14, 15 and 16, a generalized up/down counter "GCTUD_16" is implemented as follows: the count up input CU is taken from M1.4, the count down input CD is taken from M1.5, the reset input R is taken from I1.6 and the load input LD is taken from I1.7. The load count value is LV = 65510. The low limit value is L_lim = 65520, while the high limit value is H_lim = 65530.

The unique memory bit used to detect rising edge signals at the CU input is MregU,MbitU=M0.6. The unique memory bit used to detect rising edge signals at the CD input is MregD,MbitD=M0.7. num=79 and therefore the within-limits status bit of this generalized up/down counter is QL79 and the down status bit of this generalized up/down counter is QD79. The 16-bit integer variable used to count up and count down is CV_H+num&CV_L+num=CV_H+79&CV_L+79. As can be seen from **rung 15**, the state of the within-limits status bit QL79 is sent to output Q0.6. Likewise, in **rung 16**, the state of the down status bit QD79 is sent to output Q0.7.

In rung 17, a 4x1 Byte Multiplexer (Chapter 4 of the "Intermediate Concepts" book) is used to observe the low byte of current count values of one of the four 16-bit variables considered in previous rungs from Q1. The following table shows the operation of this 4x1 Byte multiplexer.

inputs			output
E	s1	s0	Y
I2.2	I2.1	I2.0	Q1
0	×	×	U
1	0	0	CV_L+19
1	0	1	CV_L+49
1	1	0	CV_L+59
1	1	1	CV_L+79

×: don't care.
U: The contents of the destination register Y(Q1) remain unchanged.

In rung 18, a 4x1 Byte Multiplexer (Chapter 4 of the "Intermediate Concepts" book) is used to observe the high byte of current count values of one of the four 16-bit variables considered in previous rungs from Q2. The following table shows the operation of this 4x1 Byte multiplexer.

inputs			output
E	s1	s0	Y
I2.2	I2.1	I2.0	Q2
0	×	×	U
1	0	0	CV_H+19
1	0	1	CV_H+49
1	1	0	CV_H+59
1	1	1	CV_H+79

×: don't care.
U: The contents of the destination register Y(Q2) remain unchanged.

About the Downloadable Files for *Hardware and Basic Concepts*

Downloadable files of this book contain source and example files defined for the basic concepts of the PIC16F1847-Based PLC project. In addition, PCB (Gerber and PDF) files are also provided in order for the reader to obtain both the CPU board and I/O extension boards produced by a PCB manufacturer. A skilled reader may produce his/her own boards by using the provided PDF files. These files are downloadable from this book's webpage under the downloads section.

The files are organised in the following folders:

PICPLC_PIC16F1847_Bsc (PLC project folder)

1. PICPLC_PIC16F1847.X (a folder containing MPLAB project files)
2. PICPLC_PIC16F1847_macros_Bsc.inc (files for PLC macros defined for the basic concepts)
3. PICPLC_PIC16F1847_main.asm (main program file)
4. PICPLC_PIC16F1847_memory.inc (memory-related definitions)
5. PICPLC_PIC16F1847_subr.inc (subroutine definitions)
6. PICPLC_PIC16F1847_user_Bsc.inc (user program files and application example files for the basic concepts)

PIC16F1847_Based_PLC_32I_32O (a folder with the following contents)

CPU_board
    PCB design files for the CPU board (Gerber files and PDF files)
    Photographs of the CPU board
    The schematic diagram of the CPU board
IO_extension_board
    PCB design files for the IO extension board (Gerber files and PDF files)
    Photographs of the IO extension board
    The schematic diagram of the IO extension board
Technical_Data_for_ICs (a folder containing technical data for some integrated circuits)

Files "PICPLC_PIC16F1847_macros_Bsc.inc" and "PICPLC_PIC16F1847_user_Bsc.inc" refer to macros and user program files of the basic concepts developed in the PIC16F1847-Based PLC project, respectively. They do not contain files related to the intermediate and advanced concepts. These two files are intended for the readers who purchased this book as a standalone book. On the other hand, when this book is purchased as part of the set of three books, all project files including basic, intermediate, and advanced concepts are put in the same directory and the reader is entitled to download and use the whole of the project files in one directory, the name of which becomes "PICPLC_PIC16F1847" instead of "PICPLC_PIC16F1847_Bsc". Therefore, in the second case the name of the file "PICPLC_PIC16F1847_macros_Bsc.inc" (and PICPLC_PIC16F1847_user_Bsc.inc, respectively) becomes "PICPLC_PIC16F1847_macros.inc" (and PICPLC_PIC16F1847_user.inc, respectively).

Index

For Product Safety Concerns and Information please contact our EU
representative GPSR@taylorandfrancis.com
Taylor & Francis Verlag GmbH, Kaufingerstraße 24, 80331 München, Germany

www.ingramcontent.com/pod-product-compliance
Ingram Content Group UK Ltd.
Pitfield, Milton Keynes, MK11 3LW, UK
UKHW021024180425
457613UK00020B/1049